Lecture Notes in Energy

Volume 44

Lecture Notes in Energy (LNE) is a series that reports on new developments in the study of energy: from science and engineering to the analysis of energy policy. The series' scope includes but is not limited to, renewable and green energy, nuclear, fossil fuels and carbon capture, energy systems, energy storage and harvesting, batteries and fuel cells, power systems, energy efficiency, energy in buildings, energy policy, as well as energy-related topics in economics, management and transportation. Books published in LNE are original and timely and bridge between advanced textbooks and the forefront of research. Readers of LNE include postgraduate students and non-specialist researchers wishing to gain an accessible introduction to a field of research as well as professionals and researchers with a need for an up-to-date reference book on a well-defined topic. The series publishes single- and multi-authored volumes as well as advanced textbooks.

Indexed in Scopus and EI Compendex The Springer Energy board welcomes your book proposal. Please get in touch with the series via Anthony Doyle, Executive Editor, Springer (anthony.doyle@springer.com)

Nedunchezhian Swaminathan · Alessandro Parente
Editors

Machine Learning and Its Application to Reacting Flows

ML and Combustion

 Springer

Editors
Nedunchezhian Swaminathan
Department of Engineering
University of Cambridge
Cambridge, UK

Alessandro Parente
Aero-Thermo-Mechanics Laboratory
École polytechnique de Bruxelles
Université Libre de Bruxelles
Brussels, Belgium

Brussels Institute for Thermal-fluid
Systems, Brussels (BRITE)
Université Libre de Bruxelles and Vrije
Universiteit Brussel
Brussels, Belgium

ISSN 2195-1284 ISSN 2195-1292 (electronic)
Lecture Notes in Energy
ISBN 978-3-031-16250-3 ISBN 978-3-031-16248-0 (eBook)
https://doi.org/10.1007/978-3-031-16248-0

This Springer imprint is published by the registered company Springer Nature Switzerland AG
The registered company address is: Gewerbestrasse 11, 6330 Cham, Switzerland

Preface

Machine learning (ML) has been around for many decades and has been explored in the past for many practical applications. Currently, ML is interpreted in a broader context and finding its way into a number of sectors such as Engineering, Health Care, Transport including Traffic Prediction and Control, driverless car, Information Technology, Big Data Analysis and Processing, Agriculture, Agronomy, etc. It has found its way also into our daily life, for example, temperature and lighting controls and information searches on the internet. In a nutshell, ML is nothing but statistical interference using data collected or knowledge gained through past targeted studies or real-life experiences. The sophistication level of ML depends on the intended application and the advanced nature of the algorithms used for statistical learning and inference. This area has attracted huge interest recently because of the advent of the computational power, technology and algorithms required for data training, verification and validation, and the readiness and availability of these algorithms for application to a wide range of fields and practical systems. Hence, it is very timely to overview the various ML techniques or algorithms for big data analyses with a specific application to combustion science and technology.

This particular topic is chosen because of the important role of combustion systems and technologies covering more than 90% of the world's total primary energy supply (TPES). Although alternative renewable energy technologies are coming up, their shares for the TPES are less than 5% currently and one needs a complete paradigm shift to replace combustion sources. Whether this is practical or not is entirely a different question and an answer to this question is likely to depend on the respondent. However, a pragmatic analysis suggests that the combustion share to TPES is likely to be more than 70% even by 2070 as discussed in the chapter "Introduction" of this book. Hence, it will be prudent to take advantage of ML techniques to improve combustion sciences and technologies to better combustion system design and development so that the emission of greenhouse gases can be curtailed along with improving overall efficiencies. The level of interest in applying ML to combustion is clearly evident from the recent surge in research activities on this topic. Hence, the aim of this volume is to bring this knowledge together and make it readily accessible for researchers and graduate students interested in this multi- and cross-disciplinary topic. We attempted to keep the discussion accessible to students and researchers interested in turbulent combustion, ML techniques, and its application to turbulence and combustion on a simple physical basis while highlighting the need for ML.

Chapter "Introduction" gives an introduction to the role of combustion technologies in the future purely based on the current practical and scientific evidence. This chapter also identifies the opportunities to use ML algorithms (MLA) while investigating turbulent combustion. The chapter "Machine Learning Techniques in Reactive Atomistic Simulations" surveys various ML techniques and discusses their application for estimating atomic potential energies, required for chemical kinetics, through molecular dynamics simulation as an example. The chapter "A Novel In Situ Machine Learning Framework for Intelligent Data Capture and Event Detection" introduces in situ training for MLA which is a useful idea as it can save considerable efforts required in the training phase while using MLA. The chapter "Machine-Learning for Stress Tensor Modelling in Large Eddy Simulation" discusses the use of ML to estimate subgrid scale stresses and fluxes needed for large eddy simulation of turbulent combustion. The application of ML for combustion chemistry is discussed in the chapter "Machine Learning for Combustion Chemistry". The turbulence-chemistry interaction is a highly nonlinear stochastic problem ideally suited for ML and chapters "Deep Convolutional Neural Networks for Subgrid-Scale Flame Wrinkling Modeling–AI Super-Resolution: Application to Turbulence and Combustion" give different perspectives on the use of ML for estimating filtered reaction rate. Data-driven approaches can also be leveraged for reduced-order modeling of turbulent combustion and this is discussed in the chapter "Reduced-Order Modeling of Reacting Flows Using Data-Driven Approaches". The use of ML for thermoacoustics is described in chapter "Machine Learning for Thermoacoustics". Some of these chapters are written in a tutorial fashion and also provide hyperlinks to access the associated computer codes. The concluding remarks and future directions are summarised in the final chapter. Each of the chapters provides ample references for further reading by curious readers.

The idea for this book came during a collaborative project, ALCHEMY (mAchine Learning for ComplEx MultiphYsics problems), between Cambridge University and ULB funded by Fondation Wiener-Anspach, ULB, Brussels. The funding from this foundation is gratefully acknowledged. We cannot understate the dedication of the contributors to this volume and we thank them for their contributions.

Cambridge, UK   Nedunchezhian Swaminathan
Brussels, Belgium   Alessandro Parente
May 2022

Contents

Contributors

Aktulga H. Michigan State University, East Lansing, USA

Blonigan P. J. Sandia National Laboratories, Livermore, CA, USA

Bode M. Jülich Supercomputing Centre, Forschungszentrum Jülich GmbH, Jülich, NRW, Germany;
Fakultät für Machinenwesen, RWTH Aachen University, Aachen, NRW, Germany

Carlson M. L. Sandia National Laboratories, Livermore, CA, USA

Chen Z. X. State Key Laboratory of Turbulence and Complex Systems, Aeronautics and Astronautics, College of Engineering, Peking University, Beijing, China;
Department of Engineering, University of Cambridge, Cambridge, UK

Chrysostomou C. The Cyprus Institute, Nicosia, Cyprus

Coussement A. Aero-Thermo-Mechanics Laboratory, École polytechnique de Bruxelles, Université Libre de Bruxelles, Brussels, Belgium;
Brussels Institute for Thermal-fluid Systems, Brussels (BRITE), Université Libre de Bruxelles and Vrije Universiteit Brussel, Brussels, Belgium

Davis IV W. L. Sandia National Laboratories, Albuquerque, NM, USA

Dunlavy D. M. Sandia National Laboratories, Albuquerque, NM, USA

Echekki T. North Carolina State University, Raleigh, NC, USA

Farooq A. King Abdullah University of Science and Technology, Thuwal, Saudi Arabia

Grama A. Purdue University, West Lafayette, USA

Iavarone S. Aero-Thermo-Mechanics Laboratory, Université Libre de Bruxelles, Brussels, Belgium;
Engineering Department, University of Cambridge, Cambridge, UK

Ihme M. Stanford University, Stanford, CA, USA

Juniper Matthew P. Engineering Department, University of Cambridge, Cambridge, UK

Karpe S. School of Aerospace Engineering, Georgia Institute of Technology, Atlanta, GA, USA

Kolla H. Sandia National Laboratories, Livermore, CA, USA

Lapeyre C. J. CERFACS, Toulouse, France

Li Z. Engineering Department, University of Cambridge, Cambridge, UK

Malik M. R. Aero-Thermo-Mechanics Laboratory, École polytechnique de Bruxelles, Université Libre de Bruxelles, Brussels, Belgium;
Brussels Institute for Thermal-fluid Systems, Brussels (BRITE), Université Libre de Bruxelles and Vrije Universiteit Brussel, Brussels, Belgium

Menon S. School of Aerospace Engineering, Georgia Institute of Technology, Atlanta, GA, USA

Minamoto Y. Department of Mechanical Engineering, Tokyo Institute of Technology, Meguro, Tokyo, Japan

Nikolaou Z. M. CORIA-CNRS, Normandie Université, INSA de Rouen, Normandy, France

Panchal A. School of Aerospace Engineering, Georgia Institute of Technology, Atlanta, GA, USA

Pandit S. University of South Florida, Tampa, USA

Parente A. Aero-Thermo-Mechanics Laboratory, École polytechnique de Bruxelles, Université Libre de Bruxelles, Brussels, Belgium;
Combustion and Robust Optimization Group (BURN), Université Libre de Bruxelles and Vrije Universiteit Brussel, Brussels, Belgium;
Brussels Institute for Thermal-fluid Systems, Brussels (BRITE), Université Libre de Bruxelles and Vrije Universiteit Brussel, Brussels, Belgium

Parish E. J. Sandia National Laboratories, Livermore, CA, USA

Ranjan R. Department of Mechanical Engineering, University of Tennessee at Chattanooga, Chattanooga, TN, USA

Ravindra V. Purdue University, West Lafayette, USA

Rizzi F. NexGen Analytics, Sheridan, WY, USA

Sarathy S. M. King Abdullah University of Science and Technology, Thuwal, Saudi Arabia

Shead T. M. Sandia National Laboratories, Albuquerque, NM, USA

Sutherland J. C. Department of Chemical Engineering, University of Utah, Salt Lake City, UT, USA

Swaminathan N. Hopkinson Laboratory, Department of Engineering, University of Cambridge, Cambridge, UK

Tencer J. Sandia National Laboratories, Albuquerque, NM, USA

Tezaur I. K. Sandia National Laboratories, Livermore, CA, USA

Vervisch L. CORIA-CNRS, Normandie Université, INSA de Rouen, Normandy, France

Xing V. CERFACS, Toulouse, France

Yang H. Engineering Department, University of Cambridge, Cambridge, UK

Zdybał K. Aero-Thermo-Mechanics Laboratory, École polytechnique de Bruxelles, Université Libre de Bruxelles, Brussels, Belgium;
Brussels Institute for Thermal-fluid Systems, Brussels (BRITE), Université Libre de Bruxelles and Vrije Universiteit Brussel, Brussels, Belgium

Introduction

N. Swaminathan and A. Parente

Abstract The annual data published by IEA is analysed to get a projection for the combustion share in total primary energy supply for the world. This projection clearly identifies that more than 60% of world total primary energy supply will come from combustion based sources even in the year of 2110 despite an aggressive shift towards renewables. Hence, improving and searching for greener combustion technologies would be beneficial for addressing global warming. Computational approaches play an important role in this search. The large eddy simulation equations are presented and discussed. Potential terms which are amenable for using machine learning algorithms are identified as a prelude to later chapters of this volume.

Combustion is a socio-economically important topic for many tens of centuries and it still remains to be so because more than 90% of the world's total primary energy supply (TPES) is met through combustion in one form or another, see IEA (2021). Even the recently proposed changes towards low carbon or carbonless fuels, including E-fuels, will involve some sort of combustion employing concepts and technologies which could be substantially different from those used currently. Figure 1 shows the share of various sources for TPES which is about 606 EJ for the year 2019. This is nearly 139% of the energy used in 1973 which suggests about 3% increase per year over the past 46 years and this is inline with an estimate of about 40% increase in the global energy consumption for the next two decades by the National Academies of Science, Engineering and Medicine, see How we use energy (2022). This projected energy demand is likely to be larger because of the widespread use of energy-hungry

N. Swaminathan (✉)
Hopkinson Laboratory, Department of Engineering, University of Cambridge, Trumpington Street, Cambridge CB2 1PZ, UK
e-mail: ns341@cam.ac.uk

A. Parente
Aero-Thermo-Mechanics Laboratory, École polytechnique de Bruxelles, Université Libre de Bruxelles, Brussels, Belgium
e-mail: alessandro.parente@ulb.be

Combustion and Robust Optimization Group (BURN), Université Libre de Bruxelles and Vrije Universiteit Brussel, Brussels, Belgium

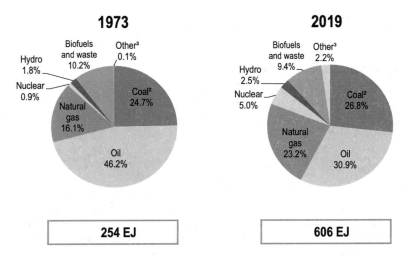

Fig. 1 World total primary energy supply, in exajoule, by source type. Adopted from IEA (2021), © IEA, 2021

consumer electronics and other technologies such as Internet of Things (IoTs), electric vehicles, etc. While these technologies bring their own advantages one cannot deny their impacts on the environment arising from their manufacturing, end-of-life treatments and more importantly higher demand for energy during their lifetime leading to global warming related issues. Indeed, the use of energy-hungry modern technologies and mitigation of global warming are at the opposite ends and bringing them together is a grand challenge requiring carefully constructed solutions.

The global temperature is expected to rise in the next 100 years according to the intergovernmental panel reports—Future climate changes, risks and impacts (2022), and as discussed by Hayhoe et al. (2017). If the emission of green house gases (GHG) follow a particular Representative Concentration Pathway (RCP 2.6) yielding Gigatons of carbon emission close to zero in the year of 2100 and the CO_2 concentration in the atmosphere is about 400 ppm then the temperature raise is expected to range from 0.3 to 1.7 °C. If the CHG emission is high following RCP 8.5 then the temperature rise may range from about 2.6 to 4.8 °C which may result in catastrophic effects.

The energy production using renewable and sustainable sources are gaining popularity and becoming wide spread in the past decade. The renewable sources include hydro, solar, wind and tidal. The nuclear energy may be considered as a renewable since the uranium deposits could provide energy for billion years (Cohen 1983) and there is no GHG emissions (Vasques 2014; Moore 2006). However, the safety issues and the concept of clean energy may exclude the nuclear energy from the renewables. Figure 1 shows that the share of this energy is 5% for the year 2019 whereas the renewables share, listed as Others, is only 2.2%. However, this substantial increase from 0.1% in 1973 is because of the advent of the renewable technologies in the recent past. The photo voltaic, both rooftop and commercial, systems become popular but the capital cost projections in Winskel et al. (2009) (see their Fig. 4.1) does not seem to be realistic (the actual cost is nearly twice the projected cost of about £1000 per kW for 2019) because the price will increase as the demand grows unless the supply is in surplus.

The levelised cost of electricity for renewable technologies at utility-scale is becoming lower than that for the traditional fossil fuels—0.038 to 0.076 USD/kWh depending on the renewables compared to 0.05 to 0.18 USD/kWh for fossil fuels (IRENA 2020)—which is an excellent progress. However, the consumer energy prices do not reflect this lower cost for the renewables yet. Perhaps, this may take some more time. Although the renewable power generation has increased by nearly 50% (a total of about 780 GW) for the year 2020 compared to 2019 (IRENA 2021), this is substantially lower than the 2019 projection of 1.5 TW for 2020 (IRENA 2019). This clearly suggests that the renewables share is growing slowly and one may have to accelerate it but the accelerated growth may have its own consequence on the environment for the reasons argued in Lørstad et al. (2022a), which are based on the data for GHG emissions and cradle-to-grave life cycle analysis (LCA) published in past studies. For example, the electric vehicles projected to have zero emission is not so in reality according to cave-to-case analysis showing that one has to drive a 110 kW size EV for about 35,000 km *without recharging* to offset the CO_2 emitted by the battery pack production alone (Alvarez 2019). This is not practical. It is likely that combustion will remain as one of the components in the energy technology mix and will play an important part for specific applications, such as transport and energy-intensive industries, requiring high energy densities but its form and type are likely to be different.

1 Combustion Technology Role

The mitigation of global warming requires solutions, targeted towards reducing GHG emissions, which arise from efforts concerted across various continents and countrywide solutions are inadequate. While a complete shift towards renewables seems attractive and achievable over longer timescales but the accelerated shift set by various governments independently does not sound pragmatic. Perhaps, this may worsen the situation because the additional energy required to achieve the accelerated shift towards renewables has to come from non-renewables. Thus, a balanced approach to meet the ever increasing energy without aggravating the global warming is needed.

Combustion technologies play important role in this respect as suggested by the results in Fig. 2 showing future projections for the combustion share of world TPES under three different scenarios (Swaminathan 2019). The inset is the actual data from the International Energy Agency (IEA 2021) showing a gradual decrease in the combustion share while a small rise in 2012 is because of the increase in coal combustion in some countries in that year. If one makes a naive projection by assuming that the progress in renewable technologies is steady and organic following the current trends

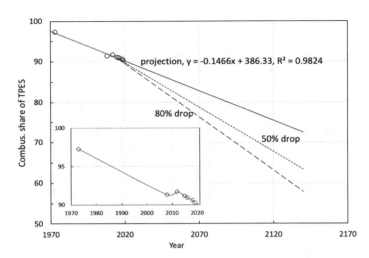

Fig. 2 Combustion share of world TPES and its future projections. Adapted from Swaminathan (2019)

then the combustion share will be more than 75% even by the year 2110 (the solid line). The slope of this curve is related to the progress and advancement of alternative energy technologies. If one keeps an optimistic view for these technologies and presumes that they are progressing at about 50% faster pace compared to the current trend then the combustion share falls to about 70% in 2110. This share decreases further to 66% for the year 2110 if one assumes that the alternative technologies progress at 80% faster pace. To achieve this, a radical paradigm shift is needed and whether this is practical or not from the economical consideration is an open question. Even the heavily accelerated shift (80% scenario) reduces the combustion share only by 40% and thus a pragmatic approach is to seek for alternative combustion concepts and technologies which can significantly reduce GHG emissions and can act as retrofits to the existing combustion systems which can also aid a quicker shift towards renewables in the longer run.

Many alternative combustion concepts such as fuel-lean and MILD (moderate, intense or low dilution) combustion emerge as potential solutions since they could deliver both low emissions and high efficiency. However, using them for practical applications bring their own challenges as discussed by Swaminathan and Bray (2011) and Lørstad et al. (2022a). Also, carbon-free and E-fuels are emerging as potential alternative solutions to mitigate CO_2 emission while catering to the ever-increasing energy demand. Specifically, hydrogen combustion seems to be gaining momentum with a view to use hydrogen as a main energy carrier. Although this solution addresses the CO_2 emission directly it brings additional challenges for its safe usage, controlled combustion for practical applications and potential increase in NO_x emissions. One of the current NO_x reduction technologies can be utilised to control this emission from hydrogen or E-fuel combustion. Nevertheless, the distribution of hydrogen from production sites to consumers is challenging which requires a complete infrastructure overhaul and the scale of economy for this cannot be underestimated adding further challenges.

Modern computational methods and approaches play significant parts in developing these alternative technologies and taking them to fruition. The use of machine learning algorithm (MLA) and techniques in computational fluid dynamics (CFD), specifically for turbulent flows and turbulent combustion are gaining renewed momentum in recent times for two reasons, *viz.,* (i) these algorithms and techniques have evolved and developed for a wide-spread use across various disciplines and (ii) to take advantage of their robustness, accuracy and computational efficiencies so that the CFD codes with MLA can be employed for quick evaluations of design changes. Before discussing the role of MLA in computational simulations of turbulent flows with chemical reactions, let us briefly review the governing principles and equations, and various computational methods used for turbulent combustion. The topic of turbulent reacting flow simulations has been discussed elaborately in many books, see for example Swaminathan and Bray (2011), Libby and Williams (1980), Poinsot and Veynante (2005), Echekki and Mastorakos (2011), Swaminathan et al. (2022b), only a brief review with detail required to fulfil the aim of this volume is discussed in the next section.

2 Governing Equations

The computational simulations of turbulent reacting flows use three numerical approaches, namely direct numerical simulation (DNS), large eddy simulation (LES) and Reynolds-Averaged Navier Stokes calculation (RANS). These approaches involve different levels of detail, approximations and modelling. The complete set of conservations equations are solved with no models using high order numerical schemes in the DNS approach and further detail can be found in many books, for example Poinsot and Veynante (2005). This approach resolves and captures the range of, from dissipative to energy containing, scales in the flow without using any modeling approximations and this range increases with turbulence Reynolds number, Re_t. The ranges of spatial and temporal scales vary as $Re_t^{3/4}$ and $Re_t^{1/2}$ respectively and thus the computational cost for using DNS at Re_t relevant for practical application in appropriate geometry is prohibitive. Hence, this approach is typically used to gain fundamental understanding of turbulence and its interaction with chemical reactions, and these knowledge are important for devising engineering models for practical use. There are many examples for this which are discussed and summarised in Swaminathan and Bray (2011), Poinsot and Veynante (2005), Echekki and Mastorakos (2011), Swaminathan et al. (2022b). Appropriately averaged conservation equations are solved in the RANS approach along with closure models and approximations, which are discussed elaborately in many past works, for example see the books edited by Libby and Williams (1980, 1994) and the works in Swaminathan and Bray (2011). The RANS equations are deterministic and do not have the stochastic aspects required for statistical inference and hence one must be cautious in using MLA for RANS calculations. However, it is possible to use some of the machine learning algorithms to address the uncertainties of RANS model parameters. LES approach is well suited to make use of MLA since there is inherent stochasticity. Before identifying the potential avenues to use MLA for LES, let us briefly review the required governing equations.

3 Equations for LES

In large eddy simulations, the low-pass filtered governing equations for mass, momentum, energy and species mass fractions are solved. The filtering, or separation of the scales, is done with a spatial filter, which is applied to the governing equations for the above quantities. The various filters and their attributes are discussed in many text books, for example see Pope (2000) and Favre-filtering, also known as density-weighted filtering, is commonly used for flows such as turbulent combustion involving strong density variations. The filtering implies that the dynamic large scales, which are larger than the filter cut-off scale, are resolved and the scales smaller than the cut-off scale, known as subgrid scales (SGS), are modelled. Hence, the computational cost for LES is much lower than that for DNS because coarser grids and larger time steps can be used for similar level of numerical fidelity.

The Favre-filtered governing equations are written as

$$\text{Mass:} \qquad \frac{\partial \overline{\rho}}{\partial t} + \nabla \cdot (\overline{\rho}\,\widetilde{\mathbf{u}}) = 0 \tag{1}$$

$$\text{Momentum:} \qquad \frac{\partial \overline{\rho}\,\widetilde{\mathbf{u}}}{\partial t} + \nabla \cdot (\overline{\rho}\,\widetilde{\mathbf{u}}\,\widetilde{\mathbf{u}}) = -\nabla \overline{p} + \nabla \cdot \overline{\tau} - \nabla \cdot \overline{\tau}^{S} \tag{2}$$

$$\text{Energy:} \quad \frac{\partial \overline{\rho}\,\widetilde{h}}{\partial t} + \nabla \cdot \left(\overline{\rho}\,\widetilde{\mathbf{u}}\,\widetilde{h}\right) = \frac{\mathrm{D}\overline{p}}{\mathrm{D}t} - \nabla \cdot \overline{\mathbf{q}} - \nabla \cdot \left(\rho \sum_{i=1}^{N_s} Y_i \mathbf{U}_i h_i\right)$$
$$+ \overline{\tau : \nabla \mathbf{u}} + \overline{Q_r} + \Pi_{\text{dil}} - \nabla \cdot \overline{\theta}^{S} \tag{3}$$

$$\text{Species:} \quad \frac{\partial \overline{\rho}\,\widetilde{Y_i}}{\partial t} + \nabla \cdot \left(\overline{\rho}\,\widetilde{\mathbf{u}}\widetilde{Y_i}\right) = \nabla \cdot \left(-\overline{\rho\,Y_i \mathbf{U}_i}\right) + \overline{\dot{\omega}_i} - \nabla \cdot \overline{\psi}_i^{S} \tag{4}$$

using standard notations and $\mathbf{U}_i$ is the diffusion velocity of species i.

The filtering procedure yields extra terms, SGS stress tensor $\overline{\tau}^{S}$, SGS enthalpy flux $\overline{\theta}^{S}$, SGS pressure-dilation Π_{dil}, and SGS species flux $\overline{\psi}_i^{S}$, given by

$$\overline{\tau}^S = \overline{\rho}\,(\widetilde{\mathbf{u}\mathbf{u}} - \widetilde{\mathbf{u}}\,\widetilde{\mathbf{u}}) \tag{5}$$

$$\overline{\theta}^S = \overline{\rho}\,(\widetilde{\mathbf{u}\mathbf{h}} - \widetilde{\mathbf{u}}\,\widetilde{h}) \tag{6}$$

$$\Pi_{\mathrm{dil}} = \overline{\mathbf{u}\cdot\nabla p} - \widetilde{\mathbf{u}}\cdot\nabla\overline{p} \tag{7}$$

$$\overline{\psi}_i^S = \overline{\rho}\,\left(\widetilde{\mathbf{u}\,Y_i} - \widetilde{\mathbf{u}}\,\widetilde{Y}_i\right) \tag{8}$$

These unknown quantities represent the influence of unresolved scales on the resolved scales and need closure models. The pressure-dilation in Eq. (7) is sometimes less important in compressible flows and therefore commonly neglected (Piomelli 1999; Martin et al. 2000). A plausible modelling for it and its limitation are explored in Langella et al. (2017). Closure models are required for all the SGS quantities in Eqs. (6) to (8), molecular diffusion related quantities in Eqs. (2) to (4) and the species reaction rate $\overline{\dot{\omega}}_i$. The molecular diffusion of momentum (viscous shear, $\overline{\tau}$), energy (heat flux, $\overline{\mathbf{q}}$), and species (diffusive flux, $-\overline{\rho\,Y_i\mathbf{U}_i}$) are modeled following classical ideas of gradient diffusion after neglecting the fluctuations in viscosity, diffusivity and heat conductivity (Piomelli 1999; Gicquel 2012). Further detail on these models and the LES governing equations are discussed in Pope (2000), Poinsot and Veynante (2005) and Garnier et al. (2009).

3.1   SGS Closures

A few common closures for the SGS terms given in Eqs. (6) to (8) are discussed briefly here. The eddy viscosity models are the most simple ones for the SGS stress in Eq. (5) and the popular of these is the classical Smagorinsky model (Smagorinsky 1963), which has been extended to the SGS kinetic energy by Yoshizawa (1986). The Smagorinsky model, in tensor notation, is

$$\tau_{ij}^S - \frac{\delta_{ij}}{3}\tau_{kk}^S = -2\,C_s^2\,\Delta^2\overline{\rho}\,|\widetilde{\mathbf{S}}|\left(\widetilde{S}_{ij} - \frac{\delta_{ij}}{3}\widetilde{S}_{kk}\right)$$

$$= -2\,\overline{\rho}\,\nu_{\mathrm{SGS}}\left(\widetilde{S}_{ij} - \frac{\delta_{ij}}{3}\widetilde{S}_{kk}\right), \quad \text{and} \tag{9}$$

$$\tau_{kk}^S = 2\,C_I\overline{\rho}\,\Delta^2|\widetilde{\mathbf{S}}|^2 \tag{10}$$

where $\widetilde{S}_{ij} = 0.5\left(\partial\widetilde{u}_i/\partial x_j + \partial\widetilde{u}_j/\partial x_i\right)$ is the resolved symmetric strain-rate tensor and $|\widetilde{\mathbf{S}}| = \sqrt{2\,\widetilde{S}_{ij}\widetilde{S}_{ij}}$. The filter width estimated typically using the local numerical cell volume is denoted as Δ. Equation (9) defines the SGS eddy viscosity, ν_{SGS}, and the symbols C_s and C_I are model constants. The τ_{kk}^S, which is twice the SGS kinetic energy, is likely to be small or negligible in low Mach number flows as noted by Martin et al. (2000) but may not be so for flows with strong heat release.

The Smagorinsky models is relatively simple and robust, but it has its limitation for near-wall and transition flows since it can give a non-vanishing eddy viscosity, which is unphysical and this can be remedied by invoking damping functions, but an alternative approach is to use a dynamical procedure to determine C_s and C_I as proposed in Moin et al. (1991). This approach is used widely by applying a second filter of typical width $\hat{\Delta} = 2\Delta$ to the resolved fields to compute the resolved stress near the filter cut-off. Assuming similarity of the stresses near the cut-off scale, Δ, this resolved stress can be used to find an expression for C_s and C_I in terms of the resolved velocity gradients, see Pope (2000), Martin et al. (2000) and Garnier et al. (2009).

The dynamic procedure allows the model to adapt itself to the local flow changes and hence ν_{SGS} naturally approaches zero near solid walls and in laminar regions which retains physical behaviour. The dynamic procedure can produce $\nu_{SGS} < 0$ implying an instantaneous reverse cascade of kinetic energy locally which may occur in turbulent flows. However, this can lead to numerical instabilities and therefore, it is common to clip C_s to avoid negative ν_{SGS} or by averaging it in either space or time.

Other algebraic approaches have also been developed in past studies to over this specific issue of ν_{SGS} not approaching zero near a wall in wall bounded flows. Details can be found in Vreman (2004), Nicoud and Ducros (1999), Nicoud et al. (2011). An alternative approach to estimate ν_{SGS} uses the SGS turbulent kinetic energy, $\tilde{k}_{SGS}$, obtained directly by using its transport equation, see Yoshizawa and Horiuti (1985) and Ghosal et al. (1995). Various approaches have also been proposed, developed and tested for the SGS stresses in many past studies and detail can be found in Zang et al. (1993), Lesieur and Métais (1996), Layton (1996), Kosovic (1997), Misra and Pullin (1997), Meneveau and Katz (1997), Armenio and Piomelli (2000), Domaradzki and Adams (2002), Chaouat and Schiestel (2005), Lucor et al. (2007).

Further to the SGS stress discussed above, the SGS fluxes needing modelling and a straightforward approach is to use an eddy diffusivity model written as

$$\overline{\psi}_i^S = \frac{-\overline{\rho}\,\nu_{SGS}}{Sc_{SGS}}\nabla\widetilde{Y}_i, \quad \text{and} \quad \overline{\theta}^S = \frac{-\overline{\rho}\,\nu_{SGS}}{Pr_{SGS}}\nabla\widetilde{h} \tag{11}$$

for species and enthalpy respectively. The symbols Sc_{SGS} and Pr_{SGS} are the SGS Schmidt and Prandtl numbers respectively. These quantities may be estimated using a static or dynamic procedure, see Martin et al. (2000), Garnier et al. (2009) and Moin et al. (1991). Many other models for the SGS stresses and fluxes have been developed and tested in past studies (Martin et al. 2000; Garnier et al. 2009; Silvis et al. 2017) and these models are introduced and discussed in later chapters, specifically in chapter "Machine-Learning for Stress Tensor Modelling in Large Eddy Simulation". The statistics obtained using these models could show some sensitivities to errors introduced by the numerical scheme, especially for second order statistics and thus some care is needed. Perhaps, one way to address these issues is to use MLA to estimate the model parameters, which is discussed in chapter "Machine-Learning for Stress Tensor Modelling in Large Eddy Simulation".

The chemical reaction rate in the species equation, Eq. (4), is important for turbulent combustion. The physical processes represented by this term typically occur at SGS level. Also, the reaction rate is a highly nonlinear function of temperature, T, and species mass fractions, Y_i, and, hence it cannot be expressed in a meaningful way using only the resolved temperature and species mass fractions. Formulating a robust yet accurate SGS closure for the reaction rate is challenging and important and this has been studied in past studies which are reviewed and summarised in many references, see for example Swaminathan and Bray (2011), Poinsot and Veynante (2005), Swaminathan et al. (2022b), Gicquel et al. (2012), Peters (2000), Pitsch (2006), Rutland (2011). Each of these approaches has their advantages and limitations in terms their predictive abilities, simplicity, ease of use, computational expenses, physical basis and these aspects are discussed in past works, for example see Swaminathan et al. (2022b). In the following, we give an brief overview on the challenges involved in LES and the role of MLA to tackle them which also helps us to articulate the objectives for this volume.

3.2   LES Challenges and Role of MLA

The SGS closures are predominantly based on the gradient flux hypothesis as discussed in the previous subsection and it is well known that in reacting flows there are processes which defy this hypothesis. Hence, modelling counter-gradient subgrid scalar fluxes are still an outstanding issue, specifically for low Reynolds number reacting flows. Despite this, LES calculations with the gradient flux models have shown good agrements between the computed and measured statistics suggesting that these models are sufficient for flows of interest to practical systems. Another challenge for LES is on the near-wall flow characteristics. It is quite well known that practical LES cannot recover the law of the wall and some special numerical treatments are required as noted by Nikitin et al. (2000) and Brasseur and Wei (2010). Recovering the law of the wall becomes important when the heat and momentum fluxes through the walls (of the combustor, for example) need to be evaluated as design variables.

It is observed generally that the numerical grids used for LES of reacting flows resolve instantaneous flame structure to some extend, which is acceptable for atmospheric pressure. High pressure flows in complex geometries are common in practical applications and thus resolving the instantaneous flame structure will likely to yield impractical grid cell counts because the flame thickness approximately scales as $\delta_{th} \sim p^{-1/2}$ (Turns 2006) and some of the important geometry detail need to be captured in the grid. Thus, the common practice of using grids having cell sizes of the order of δ_{th} is unattractive for practical LES. Consequently, SGS combustion models have to be robust and accurate in representing the relevant physical processes and machine learning algorithms can play important role here. Probably, it is useful to design or select a grid resolving most of the kinetic energy in the flow and let the SGS closures, specifically for combustion, to handle the turbulence-chemistry interactions and their intricacies for LES of reacting flows in practical systems. The guidance suggested by Pope (2000), which is $\mathcal{K} = k_{\text{sgs}}/(k_{\text{res}} + k_{\text{sgs}}) \leq 0.2$, where k_{sgs} and k_{res} are subgrid scale and resolved kinetic energies respectively, may be used. It is to be noted that this condition can only be evaluated after completing a preliminary LES of non-reacting flow in a given geometry. Alternative measures to evaluate LES grid requirement have also been suggested in past studies. However, the parameter $\mathcal{K}$ is quite practical and useful, and thus it is recommended. This requirement is to be applied for flows before igniting the flame and thus checking and satisfying this grid requirement are quite straightforward since the LES of non-reacting flow is the first step in conducting LES of turbulent combustion.

Machine learning algorithms can play a vital role in turbulent combustion calculations. These algorithms can be leveraged to build SGS models which can reduce computational requirements substantially. However, using MLA for these purposes are not common yet and there is a surge of research activities in this direction. The subgrid fluid dynamic and combustion processes and their interactions are highly non-linear stochastic events and thus MLA is well suited to infere the SGS statistics required for LES. Typically, machine learning methods are used for pattern recognition in various fields (Hinton et al. 2012; Sathiesh et al. 2016; Gogul and Sathiesh Kumar 2017) and are finding their ways into other fields such as climate modelling (Watson-Parris 2021), drug discovery (Bhati et al. 2021) and fluid mechanics (Brunton et al. 2020). Their application to reacting flows is gaining momentum although it is still at an early development and validation stage. Hence, the objective for compiling this volume is to bring together the latest developments in MLA and its application to chemically reacting flows and make it readily accessible for researchers and graduate students interested in this multi- and cross-disciplinary topic.

4   Objectives

The broad aim here is to bring together the recent developments in the field of MLA applied to reacting flow calculations. These flows in practical systems are invariably turbulent and hence there are three important aspects, *viz.,* turbulence, chemical reactions and their interactions, requiring close attention. The chemical reactions are because of molecular collisions but, at continuum level of description used commonly for turbulent reacting flow simulations, they are modelled using Arrhenius rate expressions involving kinetic parameters. These parameters, related to the atomic potential energies, are obtained typically using shock tube experiments but recent advances in ML techniques is helping to estimate these parameters using atomistic molecular dynamic simulations as described in chapter "Machine Learning Techniques in Reactive Atomistic Simulations". This chapter also gives an overview of various ML algorithms. One needs large data sets to train and validate these algorithm before using them for inferring quantities of interest and thus their robustness depends on the conditions covered in the data sets and hence these data sets can be huge. Hence one needs a clever and intelligent algorithm to detect events/patterns of interest in the data. Machine learning algorithms can come handy for this purpose as discussed in chapter "A Novel In Situ Machine Learning Framework for Intelligent Data Capture and Event Detection" suggesting an interesting idea—in situ training—to train MLA. The application of MLA to infer SGS stresses and fluxes are described in chapter "Machine-Learning for Stress Tensor Modelling in Large Eddy Simulation". The combustion chemistry is quite complex even for a simple fuel like methane or hydrogen and involves a large number of elementary reactions with disparate time and length scales. Hence integrating these reaction into numerical simulations of turbulent combustion can make the simulations to be prohibitively expensive. Machine learning can be leveraged to accelerate chemistry integration by helping us to understand combustion chemistry closely as described in chapter "Machine Learning for Combustion Chemistry". The third aspect, turbulence-chemistry interaction, of turbulent combustion noted above can be addressed using different modelling approaches which helps us to estimate the filtered reaction rate of a chemical species or a reaction progress variable depending on the modelling approach used. The application of machine learning algorithms to these approaches are discussed in chapters "Deep Convolutional Neural Networks for Subgrid-Scale Flame Wrinkling Modeling" to "AI Super-Resolution: Application to Turbulence and Combustion". Obeying constraints coming from physical conservation laws and requirements (for example species mass fractions have to positive or zero) can become an issue for machine learning methods and some extra care is required while defining the *cost function* needed in the training step for machine learning algorithms, see chapters "Machine Learning Techniques in Reactive Atomistic Simulations" and "AI Super-Resolution: Application to Turbulence and Combustion". The interaction between fluctuating heat release rate and pressure in turbulent combustion established inside a tube as in many practical combustion systems, for example gas turbines and rocket engines, will have thermoacoustic oscillations which can become an issue for safe operation of these systems if these oscillations are not controlled. Predicting these oscillations and their on-set are challenging machine learning algorithms can be applied to these problems as described in chapter "Machine Learning for Thermoacoustics". The concluding remarks are drawn in the final chapter.

Acknowledgements N. Swaminathan acknowledges the support from EPSRC through the grant EP/S025650/1.

References

Alvarez JAA (2019) Cave-to-case analysis of batteries for electric vehicles. Technical report, MPhil Thesis, University of Cambridge, Cambridge, UK

Armenio V, Piomelli U (2000) A lagrangian mixed subgrid-scale model in generalized coordinates. Flow Turbul Combust 65(1):51–81

Bhati AP, Wan S, Alfé D, Clyde AR, Bode M, Tan L, Titov M, Merzky A, Turilli M, Jha S, Highfield RR, Rocchia W, Scafuri N, Succi S, Kranzlmü D, Mathias G, Wifling D, Donon Y, Di Meglio A, Vallecorsa S, Ma H, Trifan A, Ramanathan A, Brettin T, Partin A, Xia F, Duan X, Stevens R, Coveney PV (2021) Pandemic drugs at pandemic speed: infrastructure for accelerating COVID-19 drug discovery with hybrid machine learning- and physics-based simulations on high-performance computers. Interf Focus 11:20210018

Brasseur JG, Wei T (2010) Designing large-eddy simulation of the turbulent boundary layer to capture law-of-the-wall scaling. Phys Fluids 22:021303-1-21

Brunton SL, Noack BR, Koumoutsakos P (2020) Machine learning for fluid mechanics. Ann Rev Fluid Mech 52:477-508

Chaouat B, Schiestel R (2005) A new partially integrated transport model for subgrid-scale stresses and dissipation rate for turbulent developing flows. Phys Fluids 17(6):065106-1-065106-9

Cohen BL (1983) Breeder reactors: a renewable energy source. Am J Phys 51:75-76

Domaradzki JA, Adams NA (2002) Direct modelling of subgrid scales of turbulence in large eddy simulations. J Turbul 3(24):N24

Echekki T, Mastorakos E (eds) (2011) Turbulent combustion modeling: advances. New trends and perspectives. Springer, Heidelberg

Future climate changes, risks and impacts. https://ar5-syr.ipcc.ch/topic_futurechanges.php. Accessed 01 Sept 2021

Garnier E, Adams N, Sagaut P (2009) Large eddy simulation for compressible flows. Springer

Ghosal S, Lund TS, Moin P, Akselvoll K (1995) A dynamic localization model for large-eddy simulation of turbulent flows. J Fluid Mech 286:229-255

Gicquel LYM, Staffelbach G, Poinsot T (2012) Large eddy simulations of gaseous flames in gas turbine chambers. Prog Energy Combust Sci 38:782-817

Gogul I, Sathiesh Kumar V (2017) Flower species recognition system using convolution neural networks and transfer learning. In: Proceedings of 4th international conference on signal processing, communications and networking, ICSCN-2017, Chennai, India, 2017. IEEE

Hayhoe K, Edmonds J, Kopp RE, LeGrande AN, Sanderson BM, Wehner MF, Wuebbles DJ (2017) Climate models, scenarios, and projections. In: Wuebbles DJ, Fahey DW, Hibbard KA, Dokken DJ, Stewart BC, Maycock TK (eds) Climate science special report: fourth national climate assessment, vol I. US Global Change Research Program, Washington, pp 133-160

Hinton G, Deng L, Yu D, Dahl GE, Mohamed A-R, Jaitly N, Senior A, Vanhoucke V, Nguyen P, Sainath TN, Kingsbury B (2012) Deep neural networks for acoustic modeling in speech recognition: the shared views of four research groups. IEEE Sig Proces Mag 82-97

How we use energy. http://needtoknow.nas.edu/energy/energy-use/. Accessed 10 Jan 2022

IEA (2021) Key world energy statistics. Technical report, International Energy Agency, Paris

IRENA (2019) Renewable capacity statistics 2019. Technical report, International Renewable Energy Agency, Abu Dhabi, 2019

IRENA (2020) Renewable power generation costs in 2020. Technical report, International Renewable Energy Agency, Abu Dhabi, 2021

IRENA (2021) Renewable capacity statistics 2021. Technical report, International Renewable Energy Agency, Abu Dhabi, 2021

Kosovic B (1993) Subgrid-scale modelling for the large-eddy simulation of high-reynolds-number boundary layers. J Fluid Mech 336(2):151-182

Langella I, Swaminathan N, Gao Y, Chakraborty N (2017) Large eddy simulation of premixed combustion: sensitivity to subgrid scale velocity modeling. Combust Sci Technol 189:43-78

Layton WJ (1993) A nonlinear subgrid-scale model for incompressible viscous flow problems. SIAM J Sci Comput 17(2):347-357

Lesieur M, Métais O (1996) New trends in large-eddy simulations of turbulence. Ann Rev Fluid Mech 28:45-82

Libby PA, Williams FA (eds) (1980) Turbulent Reacting Flows. Springer, New York

Libby PA, Williams FA (eds) (1994) Turbulent reacting flows. Academic Press, New York

Lucor D, Meyers J, Sagaut P (2007) Sensitivity analysis of large-eddy simulations to subgrid-scale-model parametric uncertainty using polynomial chaos. J Fluid Mech 585(6):255–279

Martin MP, Piomelli U, Candler GV (2000) Subgrid-scale models for compressible large-eddy simulations. Theoret Comput Fluid Dyn 13:361–376

Meneveau C, Katz J (1997) Scale-invariance and turbulence models for large-eddy simulation. Ann Rev Fluid Mech 32(1):1–32

Misra A, Pullin DI (1997) A vortex-based subgrid stress model for large eddy simulation. Phys Fluids 9(8):2443–2454

Moin P, Squires K, Cabot W, Lee S (1991) A dynamic subgrid-scale model for compressible turbulence and scalar transport. Phys Fluids 3(11):2746–2757

Moore P (2006) Nuclear Re-Think. IAEA Bull 48:56–58

Nicoud F, Ducros F (1999) Subgrid-scale stress modelling based on the square of the velocity gradient tensor. Flow Turbul Combust 62(3):183–200

Nicoud F, Baya Toda H, Cabrit O, Bose S, Lee J (2011) Using singular values to build a subgrid-scale model for large eddy simulations. Phys Fluids 23(8):085106

Nikitin NV, Nicoud F, Wasistho B, Squires KD, Spalart PR (2000) An approach to wall modeling in large-eddy simulations. Phys Fluids 12:1629–1632

Peters N (2000) Turbulent combustion. Cambridge University Press, Cambridge, UK

Piomelli U (1999) Large-eddy simulation: achievements and challenges. Prog Aerospace Sci 35:335–362

Pitsch H (2006) Large-eddy simulation of turbulent combustion. Ann Rev Fluid Mech 38:453–482

Poinsot T, Veynante D (2005) Theoretical and numerical combustion. RT Edwards, Inc

Pope SB (2000) Turbulent flows. Cambridge University Press, Cambridge, UK

Rutland CJ (2011) Large-eddy simulations for internal combustion engines—a review. Int J Engine Res 12:421–451

Sathiesh Kumar V, Gogul I, Deepan Raj M, Pragadesh SK, Sebastin JS (2016) Smart autonomous gardening rover with plant recognition using neural networks. Procedia Comput Sci 93:975–981

Silvis MH, Remmerswaal RA, Verstappen R (2017) Physical consistency of subgrid-scale models for large-eddy simulation of incompressible turbulent flows. Phys Fluids 29:015105

Smagorinsky J (1963) General circulation experiments with the primitive equations. I. the basic experiment. Mon Weather Rev 91:99–164

Swaminathan N (2019) Physical insights on MILD combustion from DNS. Front Mech Eng—Therm Mass Trans 5:Article 59

Swaminathan N, Bai X-S, Brethouwer G, Haugen NEL (2022a) Introduction. In: Swaminathan N, Bai X-S, Haugen NEL, Fureby C, Brethouwer G (eds) Advanced turbulent combustion physics and applications. Cambridge University Press, Cambridge, UK, pp 1–24

Swaminathan N, Bai X-S, Brethouwer G, Haugen NEL (eds) (2022b) Advanced turbulent combustion physics and applications. Cambridge University Press, Cambridge, UK

Swaminathan N, Bray KNC (Eds) (2011) Turbulent premixed flames. Cambridge University Press, Cambridge, UK

Turns SR (2006) An introduction to combustion: concepts and applications, 2nd edn. McGraw-Hill International Editions, Singapore

Vasques R (2014) Breeder reactors: a renewable energy source. Energy Res J 5:33–34

Vreman AW (2004) An eddy-viscosity subgrid-scale model for turbulent shear flow: algebraic theory and applications. Phys Fluids 16(10):3670–3681

Watson-Parris D (2021) Machine learning for weather and climate are worlds apart. Phil Trans R Soc A 379:20200098

Winskel M, Markusson N, Moran B, Jeffrey H, Anandarajah G, Hughes N, Candelise C, Clarke D, Taylor G, Chalmers H, Dutton G, Howarth P, Jablonski S, Kalyvas C, Ward D (2009) Decarbonising the UK energy system: accelerated development of low carbon energy supply technologies. Technical report, UK Energy Research Centre, Edinburgh University

Yoshizawa A (1986) Statistical theory for compressible turbulent shear flows, with the application to subgrid modeling. Phys Fluids 29(7):2152–2164

Yoshizawa A, Horiuti K (1985) A statistically-derived subgrid-scale kinetic energy model for the large-eddy simulation of turbulent flows. J Phys Soc Jpn 54(8):2834–2839

Zang Y, Street RL, Koseff JR (1993) A dynamic mixed subgrid-scale model and its application to turbulent recirculating flows. Phys Fluids 5(12):3186–3196

Machine Learning Techniques in Reactive Atomistic Simulations

H. Aktulga, V. Ravindra, A. Grama, and S. Pandit

Abstract This chapter describes recent advances in the use of machine learning techniques in reactive atomistic simulations. In particular, it provides an overview of techniques used in training force fields with closed form potentials, developing machine-learning-based potentials, use of machine learning in accelerating the simulation process, and analytics techniques for drawing insights from simulation results. The chapter covers basic machine learning techniques, training procedures and loss functions, issues of off-line and in-lined training, and associated numerical and algorithmic issues. The chapter highlights key outstanding challenges, promising approaches, and potential future developments. While the chapter relies on reactive atomistic simulations to motivate models and methods, these are more generally applicable to other modeling paradigms for reactive flows.

1 Introduction and Overview

Time-dependent reactive simulations involve complex interaction models that must be trained using experimental or highly resolved simulation data. The training process as well as data acquisition are often computationally expensive. Once trained, the coupling models are incorporated into reactive simulation procedures that involve small time-steps, and generate large amounts of data that must be effectively ana-

H. Aktulga
Michigan State University, East Lansing, USA
e-mail: hma@msu.edu

V. Ravindra · A. Grama (✉)
Purdue University, West Lafayette, USA
e-mail: ayg@cs.purdue.edu

V. Ravindra
e-mail: ravindvm@ucmail.uc.edu

S. Pandit
University of South Florida, Tampa, USA
e-mail: pandit@usf.edu

© The Author(s) 2023
N. Swaminathan and A. Parente (eds.), *Machine Learning and Its Application to Reacting Flows*, Lecture Notes in Energy 44, https://doi.org/10.1007/978-3-031-16248-0_2

lyzed for drawing scientific insights. The past few decades have witnessed significant advances in each of these facets. More recently, increasing attention has been focused on the development and application of machine learning (ML) techniques for increasing the accuracy, generalizability, and speed of such simulations.

In this chapter, we provide an overview of ML models and methods, along with their use in reactive particle simulations. We use highly resolved reactive atomistic simulations as the model problem for motivating and describing ML methods. We start by first presenting an overview of common ML techniques that are broadly used in the field. We then present the use of these techniques in training interaction models for reactive atomistic simulations. Recent work has focused on overcoming the time-step constraints of conventional reactive atomistic methods—we describe these methods and survey key results in the area. Finally we discuss the use of ML techniques in analyzing atomistic trajectories. The goal of the Chapter is to provide readers with a broad understanding of the state of the art in the area, unresolved challenges, and available methods and software for constructing simulations in diverse application domains. While we use reactive atomistics as our model problem, the discussion is broadly applicable to other particle-based/ discrete element simulation paradigms.

Reactive atomistic simulations provide understanding of chemical processes at the atomic level, which are usually not accessible through common experimental techniques. Quantum chemistry methods have come a long way in modeling electronic structures and subsequent chemical changes at the scale of a few atoms. However, if the interest is in the *thermodynamics* of chemical reactions then atomistic techniques are the methods of choice. Here, individual reactions are modeled in an approximate sense but system size (or particle number) approaches thermodynamic limit (or a suitable approximation thereof, i.e., as large as practical). One of the simplest sampling techniques used in atomistic simulations is molecular dynamics, which provides a psuedo-Newtonian trajectory of the system, and is applicable in modeling equilibrium as well as non-equilibrium problems. There are other sampling techniques such as Monte Carlo methods which are exclusively applicable to equilibrium statistical mechanical models. In this Chapter, we primarily focus on reactive molecular dynamics techniques.

1.1 Molecular Dynamics, Reactive Force Fields and the Concept of Bond Order

Molecular Dynamics (MD) is a widely adopted method for studying diverse molecular systems at an atomistic level, ranging from biophysics to chemistry and material science. While quantum mechanical (QM) models provide highly accurate results, they are of limited applicability in terms of spatial and temporal scales. MD simulations rely on parameterized force fields that enable the study of larger systems (with millions to billions of degrees of freedom) using atomistic models that are compu-

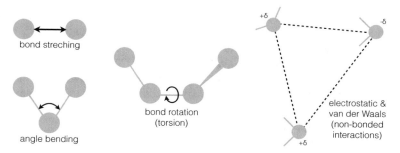

Fig. 1 Various classical force field interactions employed in atomistic MD simulations

tationally tractable and scalable on large computer systems. Typical applications of MD range from computational drug discovery to design of new materials.

MD is an active field in terms of the development of new techniques. In its most conventional form (i.e., classical MD), it relies on "Born-Oppenheimer approximation", where atomic nuclei and the core electrons together are treated as classical point particles and the interactions of outer electrons are approximated by pairwise and "many-body" terms such as bond, angle, torsion and non-bonded interactions, and additionally by using variable charge models. Each interaction is described by a parametric mathematical formula to compute relevant energies and forces. The collection of various interactions used to describe a molecular system is called a *force field*. Figure 1 illustrates interactions commonly used in various force fields. Equation 1 gives an example of a simple force field where K_b, r_0, K_a, θ_0, V_d, ϕ_0, ϵ_{ij}, v and σ_{ij} denote parameters that are specific to the types of interacting atoms (which may be a pair, triplet, or quadruplet of atoms), and ϵ denotes some global parameter.

$$V_{tot} = \sum_{bonds} K_b(r - r_0)^2 + \sum_{angles} K_a(\theta - \theta_0)^2 + \sum_{torsions} \frac{V_d}{2}[1 + cos(v\phi - \phi_0)]$$
$$+ \sum_{nonbonded} \frac{\delta_i \delta_j}{4\pi \epsilon r} + \sum_{nonbonded} 4\epsilon_{ij}[(\frac{\sigma_{ij}}{r})^{12} - (\frac{\sigma_{ij}}{r})^6] \qquad (1)$$

Classical MD models, as implemented in highly popular MD software such as Amber (Case et al. 2021), LAMMPS (Thompson et al. 2022), GROMACS (Hess et al. 2008) and NAMD (Phillips et al. 2005), are based on the assumption of static chemical bonds and, in general, static charges. Therefore, they are not applicable to modeling phenomena where chemical reactions and charge polarization effects play a significant role. To address this gap, reactive force fields (e.g., ReaxFF, Senftle et al. (2016), REBO, Stuart et al. (2000), Tersoff (1989)) have been developed. Functional forms for reactive potentials are significantly more complex than their non reactive counterparts due to the presence of dynamic bonds and charges. The development of an accurate force field (be it non-reactive or reactive) is a tedious task that relies heavily on biological and/or chemical intuition. More recently, machine learning based potentials have been proposed to alleviate the burden of force field design and fitting. Even so, the most computationally efficient way to study a large reactive molecular system, as would be necessary in a reactive flow application, is a well tuned reactive force field model. Hence, this Chapter focuses on reactive force fields and specifically on ReaxFF whenever it is necessary to discuss specific methods and results, since covering all reactive force field models would necessitate a significantly longer discussion. Nevertheless, models and methods discussed for ReaxFF are broadly applicable to other reactive force fields, as well.

Bond order is a key concept in reactive simulations; it models the overlap of electronic orbitals. This is intrinsically ambiguous in classical simulations because of approximations in assigning bond index and the bond type based on the wave function overlaps (Dick and Freund 1983). In classical reactive simulations, bond order is defined as a smooth function that vanishes with increasing distance between the atoms (van Duin et al. 2001). Clearly, such a function must depend on the environment of the atoms to correctly reproduce valencies. In non-reactive classical simulations, bond structure is maintained by either applying constraints on where a bond is expected to exist, or by assigning a large energy penalty (typically in the form of a harmonic potential, see e.g. Eq. (1)) if the atoms deviate from the expected bond length (Frenkel and Smit 2002). In either case, an improperly optimized force field can lead to divergent energies or break-down of the constraint algorithms. Reactive systems, however, have bond orders that smoothly go to zero, and usually do not have this problem but may end up with an un-physical final structure. Recently proposed ML-based approaches depend only on the atomic positions and sometimes on momenta, but do not carry information on molecular topology. Consequently, such approaches are well-suited for describing reactive simulations.

1.2 Accuracy, Complexity, and Transferability

Three key aspects must be considered when formulating simulation models: (i) *Accuracy:* A simulation is expected to reproduce structure as well as the chemical reactions and reaction rates for the model system against the target data. If a model has a sufficient number of free parameters, then, in principle, such model can accurately describe the physical system. However, the choice of model and its size depend on the availability of target training data, which are usually highly-resolved quantum chemistry calculations ranging from Density Functional Theory (DFT) to coupled cluster theory, along with a basis sets specifying the desired level of accuracy; (ii) *Complexity:* For any simulation model the complexity increases with the number of terms and free parameters in force computations (Frenkel and Smit 2002). Thus, accuracy of the model goes hand in hand with its complexity. Ideally, we would like to have a high accuracy and low complexity model. Consequently, a clever use of target data for extracting accurate results from a relatively simple model or alternately, approximations that represent minimal compromise on accuracy for significant reduction in model complexity are desirable; and (iii) *Transferability:* The models are expected to provide physical insight into the system by reproducing correct properties for different types of systems beyond the training data. This is usually achieved by breaking down the interaction terms into corresponding physical concepts, e.g., bond interaction, angle interaction, shielded 1–4 interaction, etc. Each of these interactions, although suitably abstracted, represent a physical concept that is expected to have similar interaction behavior under different conditions. Thus the total interaction can be computed as a combination of such transferable terms (Frenkel and Smit 2002). We note that the target data (usually obtained using quantum calculations) are not split into such physical abstractions. This gives rise to numerous models with similar accuracy and varying degrees of transferability. Commonly used reactive potentials such as REBO or ReaxFF are built with tranferability as a key consideration. However, even within the limited domain of atomic types and environments, these simulations rarely produce accurate results for wide variety of problems without requiring a re-tuning of the force field parameters. Unlike fixed form potential simulations, machine learnt potentials focus on tranferability of the model to similar atomic enviroments as the training datasets and optimize for higher accuracy as well as lower complexity.

In the rest of this chapter, we describe how reactive interaction models are constructed, trained, and used in accelerating simulations, in particular by making use of ML-based techniques. We begin our discussion with an overview of common ML models and methods, followed by their use in the simulation toolchain.

2 Machine Learning and Optimization Techniques

We begin our discussion with an overview of general ML techniques. This literature is vast and rapidly evolving. For this reason, we restrict ourselves to common ML techniques as they apply to reactive particle-based simulations.

ML frameworks are typically comprised of a model, a suitably specified cost function, and a training set over which the cost function is minimized. An ML model corresponds to an abstraction of the physical system—e.g., the force on an atom in its atomic context, and has a number of parameters that must be suitably instantiated. The cost function corresponds to the mismatch between the output of the model and physical (experimental or high-resolution simulated) data. Minimizing the cost function yields the necessary parametrization of the model. Training data is used to match the model output with target distribution. At the heart of ML procedures is the optimization technique used to match the model output with the target distribution.

The cost-function in typical ML applications is averaged over the training set:

$$J(\theta) = \mathbb{E}_{(x,y) \sim \hat{P}_{data}} \mathbb{L}[f(x; \theta), y] \tag{2}$$

Here, $J(.)$ represents the cost-function, P_{data} represents the empirical distribution (i.e., the training set), $L(.)$ is the loss-function that quantifies the difference between estimated and true value, and $f(.)$ is a prediction function parameterized by θ.

A key point to note here is that we operate on empirical data, and not the "true" data distribution. Hence, this approach is also called *empirical risk minimization* (Vapnik 1991). The assumption is that minimizing the loss w.r.t. empirical data will (indirectly) minimize the loss w.r.t. true data distribution, thereby allowing for generalizability (i.e., to make predictions on unseen data samples). In the rest of this section, we discuss continuous and discrete optimization strategies commonly used in ML formulations.

2.1 Continuous Optimization for Convex and Non-convex Optimization

In many applications, the objective function in Eq. 2 is continuous and differentiable. For such applications, a key consideration is whether the function is convex or non-convex (recall that a real-valued convex function is one in which the line joining any two points on the graph of the function does not lie below the graph at any point in the interval between the two points). Simple approaches to optimizing convex functions start from an initial guess, compute the gradient, and take a step along the gradient. This process is repeated until the gradient is sufficiently small (i.e., the function is close to its minima). In ML applications, the step size is determined by the gradient and the learning rate—the smaller the gradient, the lower the step size. Convex objective functions arise in models such as logistic regression and single layer neural networks.

In more general ML models such as deep neural networks, the objective function (Eq. 2) is not convex. Optimizing non-convex objective functions in high dimensions is a computationally hard problem. For this reason, most current optimizers use a gradient descent approach (or its variant) to find a local minima in the objective function space. It is important to note that a point of zero gradient may be a local minima or a saddle point. Common solvers rely on randomization and noise introduced by sampling to escape saddle points. In deep learning applications, the problem of computing the gradient can be elegantly cast as a backpropagation operator—making it computationally simple and inexpensive. Optimization methods that use the entire training set to compute the gradient are called batch or deterministic methods (Rumelhart et al. 1986). Methods that operate on small-subsets of the dataset (called minibatches) are called stochastic methods. In this context, a complete pass over the training dataset sampled in minibatches is called an epoch. Stochastic Gradient Descent (SGD) methods are workhorses for training deep neural network models.

First order methods such as SGD suffer from slow convergence, lack of robustness, and need for tuning a large number of hyperparameters. Indeed, model training using SGD-type methods incurs most of its computation cost in exploring the high-dimensional hyperparameter space to find model parametrizations with high accuracy and generalization properties (Goodfellow et al. 2014). These problems have motivated significant recent research in the development of second order methods and their variants. Second order methods scale different components of the gradient suitably to accelerate convergence. They also typically have much fewer hyperparameters, making the training process much simpler. However, these methods involve a product with the inverse of the dense Hessian matrix, which is computationally expensive. Solutions to these problems include statistical sampling, low-rank structures, and Kronecker products as approximations for the Hessian.

2.2 Discrete Optimization

In contrast to continuous optimization, in many applications, the variables and the objective function take discrete values, and thus the derivative of the objective function may not exist. This is often the case when optimizing parameters for force fields in atomistic models. Two major classes of techniques for discrete optimization are Integer Programming and Combinatorial Optimization. In Integer Programming, some (or all) variables are restricted to the space of integers, and the goal is to minimize an objective subject to specified constraints. In combinatorial optimization, the goal is to find the optimal object from a set of feasible discrete objects. Combinatorial optimization functions operate on discrete structures such as graphs and trees. The class of discrete optimization problems is typically computationally hard.

A commonly used discrete optimization procedure in optimization of force fields is genetic programming (Katoch et al. 2021; Mirjalili 2019). Genetic programming starts with a population of potentially suboptimal candidate solutions. It successively selects from this population (formally called selection) and combines them (formally called crossover) to generate new candidates. In many variants, mutations are introduced into the candidates to generate new candidates as well. A fitness function is used to screen these new candidates and the fittest candidates are retained in the population. This process is repeated until the best candidates achieve desired fitness. In the context of force-field optimization, the process is initialized with a set of parametrizations. The fitness function corresponds to the accuracy with which the candidate reproduces training data. The crossover function generates new candidates through operations such as exchange of corresponding parameters, min, max, average, and other simple operators.

3 Machine Learning Models

While the field of ML is vast, it is common to classify ML algorithms into "supervised" and "unsupervised". In supervised learning algorithms, training data contain both *features* and *labels*. The goal is to learn a function that takes as input a feature vector and returns a predicted label. Supervised learning can further be categorized into classification and regression. When labels are categorical, the learning task is commonly called "classification". On the other hand, if the task is to predict a continuous numerical value, it is called regression. In unsupervised learning algorithms, training data do not have labels. The goal of unsupervised algorithms is to analyze patterns in data without requiring annotation. Common examples of unsupervised algorithms include clustering and dimensionality reduction. We note that there are many other active areas of ML, such as reinforcement learning and semi-supervised learning that are beyond the scope of this chapter. We refer interested readers to more exhaustive sources for a comprehensive discussion (Bishop and Nasrabadi 2006; Murphy 2012; Shalev-Shwartz and Ben-David 2014; Goodfellow et al. 2016).

3.1 Unsupervised Learning

The most commonly used unsupervised learning techniques are clustering and dimensionality reduction.

3.1.1 Clustering

In *clustering*, data represented as vectors are grouped together on the basis of some inherent structures (or patterns), typically characterized by their similarities or distances (Saxena et al. 2017; Gan et al. 2020). Clustering algorithms can be categorized on the basis of their outputs into: (i) crisp versus overlapping; or (ii) hard versus soft. In crisp clustering, each data point is assigned to exactly one cluster, whereas overlapping clustering algorithms allow for multiple memberships for each data point. In hard clustering algorithms, a data-point is assigned a 0/1 membership to every cluster (a 1 corresponding to the cluster the point is assigned to). In soft clustering algorithms, each data point is assigned membership grades (typically in a 0–1 range) that indicate the degree to which data points belong to each cluster. If the grades are convex (i.e., they are positive and sum to 1), then the grades can be interpreted as probabilities with which a data point belongs to each of the classes. In the general class of fuzzy clustering algorithms (Ruspini 1969), the convexity condition is not required.

Centroid-based clustering refers to algorithms where each cluster is represented by a single, "central" point, which may not be a part of the dataset. The most commonly used algorithm for centroid-based clustering (and indeed all of clustering) is k-means algorithm of Lloyd (1982). Given a set of data-points $[\mathbf{x}_1, \mathbf{x}_2, \ldots, \mathbf{x}_n]$, and predefined number of clusters k, the objective function of k-means is given by:

$$\underset{\mathbf{C}}{\arg\min} \quad \sum_{i=1}^{k} \sum_{x \in \mathbf{C}_i} ||\mathbf{x} - \mu_i|| \tag{3}$$

where, $\mathbf{C}$ is the union of non-overlapping clusters ($\mathbf{C} = \{\mathbf{C}_1, \mathbf{C}_2, \ldots, \mathbf{C}_k\}$), and μ_i represents the mean of all data-points of belonging to cluster i. Stated otherwise, the objective of k-means clustering is to minimize the distance between data-points and their assigned clusters (as represented by the mean). The problem of k-means is NP hard, but approximation algorithms such as Lloyd's Algorithm can efficiently find local optima.

Distribution-based clustering algorithms work on the assumption that data-points belonging to the same cluster are drawn from the same distribution. Common algorithms in this class assume that data follow Gaussian Mixture Models, and typically solve the problem using the Expectation-Maximization (EM) Approach. EM does maximum likelihood estimation in the presence of latent variables. In each iteration, there are two steps. In the first step, latent variables are estimated (E-step). This is followed by the Maximization (M-step) where parameters of the models are optimized to better fit the data. In fact, the aforementioned Lloyd's algorithm for k-means clustering is a simple instance of EM.

Density-based clustering is a class of spatial-clustering algorithms, in which a cluster is modeled as a dense region in data space that is spatially separated from other clusters. Density-based spatial clustering of applications with noise (DBSCAN) by Ester et al. (1996) is the most commonly used algorithm in this class. DBSCAN requires two parameters: (i) ϵ-size of neighborhood; and (ii) Minpts—minimum number of points in each cluster. DBSCAN proceeds as follows—first, it finds all points that are in the ϵ-neighborhood of all points. Then, it designates points with more than Minpts neighbors as "core-points". Next, it finds connected components of core-points by inspecting the neighbors of each core-point. Finally, each non-core-point is assigned to the cluster if it is in an ϵ neighborhood. If a data-point is not in the neighborhood, it is identified as an outlier, or noise (Schubert et al. 2017).

Hierarchical clustering refers to a family of clustering algorithms that seeks to build a hierarchy of the clusters (Maimon and Rokach 2005). The two common approaches to build these hierarchies are bottom-up and top-down. In bottom-up (or agglomerative) clustering, each data-point initially belongs to a separate cluster. Small clusters are created on the basis of similarity (or proximity). These clusters are merged repeatedly until all data-points belong to a single cluster. The reverse process is performed in the top-down (or divisive) clustering approaches, where a single cluster is split repeatedly until each data-point is its own cluster. The main parameters to choose are the metric (i.e., the distance measures), and the linkage criterion. Commonly used metrics are L-1, L-2 norms, Hamming distance, and inner products. Linkage criterion quantifies distance between two clusters on the basis of distances between pairs of points across the clusters.

3.1.2 Dimensionality Reduction

Dimensionality reduction is an unsupervised technique common to many applications. Reducing dimensions produces a parsimonious denoised representation of data that is amenable to analysis by complex algorithms that would otherwise not be able to handle large amounts of raw data.

Linear Dimensionality Reduction Techniques

Principal component analysis (PCA) is perhaps the most commonly used linear dimension reduction technique. Principal components correspond to directions of maximum variation in data. Projecting data on to these directions, consequently, maintains dominant patterns in data. The first step in PCA is to center the data around zero mean to ensure translational invariance. This is done by computing the mean of the rows of data matrix M and subtracting it from each row to give a zero-centered data matrix M'. A covariance matrix is then computed as the nromalized form of $M'^T M'$. Note that the (i, j)th element of this covariance matrix is simply the covariance of the ith and jth rows of matrix M'. The dominant directions in this covariance matrix are then computed as the dominant eigenvectors of this matrix. Selecting the k dominant eivengectors and projecting the data matrix M to this subspace yields a k dimensional data matrix that best preserves variances in data. A common approach to selecting k is to consider the drop in magnitude of corresponding eigenvalues. PCA has several advantages: (i) by reducing the effective dimensionality of data, it reduces the cost of downstream processing; (ii) by retaining only the dominant directions of variance, it denoises the data; and (iii) it provides theroetical bounds on loss of accuracy in terms of the dropped eigenvalues.

The general class of dimensionality reduction techniques also includes other matrix decomposition techniques. In general, these techniques express matrix data M as an approximate product of two matrices $U V^T$; i.e., they minimize $||M - U V^T||$. Various methods impose different constraints on matrices U and V, leading to a general class of methods that range from dimension reduction to commonly used clustering techniques. Perhaps, the best-known technique in this class is Singular Value Decomposition (SVD) (Golub and Reinsch 1971), which is closely related to PCA, where columns of U and V are orthogonal, and rank-k for some value of k. The orthogonality of the column space of these matrices makes them hard to interpret directly in the data space.

In contrast to SVD, if matrix U is constrained to only positive entries and columns in matrix U sum to 1, we get a decomposition called archetypal analysis. In this interpretation, columns of V correspond to the corners of a convex hull of the points in matrix M, also known as pure-samples or archetypes, and all data points are expressed as convex combinations of these archetypes. A major advantage of archetypal analyses is that archetypes are directly interpretable in the data space. Another closely related decomposition is non-negative matrix factorization (NMF), which relaxes the orthogonality constraint of SVD, instead, constraining elements of matrix U to be non-negative (Gillis 2020). In doing so, it loses error norm minimization properties of SVD, but gains interpretability. All of these methods can be used to identify patterns of coherent behavior among particles in the simulation. We refer interested readers to a comprehensive survey on linear dimensionality reduction methods by Cunningham and Ghahramani (Cunningham and Ghahramani 2015).

Non-linear Dimensionality Reduction

General non-linear dimensionality reduction techniques are needed for data that resides on complex non-linear manifolds. This is commonly the case for particle datasets in reactive environments. Non-linear dimensionality reduction technqiues typically operate in three steps: (i) embedding of data onto a low-dimensional manifold (in a high-dimensional space); (ii) defining suitable distance measures; and (iii) reducing dimensionality to preserve distance measures. Among the more common non-linear dimensionality reduction technique is Isometric feature mapping. This technique first constructs a graph corresponding to the dataset by associating a node with each row of the data matrix, and edges to correspond to the k nearest neighbors of the node. This graph is then used to define distances between nodes in terms of shortest paths. Finally, techniques such as multidimensional scaling (MDS)—a generalization of PCA that can use general distance matrices, as opposed to covariance matrices used by PCA, are used to compute low-dimension representations of the matrix. An alternate approach uses the spectrum of a Laplace operator defined on the manifold to embed data points in a lower dimensional space. Such techniques fall into the general class of Laplacian eigenmaps.

An alternate approach to non-linear dimensionality reduction is the use of non-linear transformations on data in conjunction with a suitable distance measure, followed by MDS for dimensionality reduction. The first two steps of this process (non-linear transformation and distance measure computation) are often integrated into a single step through the specification of a kernel. The use of such a kernel with MDS is called kernel PCA. The key challenges in the use of these methods relate to: (i) suitable representation techniques (described in Sect. 5); (ii) kernel functions; and (iii) appropriate scaling mechanisms since distance matrices can have highly skewed distributions and the directions may be dominated by a small number of very large entries in the distance matrix. Common approaches to kernel selection rely on polynomial transformations of increasing degree until suitable spectral gap is observed. Data representations and normalization are highly application and context dependent.

Autoencoder and Deep Dimensionality Reduction

Autoencoders have been recently proposed for use in non-linear dimensionality reduction (Kramer 1991; Schmidhuber 2015; Goodfellow et al. 2016). Autoencoders are feed-forward neural networks (discussed in further detail in Sect. 3.2) that are trained to code the identity function—i.e., the output of the autoencoder neural network is the input itself. Dimensionality reduction is accomplished in this framework by having an intermediate layer with a small number of activation functions. Through this constraint, an autoencoder is trained to "encode" input data into a low-dimensional latent space, with the goal of "decoding" the input back. The output of the encoder therefore represents a non-linear reduced dimension representation of the input.

T-distributed Stochastic Neighbor Embedding (t-SNE) and Uniform Manifold Approximation and Projection (UMAP)

t-SNE (Maaten and Hinton 2008) and UMAP (McInnes et al. 2018) are commonly used non-linear dimensionality reduction techniques for mapping data to two or three dimensions—primarily for visual analysis. t-SNE computes two probability distributions—one in the high-dimensional space and one in the low-dimensional space. These distributions are constructed so that two points that are close to each other in the euclidean space have similar probability values. In the high-dimensional space, a Gaussian distribution is centered at each data point, and a conditional probability is estimated for all other data points. These conditional probabilities are normalized to generate a global probability distribution over all points. For points in the low dimensional space, t-SNE uses a Cauchy distribution to compute the probability distribution. The goal of dimensionality reduction translates to minimizing the distance (in terms of KL divergence) of these two distributions. This is typically done using gradient descent. In contrast to t-SNE, a closely related technique, UMAP assumes that the data is uniformly distributed on a locally connected Riemannian manifold and that the Riemannian metric is locally constant or approximately locally constant (https://umap-learn.readthedocs.io/en/latest/). Both of these techniques are extensively used in visualization of high-dimensional data.

3.2 Supervised Learning

The goal of supervised methods is to learn a function from input data vectors to output classes (labels) using training input-output examples. The function should "generalize" to be able to accurately predict labels for unseen inputs. The general learning procedure is as follows: first, the data is split into train and test sets. Then, the function is learnt by using the input-output training examples. The learnt function is applied to the test input to get predicted outputs. If the algorithm performs poorly on training examples, we say that the algorithm "underfits" the data. This typically occurs when the model is unable to capture the complexity of the data. When learnt functions do not perform well (say, low prediction accuracy) on test data, we say that the algorithm "overfits" to the train set. Overfitting occurs when the algorithm fits to noise, rather than true data patterns. The problem of balancing underfitting and overfitting is called the bias-variance tradeoff. Intuitively, we want the model to be sophisticated enough to capture complex data patterns, but on the other hand, we don't want to endow it with the ability to capture idiosyncrasies of the train examples.

The problem of overfitting can be controlled through a number of approaches. In cross-validation, the training set is further divided into subsets (or folds). The training procedure proceeds to learn the function by leaving out one fold in every iteration. The model is validated on the remaining fold. The parameters of the model are optimized to ensure high cross-validation accuracy. Regularization is a technique in which a penalty term is added to the error function to prevent overfitting. Tikhonov regularization is one of the early examples of regularization that is commonly used in linear regression. Early stopping is a form of regularization in which the learner uses iterative methods like gradient descent. The key idea of early stopping is to perform training until the learning algorithm continues to improve performance on external (unseen) data. It is stopped when improvement on training performance comes at the expense of test performance. Other approaches to avoid overfitting include data augmentation (increasing number of data points for training) and improved feature selection. Underfitting can be avoided by using more complex models (e.g., going from a linear to a non-linear model), increasing training time, and reducing regularization.

3.2.1 Overview of Supervised Learning Algorithms

Supervised learning algorithms are often categorized as generative or discriminative. Generative algorithms aim to learn the distribution of each class of data, whereas discriminative algorithms aim to find boundaries between different classes. Naive Bayes Classifier is a generative approach that uses the Bayes Theorem with strong assumptions on independence between the features (Rish 2001). Given a d-dimensional data vector $\mathbf{x} = [x_1, x_2, \ldots, x_d]$, naive Bayes models the probability that $\mathbf{x}$ belongs to class k as follows:

$$p(C_k|\mathbf{x}) \propto p(C_k) \prod_{i=1}^{d} p(x_i|C_k) \tag{4}$$

In practice, the parameters for the distributions of features are estimated using maximum-likelihood estimations. Despite the strong assumptions made in naive Bayes, it works well in many practical settings. Linear Discriminant Analysis (LDA) is a binary classification algorithm that models the conditional probability densities $p(\mathbf{x}|C_k)$ as normal distributions with parameters (μ_k, Σ), where $k = \{0, 1\}$ (McLachlan 2005). The simplifying assumption of homoscedasticity (i.e., the covariance matrices are the same for both classes) means that the classifier predicts class 1 if:

$$\Sigma^{-1}(\mu_1 - \mu_0) \cdot \mathbf{x} > \frac{1}{2}\Sigma^{-1}(\mu_1 - \mu_0) \cdot (\mu_1 + \mu_0) \tag{5}$$

More complex generative methods include Bayesian Networks and Hidden Markov Models.

k-Nearest Neighbor (k-NN) algorithm is an early, and still widely used discriminative algorithm used for both classification and regression. In classification, the label of a test data sample is obtained by a vote of the labels of its k-nearest neighbors. In regression, k-NN computes the predicted value of a test sample as a function of the corresponding values of its k-nearest neighbors. Logistic regression uses a logistic function (logit) to model a binary dependent variable. In the training phase, the parameters for the logit function are learnt. Logistic regression is similar to LDA, but with fewer assumptions.

Support Vector Machine (SVM) (Cortes and Vapnik 1995) is a widely used discriminative model for regression and classification. Given input data $[\mathbf{x}_1, \mathbf{x}_2, \ldots, \mathbf{x}_n]$ and corresponding labels $y_1, y_2, \ldots, y_n$, where $y_i \in \{-1, 1\}, \forall i \in \{1, 2, \ldots, n\}$, SVM aims to optimize the following objective function:

$$\text{minimize} \quad \lambda||\mathbf{w}||^2 + \sum_{i=1}^{n} \max(0, 1 - y_i(\mathbf{w} \cdot \mathbf{x} - b)) \tag{6}$$

Here, vector $\mathbf{w}$ represents the vector normal to the separating hyperplane and λ is the weight given to regularization. The $\max(.)$ term is called Hinge-loss function, which allows SVMs to work with non-linear boundaries. SVMs typically use the so-called "kernel trick". The idea is that implicit high-dimensional representation of raw data can let linear learning algorithms learn non-linear boundaries. The kernel function itself is a similarity measure. Common kernels include Fisher, Polynomial, Radial Basis Function (RBF), Gaussian, and Sigmoid functions. Other examples of discriminative methods include decision trees and random forests.

3.2.2 Neural Networks

Neural Networks are interconnected groups of units called *neurons* that are organized in layers. The first layer is called the input layer, and is typically the same dimension as the input. The final layer is called the output layer. The outputs of neural networks could be prediction of class labels, images, text, etc. Each neuron in an intermediate layer is given a number of inputs. It computes a non-linear function on a weighted sum of its input. The resulting output may be fed into a number of neurons in the next layer. The non-linear function associated with a neuron is called an *activation function*. Common examples of activation functions include hyperbolic tan (tanh), sigmoid, Rectified Linear Unit (ReLU), and Leaky ReLU, among many others.

There are two key steps to designing neural networks for specific tasks. The first step corresponds to design of the network architecture. This specifies the number of layers, connectivity, and types of neurons. The second step parametrizes weights on edges of the neural network using a suitable optimization procedure for matching the output distribution with the target distribution (as discussed earlier in Sect. 2.1).

The term *deep learning* is used to describe a family of machine learning models and methods whose architectures use neural networks as core components. The word "deep" corresponds to the the the fact that learning algorithms typically use neural network models with many layers, in contrast to shallow networks which typically have one or two intermediate (or hidden) layers (Schmidhuber 2015).

3.2.3 Convolutional Neural Networks

Convolutional Neural Networks (CNNs) are neural networks that use convolutions to quantify local pattern matches. CNNs are feed-forward networks with one or more convolution layers. CNNs are used extensively in the analysis of images, and more recently, graphs that model connected structures such as molecules. CNNs have an input layer, hidden layer(s), and output layers. The input to a CNN is a tensor of the form #*inputs* $\times$ *input height* $\times$ *input width* $\times$ *input channels*. The height and width parameters correspond to the size of the original images. The number of input channels is typically three (red, green, and blue) for images.

Each of the hidden layers can be one of: (i) convolutional layer, (ii) pooling layer, or a (iii) fully connected layer. A *convolutional layer* takes as input an image, or the output of another layer, and outputs a feature map. This produces a tensor of the form #*inputs* × *feature height* × *feature width* × *feature channels*. Each neuron of a CNN processes only a small region of the input. This region is called the *receptive field*. It convolves this input and passes it on to the next layer. *Pooling Layers* are used to reduce the dimensionality of the data. They do so by aggregating the outputs of neurons in the previous (convolutional) layer. Pooling strategies can be local (operating on a small subset of neurons), or global (operating on the entire feature map). Common pooling functions include max and average. In *fully connected layers*, outputs of neurons are connected to every single neuron in the next layer. They are often used as the penultimate layer before the output layer, where all weights are combined to compute the prediction (i.e., the output). A neural network with only fully connected layers is also called a Multiple Layer Perceptron (MLP). From this point of view, CNNs are regularized forms of MLPs.

There are a number of parameters associated with CNNs that must be tuned. Specific to convolutional layers, the common parameters are stride, depth, and padding. The depth parameter of the output volume controls the neurons in a layer that connect to the same region of the input volume. Stride controls the translation of the convolution filter. Padding allows the augmentation of input with zeros at the border of input volume. Other parameters include kernel size and pooling size. Kernel size specifies the number of pixels that are processed together, whereas pooling size controls the extent of down-sampling. Typical values for both are 2×2 in common image processing networks.

In addition to parameter tuning, regularization is also required to design robust CNNs. In addition to generic methods for regularization mentioned earlier (such as early-stopping, L1/L2 regularization), there are CNN-specific approaches. Dropout is a common measure taken to regularize neural networks. Fully-connected networks (or MLPs) are prone to overfitting, because of the large number of connections. An intuitive way to resolve this issue is to leave out individual nodes (and the corresponding inbound and outbound edges) from the training procedure. Each node is left out with a probability p (p is usually set to 0.5). During the testing phase, the expected value of the weights are computed from different versions of the dropped-out network. Other simple, CNN-specific parameter tuning techniques limit the number of units in hidden layers, number of hidden layers, and number of channels in each layer.

Commonly used CNN architectures include LeNet (LeCun et al. 1989), AlexNet (Krizhevsky et al. 2012), ResNet (He et al. 2016), Wide ResNet (Zagoruyko and Komodakis 2016), GoogleNet (Szegedy et al. 2015), VGG (Simonyan and Zisserman 2014), DenseNet (Huang et al. 2017), and Inception (v2 (Szegedy et al. 2016), v3 (Szegedy et al. 2016), v4 (Szegedy et al. 2017))).

3.2.4   Recurrent Neural Networks

A Recurrent Neural Network (RNN) is a neural network in which nodes have internal hidden states, or *memory*. RNNs can therefore process (temporal) sequences of inputs. They are typically used in the analysis of speech signals, language translation, and handwriting recognition, and more recently in prediction of atomic trajectories in molecular dynamics simulations.

A key feature of RNN is the ability to share intermediate outputs across different parts of the model. Given a sequence of inputs $[\mathbf{x}_1, \mathbf{x}_2, ..., \mathbf{x}_n]$, the state of RNN at time t is given as

$$\mathbf{h}^{(t)} = f(\mathbf{h}^{(t-1)}, \mathbf{x}^{(t)}; \theta) \tag{7}$$

where, $f(.)$ is the recurrent function, and θ is the set of shared intermediate outputs. From Eq. 7, one can see that RNNs predict the future on the basis of the past outcomes.

A generic RNN can, in theory, remember arbitrarily long-term dependencies. In practice, repeated use of back-propagation causes gradients to vanish (i.e., tend to zero), or explode (i.e., tend to infinity). *Gated RNNs* are designed to circumvent these issues. The most widely used Gated RNNs are Long Short-Term Memory (LSTM) (Hochreiter and Schmidhuber 1997; Gers et al. 2000) and Gated Recurrent Unit (GRU) (Cho et al. 2014). Recall that a regular activation neuron consists of a non-linear function applied to a linear transformation of the input. In addition to this, LSTMs have an internal cell-state (different from the hidden-state recurrence previously discussed), and a gating mechanism that controls the flow of information. In all, LSTMs have three gates—input gate, forget gate, and output gate. Specifically, the *forget gate* allows a network to forget old states that have accumulated over time, thereby preventing vanishing gradients. GRUs are similar to LSTMs, but with a simplified gating architecture. GRUs combine LSTM's input and forget gate into a reset gate. The reset gate also allows GRUs to combine hidden- and cell-states. This results in a simpler architecture that requires fewer tensor operations. The problem of exploding gradients is handled by *gradient clipping*. Two common strategies in gradient clipping are: (i) value clipping—values above and below set thresholds are set to the respective thresholds, and (ii) norm clipping—rescaling the gradient values by a chosen norm. Using CNNs and RNNs as building blocks, we can develop complex NN frameworks such as Generative Adversarial Networks (GANs).

3.2.5   Generative Adversarial Networks

A Generative Adversarial Network (GAN) is a neural network in which a zero-sum game is contested by two neural networks—the *generative network* and the *discriminative network* (Goodfellow et al. 2014). The generative network learns to map a pre-defined latent space to the distribution of the dataset, whereas a discriminative network is used to predict whether an input instance is truly from the dataset or if it is the output of the generative network. The objective of the generative network is to fool the discriminative network (i.e., increase error of the discriminative network), whereas the objective of the discriminative network is to correctly identify true data. The training procedure for a GAN is as follows: first, the discriminative network is given several instances from the dataset, so that it learns the "true" distribution. The generative network is initially seeded with a random input. From there, the generative network creates candidates with the objective of fooling the discriminative network. Both networks have separate back-propagation procedures; the discriminator learns to distinguish the two sources of inputs, even as the generative network produces increasingly realistic data.

GANs have found a number of applications in synthesis of (realistic) datasets. They have been successful in creating art, synthesizing virtual environments, generate photographs of synthetic faces, and designing animation characters. GANs are often used for the purpose of transfer learning, where knowledge obtained from one training in one application can be used in another similar, but different application.

3.2.6 Transfer Learning

Traditional machine learning is isolated, in that a model is trained in a very specific context, to perform a targeted task. The key idea in transfer-learning is that new tasks learn from the knowledge gained in a previously trained task (Weiss et al. 2016). To formally define Transfer Learning, we first define *domain* and *task*. Let $\mathcal{X}$ be a feature space, and $\mathbf{X}$ be the dataset (i.e., $\mathbf{X} = [\mathbf{x}_1, \mathbf{x}_2, \ldots, \mathbf{x}_n] \in \mathcal{X}$). Similarly, let $\mathcal{Y}$ be the label space and $Y = \{y_1, y_2, \ldots, y_n\} \in \mathcal{Y}$ be the labels corresponding to the rows of $\mathbf{X}$. Further, let $P(.)$ denote a probability distribution. A domain is defined as $\mathcal{D} = \{\mathcal{X}, P(\mathbf{X})\}$. Given a domain $\mathcal{D}$, a task $\mathcal{T}$ is defined as $\mathcal{T} = \{\mathcal{Y}, P(Y|\mathbf{X})\}$. Given source and target domains $\mathcal{D}_S$ and $\mathcal{D}_T$ and corresponding tasks $\mathcal{T}_S$ and $\mathcal{T}_T$, transfer learning aims to learn $P(Y_T|\mathbf{X}_T)$ using information from $\mathcal{D}_S$ and $\mathcal{D}_T$. In this setup, we can see that there are four possibilities: (i) $\mathcal{X}_S \neq \mathcal{X}_T$, (ii) $\mathcal{Y}_S \neq \mathcal{Y}_T$, (iii) $P(\mathbf{X}_S) \neq P(\mathbf{X}_T)$, or (iv) $P(Y_S|\mathbf{X}_S) \neq P(Y_T|\mathbf{X}_T)$. In (i), the feature spaces of the source and target domain are different. In (ii), the label space of the task are different, which happens in conjunction with (iv) where the conditional probabilities of labels are different. In (iii), the feature spaces of source and target domains are the same, while the marginal probabilities are different. Case (iii) is interesting for simulations, because the feature spaces for source (simulation) and target (reality) is typically the same, but the marginal probabilities of observations in simulation and reality can be very different.

3.3 Software Infrastructure for Machine Learning Applications

A number of software packages and libraries have been developed over the last decade in support of ML applications in different contexts. Matrix computations are often performed using NumPy (Python) (Harris et al. 2020), Eigen (et al. 2010), and Armadillo (C++) (Sanderson and Curtin 2016; Sanderson and Curtin 2020). Standard machine learning methods, including clustering such as k-means clustering and DBSCAN, classification algorithms such as SVM and LDA, regression, and dimensionality reduction are available in Python packages such as SciPy (Virtanen et al. 2020) and Theano (Theano Development Team 2016), and in C++ packages such as MLPack (Curtin et al. 2018). Deep learning approaches are often implemented using libraries such as PyTorch (Paszke et al. 2019), TensorFlow (Abadi et al. 2015), Caffe (Jia et al. 2014), Microsoft Cognitive Toolbox, and DyNet (Neubig et al. 2017). We note that a number of machine learning packages written in a source language have readily available interfaces for other languages. For example, Caffe is written in C++, with interfaces available for both Python and MATLAB. Finally, we also note that Julia has wrappers for a number of the Python and C++ libraries.

4   ML Applications in Reactive Atomistic Simulations

Building on our basic toolkit of ML models and methods, we now describe recent advances in the use of ML techniques in reactive atomistic simulations. We focus on three core challenges—use of ML techniques for training highly accurate atomistic interaction models, use of ML techniques in accelerating simulations, and use of ML methods for analysis of atomistic trajectories. Our discussion applies broadly to particle methods, however, we use reactive atomistic simulations as our model problem. In particular, we use ReaxFF as the force field for simulations.

4.1   ML Techniques for Training Reactive Atomistic Models

Optimization of force-field parameters for target systems of interest is crucial for high fidelity in simulations. However, such optimizations cannot be specific to the sets of molecules present in the target system for two reasons: (i) utility of a parameter set that only works for a particular system is marginal; and (ii) in a reactive simulation, molecular composition of a system is expected to change as a result of the reactions during the course of a simulation. For this reason, reactive force field optimizations are performed at the level of groups of atoms, e.g. Ni/C/H, Si/O/H, etc. Nevertheless, the behaviour of a given group of atoms may show variations in different contexts such as combustion, aqueous systems, condensed matter phase systems, and biochemical processes. Therefore, it may be desirable to create parameter sets optimized for different contexts (Senftle et al. 2016).

Reactive force fields such as ReaxFF are complex, with a large number of parameters that can be grouped by charge equilibration parameters, bond order parameters, and parameters based on N-body interaction (e.g., single-body, two-body, three-body, four-body and non-bonded) in addition to the system-wide global parameters. As the number of elements in a parameter set increases, force field optimization quickly becomes a challenging problem due to the high dimensionality and discrete nature of the problem. Several methods and software systems have been developed for force field optimization over the years, starting with more traditional methods early on and moving to ML-based methods more recently. After giving an overview of the force field optimization problem, we briefly review traditional methods first and then discuss the ML-based techniques, which mainly draw upon Genetic Algorithms (see Sect. 2.2) as well as the extensive ML software infrastructure that has been built recently (see Sect. 3.3).

4.1.1 Training Data and Validation Procedures

Training procedures for typical force fields require three inputs: (i) model parameters to be optimized; (ii) *geometries*, a set of atom clusters that describe the key characteristics of the system of interest (e.g., bond stretching, angle and torsion scans, reaction transition states, crystal structures, etc.); and (iii) *training data*, chemical and physical properties associated with these atom clusters (such as energy minimized structures, relative energies for bond/ angle/ torsion scans, partial charges and forces), which are typically obtained from high-fidelity quantum mechanical (QM) models or sometimes experiments, along with a function that combines these different types of training items into a quantifiable fitness value:

$$\text{Error}(m) = \sum_{i=1}^{N} \left(\frac{x_i - y_i}{\sigma_i} \right)^2.$$

(8)

In Eq. 8, m represents the model with a given set of force field parameter values, x_i is the predicted training data value calculated using the model m, y_i is the ground truth value of the corresponding training data item, and σ_i^{-1} is the weight assigned to each training item.

Table 1 summarizes commonly used training data types and provides some examples. An energy-based training data item uses a linear relationship of different molecules (expressed through their identifiers) because relative energies rather than the absolute energies drive the chemical and physical processes. For structural items, geometries must be energy minimized as accurate prediction of the lowest energy states is crucial. For other training item types, energy minimization is optional, but usually preferred.

4.1.2 Global Methods for Reactive Force Field Optimization

The earliest ReaxFF optimization tool is the sequential parabolic parameter interpolation method (SOPPI) (van Duin et al. 1994). SOPPI uses a one-parameter-at-a-time approach, where consecutive single parameter searches are performed until a certain

Table 1 Examples for commonly used training items. Identifiers (e.g., ID1) refer to structures/molecules

Type	Training item	Target	Description
Charge	ID1 1	0.5	Charge for atom 1 (in elementary charge)
Energy	ID1–ID2/2–ID3/3	50	Energy difference (in kcal/mol)
	ID1	−150	
	ID3/2–ID1/3	30	
Geometry	ID1 1 2	1.25	Distance between atom 1 and 2 (in Å)
	ID2 1 2 3	120	Valence angle between atom 1, 2 and 3 (in degree)
	ID3 1 2 3 4	170	Torsion angle between atom 1, 2, 3 and 4 (in degree)
Force	ID1 1	0.5 0.5 0.5	Forces on atom 1 (in kcal/mol Å)
	ID2	1.0	RMSG (in kcal/mol Å)

convergence criteria is met. The algorithm is simple, but as the number of parameters increases, the number of one-parameter optimization steps needed for convergence increases drastically. Furthermore, the success of this method is highly dependent on the initial guess and the order of the parameters to be optimized.

Due to the drawbacks of SOPPI, various global methods such as genetic or evolutionary algorithms (Dittner et al. 2015; Jaramillo-Botero et al. 2014; Larsson et al. 2013; Trnka et al. 2018), simulated annealing (SA) (Hubin et al. 2016; Iype et al. 2013) and particle swarm optimization (PSO) (Furman et al. 2018) have been investigated for force field optimization. We discuss some of the promising techniques below.

Genetic Algorithms (GA) often work well for global optimization because via crossover they can exploit (partial) separability of the optimization problem even in the absence of any explicit knowledge about its presence. They are also able to make long-range "jumps" in search space. Due to the continuous presence of multiple individuals that have survived several selection rounds it is ensured that these "jumps," based on information interchange between individuals, have a high probability of landing at new, promising locations. Last but not least, by admitting operators other than the classic crossover and mutation steps, it is possible to extend GAs within this abstract meta-heuristic framework with desirable features of other global optimization strategies, too. GAs are especially useful when dealing with challenging and time-critical optimization problems. The straightforward parallelism and intrinsic high scalability property of GAs provide an advantage over other strategies that are either serial in nature or where parallelization facilitates decoupled or only loosely coupled task-level parallelism. An efficient and scalable implementation of GAs for ReaxFF is provided in the ogolem-spuremd software (Dittner et al. 2015), where the authors demonstrate convergence to fitness values similar to or better than those reported in the literature in a matter of a few hours of execution time through effective use of high-performance computers and advanced GA techniques.

Recently, other population-based global ReaxFF optimization methods have been proposed, such as the particle swarm optimization algorithm RiPSOGM (Furman et al. 2018), covariance matrix adaptation evolutionary strategy (CMA-ES) (Shchygol et al. 2019), and the KVIK optimizer (Gaissmaier et al. 2022). Shchygol et al. (2019) explore different optimization choices for the CMA-ES method, the ogolem-spuremd software, as well as a Monte-Carlo force field optimizer (MCFF), and they systematically compare these techniques using three training sets from literature. Their CMA-ES method is an implementation of the stochastic gradient-free optimization algorithm proposed by Hansen (2006), where the main idea is to iteratively improve a multi-variate normal distribution in the parameter space to find a distribution whose random samples minimize the objective function starting from a user provided initial guess. The MCFF technique is based on the simulated annealing algorithm to optimize a given set of parameters. In every iteration, MCFF makes a small random change to the parameter vector and computes the corresponding change in the error function. Any change that reduces the error is accepted; changes that increase the error are accepted with a predetermined probability. With sufficiently small random changes and acceptance rates, MCFF can become a rigorous global optimization method, but at very high computational cost. Through extensive benchmarking, Shchygol et al. conclude that while CMA-ES can often converge to the lowest error rates, it cannot do this on a consistent basis. The GA method employed by ogolem-spuremd can produce consistently good (but not necessarily the lowest) error rates, but at higher computational costs compared to CMA-ES. Overall, they have found MCFF to underperform compared to CMA-ES and GA for similar computational costs.

4.1.3 Machine Learning Based Search Methods

While global methods have been proven to be successful for force-field optimization, due to the absence of any gradient information, these global search methods require a large number of potential energy evaluations, as such they can be very costly. With the emergence of advanced tools to calculate the gradients of complex functions automatically, machine learning based techniques for optimization of force fields have attracted interest.

iReaxFF: One of the earliest such attempts is the Intelligent-ReaxFF, iReaxFF, software (Guo et al. 2020). iReaxFF uses the TensorFlow library for automatically calculating gradient information and use local optimizers such as Adam or BFGS. An additional benefit of the Tensorflow implementation is that iReaxFF can automatically leverage GPU acceleration. However, iReaxFF does not have the expected flexibility in terms of the training data as it can only be trained to match the ReaxFF energies to the absolute energies from Density Functional Theory (DFT) computations on the training data; relative energies, charges or geometry optimizations cannot be used in the training, essentially limiting its usability. As iReaxFF tries to exactly match the energies of the training data, the transferability of force fields generated by iReaxFF is also limited. While it is not clearly stated what kind of gradient information is calculated using Tensorflow, their definition of the loss function (which is the sum of the squared differences between absolute DFT and ReaxFF energies) suggests that their gradients are calculated with respect to atomic positions, which essentially amounts to performing a force matching based force field optimization. The number of iterations required to reach the desired accuracies for their test cases is rather large, on the order of tens to hundreds of iterations. Even with GPU acceleration, the training time for a test case reportedly takes several days. This is partly because iReaxFF does not filter out the unnecessary 2-body, 3-body and 4-body interactions before the optimization step.

JAX-ReaxFF: Another recent effort that utilizes the Tensorflow framework is the JAX-ReaxFF software (Kaymak et al. 2022). JAX is an auto-differentitation software by Google that is built on top of Tensorflow for high performance machine learning research (Bradbury et al. 2020), it can automatically differentiate native Python and NumPy functions. Leveraging this capability, JAX-ReaxFF automatically calculates the derivative of a given fitness function with respect to the set of force field parameters to be optimized from Python-based implementation of the ReaxFF potential energy terms. By learning the gradient information of the high dimensional optimization space (which generally includes tens to over a hundred parameters), JAX-ReaxFF can employ highly effective local optimization methods such as the Limited Memory Broyden–Fletcher–Goldfarb–Shanno (L-BFGS) algorithm (Zhu et al. 1997) and Sequential Least Squares Programming (SLSQP) (Kraft et al 1988) optimizer. The gradient information alone is obviously not sufficient to prevent local optimizers from getting stuck in a local minima, but when combined with a multistart approach, JAX-ReaxFF can greatly improve the training efficiency (measured in terms of the number of fitness function evaluations) performed. As they demonstrate through a diverse set of systems such as cobalt, silica, and disulfide, which were also used in other related work, they can reduce the number of optimization iterations from tens to hundreds of thousands (as in CMA-ES, ogolem-spuremd or iReaxFF) down to only a few tens of iterations.

Another important advantage of JAX is its architectural portability enabled by the XLA technology (Sabne 2020) used under the hood. Hence, JAX-ReaxFF can run efficiently on various architectures, including graphics processing units (GPU) and tensor processing units (TPU), through automatic thread parallelization and vector processing. By making use of efficient vectorization techniques and carefully trimming the 3-body and 4-body interaction lists, JAX-ReaxFF can reduce the overall training time by up to three orders of magnitude (down to a few minutes on GPUs) compared to the existing global optimization schemes, while achieving similar (or better) fitness scores. The force fields produced by JAX-ReaxFF have been validated by measuring the macroscale properties (such as density and radial distribution functions) of their target systems.

Beyond speeding up force field optimization, the Python based JAX-ReaxFF software provides an ideal sandbox environment for domain scientists, as they can move beyond parameter optimization and start experimenting with the functional forms of the interactions in the model, add new types of interactions or remove existing interactions as desired. Since evaluating the gradient of the new functional forms with respect to atom positions gives forces, scientists are freed from the burden of coding the lengthy and error-prone force calculation parts. Through automatic differentiation of the fitness function as explained above, parameter optimization for the new set of functional forms can be performed without any additional effort by the domain scientists. After parameter optimization, they can readily start running MD simulations to test the macro-scale properties predicted by the modified set of functional forms as a further validation test before production-scale simulations, or go back to editing the functional forms if desired results cannot be confirmed in this sandbox environment provided by JAX-ReaxFF.

4.2 Accelerating Reactive Simulations

We now discuss how ML techniques can be directly used to accelerate reactive simulations and to improve their accuracy in different application contexts.

4.2.1 Machine Learning Potentials

At a high level, ML based potentials can be defined as follows (Behler 2016):

1. The potential must establish a direct functional relation between atomic configuration and the corresponding energy, where the functional must be based on an ML model. As an example, a forward propagating deep neural network may serve as a functional, where input is the atomic configuration and output is the energy.
2. Any physical approximations or theoretically grounded constraints are explicitly incorporated into the training data and are not part of the energy functional.

The second requirement in the definition distinguishes traditional fixed form potentials from the ML potentials. It also ensures that for a "sufficiently complex" energy functional and "sufficiently large and diverse" training set, an ML based potential can produce arbitrarily accurate model predictions. Often it is expected that the training data are generated using a consistent and specific set of methods. It has been observed that mixing data from different QC techniques or experiments lead to poor learning outcomes. Sizes of the training sets depend on the computational cost of the training sets and the desired accuracy expected out of the ML model.

As with most traditional fixed form potentials, ML potential energy is expressed a sum of local energies:

$$E_{\mathrm{ML}} = \sum_{i=1}^{N} E_i^{\mathrm{nbd}},$$

where the local energy corresponds to the ML energy, which depends on the local neighborhood of the ith atom. Chemical environment of an atom is primarily decided by short range interactions (Kohn 1996). The long range interactions, which decay slower than r^{-2}, are usually either approximated at cutoff distance R_c as zero or smoothly reduced to zero using tapering functions. As an example, polynomial tapering functions are used in the ReaxFF. The accuracy of such model depends on the cutoff distance R_c – larger values of R_c lead to better approximation of long range interactions. However, larger R_c implies larger atomic neighborhood (which grows as R_c^3), which means that more sample points are required in the training set. Thus R_c must be chosen appropriately to provide better long range approximation while keeping the neighborhood size tractable.

4.2.2 Training Considerations

ML potentials, like fixed form potentials require training. Here we briefly explore the steps and potential issues with the design and training of ML potentials (See e.g. (Unke et al. 2021)).

Choice of quantum methods used in generation of training data: Typically ML based simulations are orders of magnitude slower than the equivalent fix form potential simulations (Brickel et al. 2019). However, unlike the fixed form potentials, ML potentials may offer accuracy similar to that of an ab initio method (Sauceda et al. 2020). Thus it is essential to choose an appropriate ab initio method. On one hand if the ab inito method is very fast and/or less accurate, it defeats the purpose of further approximating these data into a machine learnt model. On the other hand a method such as CCSD(t), that are computationally expansive makes it difficult to generate enough training data for ML models.

How much data? The amount of data needed depends on the size of the ML model, the desired accuracy, and the sampling technique used in producing the data set.

Sampling: Sampling of training data over the domain of atomic configurations is crucial in achieving good training of the potentials. For the models designed to simulate equilibrium problems, one can potentially rely on samples that are output of an ab initio molecular dynamics simulation. Depending on the desired accuracy, generating such samples can become prohibitively expensive. Another alternative is to use meta dynamics type sampling techniques and generate samples that are in the vicinity of the free energy minima of the system. However, if the model is intended to address chemical reactions or transition states, then a more uniform sampling is required where "rare events" are also sampled with relatively higher frequencies. The framework provided by an ML model does not include any "physics" of the problem, thus the training data must sample the configuration space sufficiently to include the relevant "physics" in the problem.

Training/validation and testing: In usual ML methodology, models are trained and tested against similarly structured but disjoint data sets. In this case, the training and the validation is performed on the data sets that are similarly sampled but distinct. However, the testing of the model is usually performed against bulk or physically measurable quantities computed using the trained models. Often the ML potential frameworks have hyperparameters that require a second step of optimizations. The testing phase must be repeated for ifferent hyperparameter values.

4.2.3 Descriptors

Unique description of atomic neighborhood is a central issue in structure–function prediction problems in biophysics and materials science (Ghiringhelli et al. 2015; Deviller and Balaban 1999; Valle and Oganov 2010). For ML systems, such uniqueness is crucial for effective training. Thus, one must express any atomic neighborhood in a representation that is invariant with respect to the action of the symmetry group of the system. In case of three dimensional atomistic systems, we have a group of Galilean transformations and discrete group of atomic permutations. We summarize commonly used descriptors, noting that the state of the art in this context is continually evolving.

Atom Centered Symmetry Function (ACSF)

This descriptor expresses the environment of ith atom in terms of a Gaussian basis of varying widths and angular basis at different resolution. It uses a cosine taper function given by:

$$T_{R_c}(r_{ij}) = \begin{cases} \frac{1}{2}\left(\cos\left(\frac{\pi r_{ij}}{R_c}\right) + 1\right) & \text{for } r_{ij} \leq R_c \\ 0 & \text{for } r_{ij} > R_c, \end{cases} \tag{9}$$

where r_{ij} is the distance between ith and jth particles. This ensures that, when multiplied, the quantity goes smoothly to zero as r_{ij} approaches R_c from below. Using this taper function, an atom centered descriptor can be written with radial and angular parts as:

$$G_i^r(\eta, \mu) = \sum_{j=1}^{n} e^{-\eta(r_{ij}-\mu)^2} \cdot T_{R_c}(r_{ij}) \tag{10}$$

$$G_i^\theta(\eta, \zeta, \lambda) = 2^{1-\zeta} \sum_{j,k \neq i}^{n} \left(1 + \lambda \cos\theta_{ijk}\right)^\zeta e^{-\eta\left(r_{ij}^2 + r_{ik}^2 + r_{jk}^2\right)}$$
$$\cdot T_{R_c}(r_{ij}) \cdot T_{R_c}(r_{ik}) \cdot T_{R_c}(r_{jk}), \tag{11}$$

where n is the number of neighbors in cutoff distance R_c, $\lambda = \pm 1$. The descriptor vector is generated by sampling the parameters η, ζ, μ, and λ. By design, ACSF produces a description that is invariant under translation and rotation. We note that the number of symmetry functions needed does not depend on n. However, the number of symmetry functions grow very rapidly. Typically for an atom 50–100 symmetry functions are used with various values of parameters (Behler 2016). Further the number of functions required grows quadraticaly with respect to the number of types of atoms used in the model. ACSF can be generalized with additional weight functions to improve resolution and complexity (Gastegger et al. 2017).

Coulomb Matrix (CM)

An alternate descriptor uses the Fourier transform of the Coulomb matrix (Rupp et al. 2012), which is defined as:

$$M_{ij} = \begin{cases} \frac{1}{2}Z_i^{2.4} & i = j \\ \frac{Z_i Z_j}{|\mathbf{r}_i - \mathbf{r}_j|} & i \neq j, \end{cases} \tag{12}$$

where Z_i is the chanrge on the ith particle. This descriptor is invariant under the transformations listed, however, it is computationally expensive unless restricted to a local coulomb matrix (Rupp et al. 2012). The descriptor can be further generalized to include Ewald matrix instead of Coulomb matrix (Faber et al. 2015).

Bispectral Coefficients (BC)

In this descriptor, the atomic environment is represented as a local density that is expressed in terms spherical harmonics on a four dimensional sphere. The density is written as superposition of delta function densities using the taper function from Eq. (9) as:

$$\rho_i(\mathbf{r}) = \delta(\mathbf{r}_i) + \sum_{r_{ij} < R_c} T_{R_c}(r_{ij})\omega_j \delta(\mathbf{r} - \mathbf{r}_{ij}), \tag{13}$$

where the dimensionless parameter ω_j represents atom type or other internal properties of the jth atom. Angular part of such density can be expanded in spherical harmonics basis and radial part can be expanded in terms of a linear basis. The radial part is transformed into an additional angle, converting the basis to spherical harmonics on 3-sphere. Let $U_{m,m'}^j$ be these hyper-spherical harmonics, then one can express the local density as:

$$\rho = \sum_{j=0}^{\infty} \sum_{m,m'=-j}^{j} c_{m,m'}^j U_{m,m'}^j, \tag{14}$$

where $c_{m,m'}^j$ are the coefficients of expansion. The $c_{m,m'}^j$ are computed by evaluating the inner product $\langle U_{m,m'}^j | \rho \rangle$. The BC are then computed using the mixing rules as:

$$B_{j_1,j_2,J} = \sum_{m_1,m_1'=-j_1}^{j_1} \sum_{m_2,m_2'=-j_2}^{j_2} \sum_{m,m'=-j}^{j} c_{m,m'}^{j}$$
$$\times C_{j_1 m_1 j_2 m_2}^{jm} C_{j_1 m_1' j_2 m_2'}^{jm'} c_{m_1',m_1}^{j_1} c_{m_2',m_2}^{j_2}, \tag{15}$$

where $C_{j_1 m_1 j_2 m_2}^{jm}$ are the Clebsch–Gordon coefficients of mixing. These descriptors also satisfy the required invariance properties. One key advantage of BC over ACSF is that BCs can be systematically expanded or truncated based on accuracy versus complexity trade offs of the model (Thompsona et al. 2015).

Smooth Overlap of Atomic Positions (SOAP)

In SOAP descriptor local density is generated by smoothing delta functions into a Gaussian as (Albert et al. 2013)

$$\rho_{\text{SOAP}}(\mathbf{r}) = \sum_{j=1}^{N_i} e^{-\alpha(\mathbf{r}-\mathbf{r_j})^2}.$$

This density can be expanded in term of radial and angular basis as

$$\rho_{\text{SOAP}}(\mathbf{r}) = \sum_{j=1}^{N_i} \sum_{n,l,m} c_{n,l,m}^{j} g_n(r) Y_{l,m}(\theta, \phi),$$

where $Y_{l,m}(\theta, \phi)$ are spherical harmonics basis, and $g_n(r)$ is a radial basis set chosen based on specific model. Thus the descriptor for atom i is written as an appropriately normalized power spectrum

$$p_{n,k,l}(i) = \sum_{m} c_{n,l,m}^{i} \left(c_{k,l,m}^{i}\right)^*.$$

4.2.4   Energy Functionals

The input to the ML model is a descriptor using one of the models described above. The output of the ML model is an energy functional. We describe common forms of the energy functional here.

Feed Forward Neural Network Based Energy Functional

One of the common ML energy functionals is based on feed forward neural networks (FFNN) (see e.g. Blank et al. (1995), Gassner et al. (1998), Lorenz et al. (2004), Manzhos and Carrington (2006), Behler et al. (2007), Geiger and Dellago (2013), Behler (2014), Behler (2015)). These networks typically use descriptor as input and produce an energy value as output. One can write the energy as:

$$E_i = g_m \circ g_{m-1} \circ \cdots \circ g_2 \circ g_1 \left(\mathbf{b}_1 + \mathbf{W}_{0,1} \cdot \mathbf{G}_i \right)$$
$$h_{k+1} = g_k(h_k) = f_k(\mathbf{b}_k + \mathbf{W}_{k-1,k} \cdot \mathbf{h}_{k-1}),$$

where the neural network has m layers, $\mathbf{W}_{k-1,k}$, $\mathbf{b}_k$ are the weights and the bias values associated with the kth layer respectively, and f_k are the nonlinear activation functions associated with the kth layer. Forces are computed as the negative gradients of the energy functional. Thus we expect the activation functions f_k to be differentiable functions.

Gaussian Approximation Potential (GAP)

This approximation establishes a mapping between the environment of an atom and the corresponding energy using a Gaussian kernel function.

$$E_i = \sum_n^{N_i} \alpha_n G(\mathbf{b}, \mathbf{b}_n)$$
$$= \sum_n^{N_i} \alpha_n e^{-\frac{1}{2} \sum_l^L \left(\frac{b_l - b_{n,l}}{\theta_l} \right)^2},$$

where L is the number of truncated bispectrum components, $\mathbf{b}$ are the BCs. The determination of the coefficients α_n is computationally expensive, since it grows as N^3 (Li et al. 2015).

Spectral Neighbour Analysis Potential (SNAP)

SNAP simplifies the computation of α_i in GAP by changing problem of Gaussian regression to a linear regression. Thus now the energy functional is given by (Thompsona et al. 2015)

$$E_i = \beta_0^{\omega_i} + \sum_{k=1}^M \beta_k^{\omega_i} \cdot B_k^i,$$

where M is the number of bispectrum coefficients used in an approximation. Most important advantage of SNAP over GAP is the simplification of computation due to linear regression.

4.2.5 Accelerating Time-stepping Using Deep Networks

We have previously described the use of ML potentials to increase the accuracy and scope of modeled interactions. An important bottleneck in reactive atomistic simulations is the need for small timesteps (sub-femtoseconds in typical applications), whose sequential nature limits the temporal scope of simulations. There have been some recent efforts aimed at ML techniques for long-timestep integration. Conventional time-stepping schemes use the current atomic state (and in some cases, the few states leading up to the current state), combined with the force (derived from energy) to advance system state to the next step. The goal of ML-based time integrators is to use a sequence of past atomic states, along with the energy, to predict system state over longer timesteps (e.g., three orders of magnitude longer than conventional integrators).

The use of multiple past states in predicting the next state motivates the use of Recurrent Neural Networks (RNNs) for this task. Recall that RNNs use internal states to process time-series data. To address the 'vanishing gradient' problem discussed in Sect. 3.2.2, RNN variants such as Long Short-Term Memory (LSTM) networks are used for this purpose. There are three key issues in the use of LSTMs in long time-step integrators: (i) specification of input states for the deep network; (ii) the network architecture; and (iii) training process. The input to a LSTM-based time integrator is typically limited to a finite region around the atom for which the trajectory is predicted. Larger neighborhoods require significantly larger number of degrees of freedom in the network. While in theory, this would improve accuracy, the need for large amounts of training data and training error typically negate this improvement in accuracy. The network architecture is determined by the complexity of the energy functional and specific domain properties. In current practice, even simple energy terms (Lennard-Jones interactions) require large networks (100K parameters) for ensembles of as few as 16 particles. The need for training data and associated training cost for these is significant. However, such integrators are shown to be capable of timesteps three orders of magnitude longer than conventional Verlet integrators (Kadupitiya et al. 2020).

In current proposals, which are in relative infancy, the training procedures for the LSTMs use simulation data generated from the specific potential, with well specified boundary conditions (e.g., periodic boundaries). Even in these simple systems, a large amount of training data is needed to accurately predict trajectories. It is observed that for more complex potentials (with multiple terms) and diverse atomic contexts, the need for training data increases substantially.

We note that the use of deep networks for particle dynamics is in relative infancy. There has been significant interest in the use of deep networks for time-integrating ODEs since the recent work of Chen et al. (2018). Recent advances include symplectic ODE-Nets for learning the dynamics of Hamiltonian systems (Zhong et al. 2019), and associated deep learning architectures (Rusch and Mishra 2021).

5 Analyzing Results from Atomistic Simulations

A key use of machine learning techniques is in the analysis of large amounts of data generated from time-dependent simulations. This data generally takes the form of snapshots of trajectories—with each snapshot corresponding to system state comprised of degrees of freedom (position, momentum, etc.) associated with particles, and in the case of reactive simulations, bond information. Complex simulations scale to millions of particles and beyond, over billions of time-steps—leading to datasets that are in excess of terabytes. A number of techniques are deployed to deal with this data volume, including subsampling for reducing storage, indexing for fast access, and compression. While these techniques facilitate storage and access, the focus of this section is primarily on analysis techniques that abstract and extract useful information from trajectories.

We note that ML techniques for analyses of time-dependent simulation is an active area of research. This section summarizes the rich state of the art in the area—for a more detailed recent summary, we refer readers to excellent reviews by Glielmo et al. (2021), Sidky et al. (2020), and Noé et al. (2020).

5.1 Representation Techniques

We consider a general class of simulations that result in a set of T snapshots of data—each snapshot S_i, $i = 0 \ldots T - 1$, stored as a D dimensional vector, in a matrix M of dimension $T \times D$. The first challenge we face is to suitably encode system state at time t_i into a corresponding vector S_i. This poses challenges w.r.t. different data structures and their consistent encoding. We consider two common data structures and associated representation techniques:

Vector Fields

The most common data associated with particles is in vector fields. This includes position data, momentum, and other particle properties. The first step in representing these vector fields is to account for underlying invariants. For instance, a particle aggregate (e.g., a molecule) may be invariant under rotation and translation. To account for this invariance, these aggregates must be represented in a canonical framework so that two aggregates in different orientations can be viewed as being identical under affine transformations. The most common technique relies on aligning particle aggregates with known reference aggregates (e.g., reference geometries of molecules) and to store them as deviations from these reference molecules under affine transformations. Such transformations can easily be computed through local formulations solved using Shapelets or global formulations such as the Orthogonal Procrustes Problem, which has an optimal solution due to Kabsch (1976). Once suitable alignments have been computed, the particle aggregates are stored as suitable vectors of deviations from reference aggregates. When reference aggregates are unavailable, canonical representations can be derived through suitable internal representations, for example, in the form of internal distances between reference particles (e.g., distance between pairs of marked atoms in a molecule). This vector of distances provides a canonical representation.

Network Models

Reactive simulations often store bond structure of molecules within snapshots S_i. These structures are invariant to within an isomorphism; i.e., any relabeling of atoms in the molecule should be treated identically. Canonical labelings are challenging because there exist an exponential number of permutations, and corresponding labelings. Deriving canonical labelings to represent graphs corresponding to molecular structures as vectors require solution of the graph isomorphism problem. For small molecules, this can be done by enumeration; however, for larger molecules, this is more computationally expensive. One solution to this problem relies on a diffusion kernel to derive canonical labelings. The Laplacian of the given graph structure is used to simulate a diffusion process on the graph. The stationary probabilities associated with the diffusion process are used to represent the graph in a canonical vector form. One may also view this vector in terms of the spectra of the graph. Other approaches to canonical labelings rely on graph neural networks (GNNs). These networks are trained to input a given graph and to generate canonical labels as output. This training procedure for GNNs associates the identical labelings for isomorphic graphs.

5.2 Dimensionality Reduction and Clustering

Using suitable representation techniques, state S_i at timestep i is represented as a vector v_i in dimension D_n. We use subscript n to denote the native dimension of the representation. The next step in typical analyses is to reduce the native dimension D_n to a lower (reduced) dimension D_r. This facilitates downstream analyses by denoising data (filtering dimensions that are less important), while simultaneously reducing computational cost. Dimensionality reduction is accomplished through the linear (PCA, SVD, NMF, AA) or non-linear techniques (Kernel PCA, Autoencoders) described in Sect. 3.1.

5.3 Dynamical Models and Analysis

Molecular systems evolve through a dynamical operator acting on successive system states. This motivates the natural observation that the data-points associated with temporal snapshots are not independent; rather, they have temporal correlations that reveal interesting aspects of underlying systems. Identification of temporally coherent subdomains is an important analysis task. The starting point for such analysis is a time-lagged covariance matrix, which is computed as the distance (normalized dot product) of a state descriptor at time t with that at time $t + \delta t$, for a suitably selected lag δt. A commonly used method, Time Lagged Independent Component Analysis (TL-ICA) uses this time-lagged covariance matrix, along with the covariance matrix at current state to define a generalized eigenvalue problem. The eigenvectors derived from this generalized eigenvalue problem correspond to the slow modes in the underlying dynamics in the system. We refer to the work of Naritomi and Fuchigami (2013) for a detailed description of this method and its use in analysing atomic trajectories. These approaches are generalized into a variational framework that aims to characterize the dominant eigenpairs of the propagation operator corresponding to the dynamical system. This is achieved by first computing a discrete approximation to the propagation operator, which uses abstractions of the self and time-lagged covariance matrices to compute transition probabilities for each state at time t to a state at time $t + \delta t$. The eigenvectors of this operator correspond to the dominant modes in the system. This general variational model is equivalent to TL-ICA if data points are represented through a linear basis. However, the variational model admits a more general basis, through the use of higher-order kernels and the underlying optimization problem is solved using conventional gradient-descent type methods.

5.4 Reaction Rates and Chemical Properties

Reactive simulations often produce diverse chemical constituents. Some of these compounds are transient, however these still require careful analysis and classification. In the simple case of two component Silica–Water system, the molecular components observed at the end of the simulations include Si–O, Si–O$_2$, OH, H$_2$ etc. (Fogarty et al. 2010). Identifying all the molecular components and corresponding chemical reaction is a difficult problem.

In order to enumerate all the molecular components, one can treat a simulation time step as a colored graph with atom type as color on the node and the existence of an edge between two atoms is decided by the bond order between the pair being greater than a cutoff value. Further the enumeration requires identification of all the distinct classes of isomorphic subgraphs of atoms. Each such class entry is either a molecule or molecular fragment present in a simple time frame. Then a hash table of such fragments is constructed to label the frequency of occurrence of reactant or product in a single time frame.

For the most common molecular fragments, often it is possible to identify reactions of kind, $A + B \rightleftharpoons AB$. Such reactions can be modeled using first order differential equations, which can be solved as:

$$N_{AB}(t) = \frac{K_f \cdot N}{K_f + K_b} \left(1 - \exp\left[-(K_f + K_b)(t - t_0)\right]\right), \qquad (16)$$

where N is total number of molecules of type A and B, N_{AB} is the number of molecules of AB; K_f, K_b are forward and backward reaction rates respectively (Saunders et al. 2022). Within simulations the computed number of molecular types can be fitted to Eq. (16) as a function of time, giving the reaction rates and equilibrium concentrations of various chemical components.

6   Concluding Remarks

In this chapter, we presented an overview of common ML techniques and formulations. We discussed how computationally expensive components of reactive atomistic simulations are formulated in ML frameworks, considerations for training ML models, tradeoffs of accuracy, need for training data, transferrability, and computational cost. While we primarily focused on reactive atomistic simulations, the models and methods discussed apply more generally to discrete element models.

The area of ML techniques for reactive simulations is extremely active and fluid. There is tremendous potential for significant new developments in the area, enabling simulation scales and scope far beyond those currently accessible. In doing so, these techniques hold the promise of new applications and domains.

Acknowledgements This work is supported by the US National Science Foundation through grants OAC 1807622, OAC 1908691 and CCF 2019263, as well as the National Institutes of Health through the grant 5R01GM130641.

References

Abadi M, Agarwal A, Barham P, Brevdo E, Chen Z, Citro C, Corrado GS, Davis A, Dean J, Devin M, Ghemawat S, Goodfellow I, Harp A, Irving G, Isard M, Jia Y, Jozefowicz R, Kaiser L, Kudlur M, Levenberg J, Mané D, Monga R, Moore S, Murray D, Olah C, Schuster M, Shlens J, Steiner B, Sutskever I, Talwar K, Tucker P, Vanhoucke V, Vasudevan V, Viégas F, Vinyals O, Warden P, Wattenberg M, Wicke M, Yu Y, Zheng X (2015) TensorFlow: large-scale machine learning on heterogeneous systems. Software available from tensorflow.org

Albert B, Kodor PG, Csányi R (2013) On representing chemical environments. Phys Rev B 87:184115

Behler J (2014) Representing potential energy surfaces by high-dimensional neural network potentials. J Phys: Condensed Matter 26:183001

Behler J (2016) Perspective: machine learning potentials for atomistic simulations. J Chem Phys 145:170901

Behler J (2015) Constructing high-dimensional neural network potentials: a tutorial review. Int J Quant Chem 115(16):1032–1050

Behler J, Lorenz S, Reuter K (2007) Representing molecule-surface interactions with symmetry-adapted neural networks. J Chem Phys 127:014705

Bishop CM, Nasrabadi NM (2006) Pattern recognition and machine learning, vol 4. Springer

Blank TB, Brown SD, Calhoun AW, Doren DJ (1995) Neural network models of potential energy surfaces. J Chem Phys 103:4129

Bradbury J, Frostig R, Hawkins P, Johnson MJ, Leary C, Maclaurin D, Wanderman-Milne S (2020) Jax: composable transformations of python+ numpy programs, p 18. http://github.com/google/jax

Brickel S, Das AK, Unke OT, Turan HT, Meuwly M (2019) Reactive molecular dynamics for the [cl-ch3-br]-reaction in the gas phase and in solution: a comparative study using empirical and neural network force fields. Elect Struc 1:024002

Case DA, Aktulga HM, Belfon K, Ben-Shalom I, Brozell SR, Cerutti DS, Cheatham TE III, Cruzeiro VWD, Darden TA, Duke RE et al (2021) Amber 2021. University of California, San Francisco

Chen TQ, Rubanova Y, Bettencourt J, Duvenaud D (2018) Neural ordinary differential equations. arxiv:1806.07366

Cho K, Merriënboer BV, Bahdanau D, Bengio Y (2014) On the properties of neural machine translation: encoder-decoder approaches. arXiv:1409.1259

Cortes C, Vapnik V (1995) Support-vector networks. Mach Learn 20(3):273–297

Cunningham JP, Ghahramani Z (2015) Linear dimensionality reduction: survey, insights, and generalizations. J Mach Learn Res 16(1):2859–2900

Curtin RR, Edel M, Lozhnikov M, Mentekidis Y, Ghaisas S, Zhang S (2018) mlpack 3: a fast, flexible machine learning library. J Open Source Softw 3(26):726

Deviller J, Balaban AT (eds) (1999) Topological indices and related descriptors in QSAR and QSPR. Gordon and Breach Science Publishers

Dick B, Freund H-J (1983) Analysis of bonding properties in molecular ground and excited states by a cohen-type bond order. Int J Quant Chem 24:747–765

Dittner M, Müller J, Aktulga HM, Hartke B (2015) Efficient global optimization of reactive forcefield parameters. J Comput Chem 36(20):1550–1561

Ester M, Kriegel H-P, Sander J, Xu X et al (1996) A density-based algorithm for discovering clusters in large spatial databases with noise. In KDD 96:226–231

Faber F, Lindmaa A, von Lilienfeld OA, Armiento R (2015) Crystal structure representations for machine learning models of formation energies. Int J Quant Chem 115(16):1094–1101

Fogarty JC, Aktulga HM, Grama AY, van Duin ACT, Pandit SA (2010) A reactive molecular dynamics simulation of the silica-water interface. J Chem Phys 132(17):174704

Frenkel D, Smit B (2002) Understanding molecular simulation from algorithms to applications. Academic Press

Furman D, Carmeli B, Zeiri Y, Kosloff R (2018) Enhanced particle swarm optimization algorithm: efficient training of reaxff reactive force fields. J Chem Theory Comput 14(6):3100–3112

Gaissmaier D, van den Borg M, Fantauzzi D, Jacob T (2022) Kvik optimiser—an enhanced reaxff force field training approach, ChemRxiv

Gan G, Ma C, Wu J (2020) Data clustering: theory, algorithms, and applications. SIAM

Gassner H, Probst M, Lauenstein A, Hermansson K (1998) Representation of intermolecular potential functions by neural networks. J Phys Chem A 102(24):4596–4605

Gastegger M, Behler J, Marquetand P (2017) Machine learning molecular dynamics for the simulation of infrared spectra. Chem Sci 8(10):6924–6935

Geiger P, Dellago C (2013) Neural networks for local structure detection in polymorphic systems. J Chem Phys 139:164105

Gers FA, Schmidhuber J, Cummins F (2000) Learning to forget: continual prediction with LSTM. Neural Comput 12(10):2451–2471

Ghiringhelli LM, Vybiral J, Levchenko SV, Draxl C, Scheffler M (2015) Big data of materials science: critical role of the descriptor. Phys Rev Lett 114:105503

Gillis N (2020) Nonnegative matrix factorization. SIAM

Glielmo A, Husic BE, Rodriguez A, Clementi C, Noé F, Laio A (2021) Unsupervised learning methods for molecular simulation data. Chem Rev 121(16):9722–9758

Golub GH, Reinsch C (1971) Singular value decomposition and least squares solutions. In: Linear algebra. Springer, pp 134–151

Goodfellow I, Bengio Y, Courville A (2016) Deep learning. MIT Press

Goodfellow I, Pouget-Abadie J, Mirza M, Xu B, Warde-Farley D, Ozair S, Courville A, Bengio Y (2014) Generative adversarial nets. Advances in neural information processing systems, vol 27

Guennebaud G, Jacob B et al (2010) Eigen v3. http://eigen.tuxfamily.org

Guo F, Wen Y-S, Feng S-Q, Li X-D, Li H-S, Cui S-X, Zhang Z-R, Hu H-Q, Zhang G-Q, Cheng X-L (2020) Intelligent-reaxff: evaluating the reactive force field parameters with machine learning. Comput Mat Sci 172:109393

Hansen N (2006) Towards a new evolutionary computation. Stud Fuzziness Soft Comput 192:75–102

Harris CR, Millman KJ, van der Walt SJ, Gommers R, Virtanen P, Cournapeau D, Wieser E, Taylor J, Berg S, Smith NJ, Kern R, Picus M, Hoyer S, van Kerkwijk MH, Brett M, Haldane A, del Río JF, Wiebe M, Peterson P, Gérard-Marchant P, Sheppard K, Reddy T, Weckesser W, Abbasi H, Gohlke C, Oliphant TE (2020) Array programming with NumPy. Nature 585(7825):357–362

Hess B, Kutzner C, Spoel DV-D, Lindahl E (2008) Gromacs 4: algorithms for highly efficient, load-balanced, and scalable molecular simulation. J Chem Theory Comput 4(3):435–447

He K, Zhang X, Ren S, Sun J (2016) Deep residual learning for image recognition. In: Proceedings of the IEEE conference on computer vision and pattern recognition, pp 770–778

Hochreiter S, Schmidhuber J (1997) Long short-term memory. Neural Comput 9(8):1735–1780

Huang G, Liu Z, Maaten LVD, Weinberger KQ (2017) Densely connected convolutional networks. In: Proceedings of the conference on computer vision and pattern recognition, pp 4700–4708

Hubin PO, Jacquemin D, Leherte L, Vercauteren DP (2016) Parameterization of the reaxff reactive force field for a proline-catalyzed aldol reaction. J Comput Chemistry 37(29):2564–2572

Iype E, Hütter M, Jansen APJ, Nedea SV, Rindt CCM (2013) Parameterization of a reactive force field using a monte carlo algorithm. J Comput Chemistry 34(13):1143–1154

Jaramillo-Botero A, Naserifar S, Goddard WA III (2014) General multiobjective force field optimization framework, with application to reactive force fields for silicon carbide. J Chem Theory Comput 10(4):1426–1439

Jia Y, Shelhamer E, Donahue J, Karayev S, Long J, Girshick R, Guadarrama S, Darrell T (2014) Caffe: convolutional architecture for fast feature embedding. In: Proceedings of the 22nd ACM international conference on multimedia, pp 675–678

Kabsch W (1976) A solution for the best rotation to relate two sets of vectors. Acta Crystallographica Sect A 32:922–923

Kadupitiya JCS, Fox GC, Jadhao V (2020) Solving newton's equations of motion with large timesteps using recurrent neural networks based operators

Katoch S, Chauhan SS, Kumar V (2021) A review on genetic algorithm: past, present, and future. Multi Tools App 80(5):8091–8126

Kaymak MC, Rahnamoun A, O'Hearn KA, van Duin ACT, Merz KM Jr, Aktulga HM (2022) Jax-reaxff: a gradient based framework for extremely fast optimization of reactive force fields. ChemRxiv

Kohn W (1996) Density functional and density matrix method scaling linearly with the number of atoms. Phys Rev Lett 76:3168

Kraft D et al (1988) A software package for sequential quadratic programming. DFVLR Obersfaffeuhofen, Germany

Kramer MA (1991) Nonlinear principal component analysis using autoassociative neural networks. AIChE J 37(2):233–243

Krizhevsky A, Sutskever I, Hinton G (2012) Imagenet classification with deep convolutional neural networks. Neural Info Proc Sys 25:01

Larsson HR, van Duin ACT, Hartke B (2013) Global optimization of parameters in the reactive force field reaxff for sioh. J Comput Chem 34(25):2178–2189

LeCun Y, Boser B, Denker JS, Henderson D, Howard RE, Hubbard W, Jackel LD (1989) Backpropagation applied to handwritten zip code recognition. Neural Comput 1(4):541–551

Li Z, Kermode JR, Vita AD (2015) Molecular dynamics with on-the-fly machine learning of quantum-mechanical forces. Phys Rev Lett 114:096405

Lloyd S (1982) Least squares quantization in PCM. IEEE Trans Info Theory 28(2):129–137

Lorenz S, Groß A, Scheffler M (2004) Representing high-dimensional potential-energy surfaces for reactions at surfaces by neural networks. Chem Phys Lett 395:4–6

Maaten LV, Hinton G (2008) Visualizing data using t-sne. J Mach Learn Res 9(11)

Maimon O, Rokach L (2005) Data mining and knowledge discovery handbook. Springer

Manzhos S, Carrington T Jr (2006) A random-sampling high dimensional model representation neural network for building potential energy surfaces. J Chem Phys 125:084109

McInnes L, Healy J, Melville J (2018) Umap: uniform manifold approximation and projection for dimension reduction. arXiv:1802.03426

McLachlan GJ (2005) Discriminant analysis and statistical pattern recognition. Wiley

Mirjalili S (2019) Genetic algorithm. In: Evolutionary algorithms and neural networks. Springer, pp 43–55

Murphy KP (2021) Machine learning: a probabilistic perspective. MIT Press

Naritomi Y, Fuchigami S (2013) Slow dynamics of a protein backbone in molecular dynamics simulation revealed by time-structure based independent component analysis. J Chem Phys 139:215102

Neubig G, Dyer C, Goldberg Y, Matthews A, Ammar W, Anastasopoulos A, Ballesteros M, Chiang D, Clothiaux D, Cohn T, Duh K, Faruqui M, Gan C, Garrette D, Ji Y, Kong L, Kuncoro A, Kumar G, Malaviya C, Michel P, Oda, M. Richardson Y, Saphra N, Swayamdipta S, Yin P (2017) Dynet: the dynamic neural network toolkit. arXiv:1701.03980

Noé F, Tkatchenko A, Müller K-R, Clementi C (2020) Machine learning for molecular simulation. Ann Rev Phys Chem 71(1):361–390 PMID: 32092281

Paszke A, Gross S, Massa F, Lerer A, Bradbury J, Chanan G, Killeen T, Lin Z, Gimelshein N, Antiga L, Desmaison A, Kopf A, Yang E, DeVito Z, Raison M, Tejani A, Chilamkurthy S, Steiner B, Fang L, Bai J, Chintala S (2019) Pytorch: An imperative style, high-performance deep learning library. In: Wallach H, Larochelle H, Beygelzimer A, d' Alché-Buc F, Fox E, Garnett R (eds) Advances in neural information processing systems, vol 32. Curran Associates, Inc, pp 8024–8035

Phillips JC, Braun R, Wang W, Gumbart J, Tajkhorshid E, Villa E, Chipot C, Skeel RD, Kale L, Schulten K (2005) Scalable molecular dynamics with NAMD. J Comput Chem 26(16):1781–1802

Rish I (2001) An empirical study of the naive bayes classifier. In: IJCAI 2001 workshop on empirical methods in artificial intelligence, vol 3, pp 41–46

Rumelhart DE, Hinton GE, Williams RJ (1986) Learning representations by back-propagating errors. Nature 323(6088):533–536

Rupp M, Tkatchenko A, Müller K-R, von Lilienfeld OA (2012) Fast and accurate modeling of molecular atomization energies with machine learning. Phys Rev Lett 108:058301

Rusch TK, Mishra S (2021) Unicornn: a recurrent model for learning very long time dependencies. arxiv:2103.05487

Ruspini EH (1969) A new approach to clustering. Info Control 15(1):22–32

Sabne A (2020) XLA: compiling machine learning for peak performance. Google Res

Sanderson C, Curtin R (2016) Armadillo: a template-based c++ library for linear algebra. J Open Source Softw 1(2):26

Sanderson C, Curtin R (2020) An adaptive solver for systems of linear equations. In: 2020 14th international conference on signal processing and communication systems (ICSPCS). IEEE, pp 1–6

Sauceda HE, Gastegger M, Chmiela S, Müller K-R, Tkatchenko A (2020) Molecular force fields with gradient-domain machine learning (gdml): comparison and synergies with classical force fields. J Chem Phys 153:124109

Saunders M, Wineman-Fisher V, Jakobsson E, Varma S, Pandit SA (2022) High-dimensional parameter search method to determine force field mixing terms in molecular simulations. Langmuir

Saxena A, Prasad M, Gupta A, Bharill N, Patel OP, Tiwari A, Er MJ, Ding W, Lin C-T (2017) A review of clustering techniques and developments. Neurocomputing 267:664–681

Schmidhuber J (2015) Deep learning in neural networks: an overview. Neural Netw 61:85–117

Schubert E, Sander J, Ester M, Kriegel HP, Xu X (2017) Dbscan revisited, revisited: why and how you should (still) use dbscan. ACM Trans Database Syst (TODS) 42(3):1–21

Senftle TP, Hong S, Islam MM, Kylasa SB, Zheng Y, Shin YK, Junkermeier C, Engel-Herbert R, Janik MJ, Aktulga HM et al (2016) The reaxff reactive force-field: development, applications and future directions. NPJ Comput Mater 2(1):1–14

Shalev-Shwartz S, Ben-David S (2014) Understanding machine learning: from theory to algorithms. Cambridge University Press

Shchygol G, Yakovlev A, Trnka T, van Duin ACT, Verstraelen T (2019) Reaxff parameter optimization with monte-carlo and evolutionary algorithms: Guidelines and insights. J Chem Theory Comput 15(12):6799–6812

Sidky H, Chen W, Ferguson AL (2020) Machine learning for collective variable discovery and enhanced sampling in biomolecular simulation. Mol Phys 118(5):e1737742

Simonyan K, Zisserman A (2014) Very deep convolutional networks for large-scale image recognition. arXiv:1409.1556

Stuart SJ, Tutein AB, Harrison JA (2000) A reactive potential for hydrocarbons with intermolecular interactions. J Chem Phys 112(14):6472–6486

Szegedy C, Ioffe S, Vanhoucke V, Alemi AA (2017) Inception-v4, inception-resnet and the impact of residual connections on learning. In: 31 AAAI conferences on artificial intelligence

Szegedy C, Liu W, Jia Y, Sermanet P, Reed S, Anguelov D, Erhan D, Vanhoucke V, Rabinovich A (2015) Going deeper with convolutions. In: Proceedings of the IEEE conference on computer vision and pattern recognition, pp 1–9

Szegedy C, Vanhoucke V, Ioffe S, Shlens J, Wojna Z (2016) Rethinking the inception architecture for computer vision. In: Proceedings of the conference on computer vision and pattern recognition, pp 2818–2826

Tersoff J (1989) Modeling solid-state chemistry: interatomic potentials for multicomponent systems. Phys Rev B 39(8):5566

Theano Development Team. Theano: a python framework for fast computation of mathematical expressions, May 2016. arxiv:abs/1605.02688

Thompson AP, Aktulga HM, Berger R, Bolintineanu DS, Brown WM, Crozier PS, Veld PJ, Kohlmeyer A, Moore SG, Nguyen TD et al (2022) Lammps-a flexible simulation tool for particle-based materials modeling at the atomic, meso, and continuum scales. Comput Phys Commun 271:108171

Thompsona A, Swilerb P, Trottc R, Foilesd M, Tucker J (2015) Spectral neighbor analysis method for automated generation of quantum-accurate interatomic potentials. J Comput Phys 285:316–330

Trnka T, Tvaroska I, Koca J (2018) Automated training of reaxff reactive force fields for energetics of enzymatic reactions. J Chem Theory Comput 14(1):291–302

Unke OT, Chmiela S, Sauceda HE, Gastegger M, Poltavsky I, Schütt KT, Tkatchenko A, Muller K-R (2021) Machine learning force fields. Chem Rev 121:10142–10186

Valle M, Oganov AR (2010) Crystal fingerprint space—a novel paradigm for studying crystal-structure sets. Acta Crystallographica A A66:507–517

van Duin ACT, Baas JMA, Graaf BVD (1994) Delft molecular mechanics: a new approach to hydrocarbon force fields. Inclusion of a geometry-dependent charge calculation. J Chem Soc Faraday Trans 90(19):2881–2895

van Duin ACT, Dasgupta S, Lorant F, Goddard WA (2001) Reaxff: a reactive force field for hydrocarbons. J Phys Chem A 105(41):9396–9409

Vapnik V (1991) Principles of risk minimization for learning theory. Advances in neural information processing systems, vol 4

Virtanen P, Gommers R, Oliphant TE, Haberland M, Reddy T, Cournapeau D, Burovski E, Peterson P, Weckesser W, Bright J, van der Walt SJ, Brett M, Wilson J, Millman KJ, Mayorov N, Nelson ARJ, Jones E, Kern R, Larson E, Carey CJ, Polat İ, Feng Y, Moore EW, VanderPlas J, Laxalde D, Perktold J, Cimrman R, Henriksen I, Quintero EA, Harris CR, Archibald AM, Ribeiro AH, Pedregosa F, van Mulbregt P, SciPy 1.0 contributors. SciPy 1.0: fundamental algorithms for scientific computing in python. Nat Methods 17:261–272

Weiss K, Khoshgoftaar TM, Wang D (2016) A survey of transfer learning. J Big Data 3(1):1–40

Zagoruyko S, Komodakis N (2016) Wide residual networks. arXiv:1605.07146

Zhong YD, Dey B, Chakraborty A (2019) Symplectic ode-net: learning hamiltonian dynamics with control. arxiv:1909.12077

Zhu C, Byrd RH, Lu P, Nocedal J (1997) Algorithm 778: L-bfgs-b: fortran subroutines for large-scale bound-constrained optimization. ACM Trans Math Softw (TOMS) 23(4):550–560

A Novel In Situ Machine Learning Framework for Intelligent Data Capture and Event Detection

T. M. Shead, I. K. Tezaur, W. L. Davis IV, M. L. Carlson, D. M. Dunlavy, E. J. Parish, P. J. Blonigan, J. Tencer, F. Rizzi, and H. Kolla

Abstract We present a novel framework for automatically detecting spatial and temporal events of interest in situ while running high performance computing (HPC)

T. M. Shead
Sandia National Laboratories, Albuquerque, NM, USA
e-mail: tshead@sandia.gov

I. K. Tezaur
Sandia National Laboratories, Livermore, CA, USA
e-mail: ikalash@sandia.gov

W. L. Davis IV
Sandia National Laboratories, Albuquerque, NM, USA
e-mail: wldavis@sandia.gov

M. L. Carlson
Sandia National Laboratories, Livermore, CA, USA
e-mail: maxcarl@sandia.gov

D. M. Dunlavy
Sandia National Laboratories, Albuquerque, NM, USA
e-mail: dmdunla@sandia.gov

E. J. Parish
Sandia National Laboratories, Livermore, CA, USA
e-mail: ejparis@sandia.gov

P. J. Blonigan
Sandia National Laboratories, Livermore, CA, USA
e-mail: pblonig@sandia.gov

J. Tencer
Sandia National Laboratories, Albuquerque, NM, USA
e-mail: jtencer@sandia.gov

F. Rizzi
NexGen Analytics, Sheridan, WY, USA
e-mail: francesco.rizzi@ng-analytics.com

H. Kolla (✉)
Sandia National Laboratories, Livermore, CA, USA
e-mail: hnkolla@sandia.gov

© The Author(s) 2023
N. Swaminathan and A. Parente (eds.), *Machine Learning and Its Application to Reacting Flows*, Lecture Notes in Energy 44, https://doi.org/10.1007/978-3-031-16248-0_3

simulations. The new framework – composed from *signature*, *measure*, and *decision* building blocks with well-defined semantics – is tailored for parallel and distributed computing, has bounded communication and storage requirements, is generalizable to a variety of applications, and operates in an unsupervised fashion. We demonstrate the efficacy of our framework on several cases spanning scientific domains and applications of event detection: optimized input/output (I/O) in computational fluid dynamics simulations, detecting events that can lead to irreversible climate changes in simulations of polar ice sheets, and identifying optimal space-time subregions for projection-based model reduction. Additionally, we demonstrate the scalability of our framework using a HPC combustion application on the Cori supercomputer at the National Energy Research Scientific Computing Center (NERSC).

1   Introduction

Scientific investigations – whether computational, experimental or observational – are ever expanding to include larger sets of coupled physics spanning broader ranges of scales, and the volumes of data generated from these investigations consistently outpace the growth of computational and data storage resources. As a consequence, specifically in the area of HPC modeling and simulation, the process of mining scientific data to glean insight is shifting from one of a posteriori to one of in situ analysis, i.e., analysis performed simultaneously with a simulation while sharing resources with it. Capturing events of interest to scientists in complex, high-fidelity HPC simulations is difficult because it is rarely feasible to export the entire simulation state at every timestep. Crucial stages in the development of events can be lost between checkpoints, and ephemeral events can be missed altogether, making a posteriori event detection problematic. Identifying events in situ is equally challenging, as traditional analysis algorithms that assume global access to data require excessive communication bandwidth.

Machine learning (ML) is being applied to scientific data for various purposes, including establishing constitutive laws, developing mathematically and statistically compact models of governing physics, identifying embedded patterns, dimensionality reduction, parameter importance and sensitivity analysis, and uncertainty quantification (UQ). In this work we focus on one specific application of ML: in situ *event detection*. Specifically, we seek to develop event detection algorithms that are:

- **Generalizable**: deployable in a variety of different scientific computing domains without the need for application-specific tuning;
- **Unsupervised**: able to operate without labeled examples defining events of interest;
- **Low Overhead**: requiring minimal communication between processors;
- **Online**: able to make predictions with minimal retention of data from prior timesteps.

To motivate the main contributions of this chapter, we first provide a brief overview of related past work.

1.1 Overview of Related Work

Event detection is related to anomaly detection, since the purpose of each is to detect behavior that is locally different. There has been substantial previous research on developing streaming anomaly detection algorithms for HPC simulation data. However, many of these algorithms require significant communication between processors. For example, Wu et al. (2014) proposed the Random Subspace Forest (RS-Forest) algorithm in which decision trees with random splits and random thresholds are used to construct a density estimate over the data observations in a continuous feature space. While this algorithm is very fast for local or shared memory applications, it is not communication efficient in this context because it requires sharing the entire RS-Forest data model across all processors. Similarly, Kernel Density Estimation (KDE) has been proposed for online anomaly detection (Ahmed 2009), but also requires significant communication between processors.

Some anomaly detection methods have been designed for parallel implementation with low communication overhead. Zhao et al. (2009) proposed a parallel framework for k-means clustering that could be adapted for anomaly detection. However, k-means clustering requires a user-defined number of clusters k, and performance is often strongly dependent on the selected value of this variable. Such sensitivity to algorithm parameters is undesirable for unsupervised in situ event detection.

Application-specific event detectors have also been developed. These include detectors to flag when ignition has occurred in combustion simulations (Bennett et al. 2016) and tropical cyclone trackers for climate simulations (Ullrich and Zarzycki 2016; Zhao et al. 2009). These algorithms make use of significant domain knowledge and are only applicable in the specific field for which they were developed, which is contrary to our goal of developing generalizable algorithms.

Ensemble anomaly detection techniques, such as iForest (Liu et al. 2012) and iNNE (Bandaragoda et al. 2014), are often considered to be robust and highly generalizable. Furthermore, these techniques have been shown to be compatible with data sub-sampling. The disadvantage of these methods is that they require communication to share the ensemble model between processors. For large ensembles this overhead can be prohibitively high.

Finally, it is not clear that conventional anomaly detection algorithms are well-suited for event detection in simulations. Because simulations often make use of highly refined meshes to resolve complex physical phenomena, an event of interest could occur over tens of thousands of mesh points, making it well-represented in the data, and therefore not anomalous. Moreover, comparisons to previous timesteps also are not straightforward, since many simulations exhibit significant drift over time: what is unusual at one timestep might become the norm later in time.

1.2   Contributions and Organization

We present herein a novel framework for applying ML to detect events of interest in situ in HPC simulation data. In this context, "events of interest" can be defined as any local dynamics in a region that differ significantly from the dynamics of other regions or timesteps. Our framework is tailored for parallel and distributed computing with the data typically representing a space-time domain of interest, with the spatial domain distributed across computing resources (processors/nodes) and data along the time dimension arriving in a streaming manner.

Consider a region handled by a single processor exhibiting behavior that differs significantly from the regions on other processors. Such a region could be considered interesting even if the behavior persists over multiple timesteps. An example of this type of event could be a tropical cyclone that persists over many timesteps in a weather simulation but is geographically localized. We refer to events of this type as *spatial* events of interest. Conversely, a sudden change across all processors from one timestep to the next could also be considered interesting. An example of this type of event could be simultaneous ignition across an entire domain in a combustion simulation. We refer to these as *temporal* events of interest.

This research presents a framework for developing in situ spatial and temporal event detection algorithms with tightly bounded communication and storage requirements, composed from *signature*, *measure*, and *decision* building blocks with well-defined semantics. The goal of this framework is to facilitate event detection in a computationally scalable and efficient manner, while allowing the flexibility to compose a learning workflow best suited for the scientific domain and problem at hand. The proposed framework can be used not only to optimize I/O within an HPC simulation (by flagging the locations where events of interest occur so that only a subset of the simulation state is stored to disk), but also to detect scientifically meaningful phenomena within HPC simulations and even to improve a simulation's accuracy/efficiency. A detected event can be used as a trigger for mesh and/or timestep refinement, e.g., Adaptive Mesh Refinement (AMR) (Berger and Oliger 1984).

The remainder of this chapter is organized as follows. The specific components of the proposed event detection framework are detailed in Sect. 2. In Sect. 3 we present results from three use cases that demonstrate the versatility and composability of the framework. The use cases span different scientific domains and different applications of event detection: optimized I/O in fluid flow simulations (Sect. 3.1), detecting events that are scientifically interesting in ice sheet simulations (Sect. 3.2), and identifying optimal space-time sub-regions for projection-based model reduction (Sect. 3.3). Section 3.4 presents results, using an exemplar turbulent combustion simulation, that demonstrate the scalability and computational efficiency of the framework when deployed in parallel computing simulations. Finally, conclusions are provided in Sect. 4.

2 Approach

Our framework for event detection is as follows. First, we assume a simulation domain with any number of dimensions. We further assume that the domain is divided into a set of P analysis partitions, where each analysis partition $p_i = 0, \ldots, P - 1$ is a spatially-contiguous subset of mesh points of the simulation domain. Each partition is always associated with a single processor so that analysis partitions never straddle processor boundaries or migrate from one processor to another throughout the simulation. Thus, a single processor will be responsible for one-to-many analysis partitions, with the size and number of partitions chosen based on the problem domain (Fig. 1).

Next, we execute the following workflow at each timestep of the running simulation. For each analysis partition p_i we compute a *signature* s_i, a fixed-length vector representing the simulation state within that partition where $|s_i| \ll |p_i|$ (Fig. 2). Conceptually, signatures are compressed, low-dimensional representations of an analysis partition's content, and our intent is that the signature should contain crucial aspects

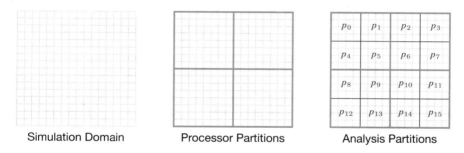

Simulation Domain   Processor Partitions   Analysis Partitions

Fig. 1 Example simulation domain (gray), split across processors (green), and divided into analysis partitions (blue)

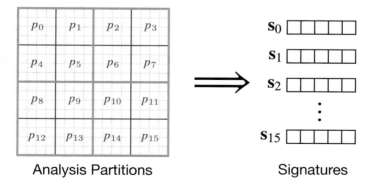

Analysis Partitions   Signatures

Fig. 2 Each analysis partition is represented by a low-dimensional signature

Table 1 Signature functions

Signature	Description
fieda	Vector of feature importance values described in Ling et al. (2017)
fmm	Vector based on the Feature Moment Metric algorithm in Konduri et al. (2018)
mean	Vector of mean values for each simulation feature
minimax	Vector of minimum and maximum values for each simulation feature
quartile	Vector of quartile boundaries for each simulation feature (a generalization of minimax)
svd	Vector of singular values computed using an SVD on the flattened partition state matrix

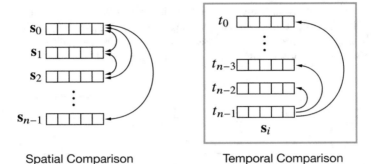

Spatial Comparison   Temporal Comparison

Fig. 3 Signatures can be compared all-to-all across analysis partitions to identify spatial events (left), and current signatures can be compared to previous signatures within partitions to identify temporal events (right)

of the state of the simulation within that partition, stored in such a way that changes across space or time can be detected by subsequent analysis of that representation.

As an example, for a simulation with state variables $F \in \mathbb{R}^n$, a signature could be vector of size $2|F|$ containing the minimum and maximum value for each variable $f \in F$ within the partition. Of course, this is only one possible signature type among many (we call this type *minimax*); we provide the subset of signature functions used in our experiments in Table 1. Note that, because analysis partitions are always associated with a single processor, computing signatures can be a purely local operation. Further, because signatures are a small, fixed size relative to the partitions they represent, they can be broadcast to other processors for spatial (partition-to-partition) comparisons and stored between timesteps for temporal (timestep-to-timestep) comparisons (Fig. 3). The user can choose, based on domain knowledge and the problem specifics, the set of features used to compute signatures. This set could consist of all of the state variables, a subset, derived variables, or any combination thereof; the only requirement is that the same set of features be used across all analysis partitions.

Given a set of signatures, we can compute spatial or temporal *measures* to identify events. Measures are functions applied to signatures that detect changes across space

Table 2 Spatial measures

Measure	Description
dbscan	Uses DBSCAN (Ester et al. 1996) to flag outlier signatures as events
m1	M1 metric described in Ling et al. (2017)
m1-hellinger	Modified version of the Hellinger distance used as a spatial metric, described in Konduri et al. (2018)
msd	Compares the distance between one signature and the mean signature for all partitions
sigscal	Normalizes the signature matrix using the product of the inverse of each signature and measures the ability of each signature to drive values in the product to zero

Table 3 Temporal measures

Measure	Description
changefreq	Counts the number of changes (dramatic increases or decreases across any two timesteps within a temporal window)
maxchange	Uses the maximum change across any timesteps within the current temporal window
mse	Based on mean squared error between the current temporal block ($S \times T_{window}$ matrix) and previous temporal block
psd	Estimates power spectral density (power spectrum) for each feature within the temporal block ($S \times T_{window}$ matrix) using Welch's method (Welch 1967)
svd	Ratio of the largest non-zero singular value to the smallest non-zero singular value

or time. Spatial measures compare signatures across analysis partitions to identify spatial events; typically, they compare the signature for a given partition to every other partition's signature, which requires communication. Temporal measures compare an analysis partition's current signature to its past signatures and are thus completely local, requiring only storage of a finite number of signatures from previous timesteps. In both cases, the output of the measure is a per-analysis-partition continuous scalar value indicating how interesting the partition's state is at the current timestep. We list representative spatial and temporal measures implemented for our experiments in Tables 2 and 3, respectively.

Finally, we use *decision* functions to convert continuous per-analysis-partition measures into boolean values to indicate whether the partitions should be flagged as containing events of interest for the current timestep. Decision functions are purely local, requiring no communication. Table 4 describes the decision functions that we used in our experiments.

We refer to a combination of signature, measure, and decision functions as an *algorithm* for in situ event detection; because we have many instances of each type, and they can be combined almost without exception, there are many possible algo-

Table 4 Decision functions

Decision	Description
threshold	Flag a partition when its measure exceeds a fixed threshold
percentile	Flag a partition when its measure exceeds the nth percentile of the measure for all partitions
memory	Decision modifier which continues to flag a partition for a fixed number of timesteps after another decision function has flagged it

rithms that can be created with just a few components (and the set of components continues to grow as we explore new ideas). The few incompatibilities tend to be driven by the expected inputs for a component. For example, it makes little sense to combine the *dbscan* measure with the *percentile* decision, since the former only produces binary values as output, and the latter is only useful with a continuous distribution as input.

3 Results

In this section we demonstrate our methodology on three important use cases for in situ machine learning: data capture for optimizing I/O (Sect. 3.1), detection of interesting physical events (Sect. 3.2), and facilitating reduced order model construction (Sect. 3.3). The use cases represent different scientific domains, but have similarities with reacting flows: Sect. 3.1 pertains to low speed non-reacting turbulent flows with passive tracers; Sect. 3.2 pertains to an incompressible fluid flow (glacier ice) solved using Stokes flow equations; Sect. 3.3 pertains to supersonic flow with shock. The purpose behind choosing such different use cases is to illustrate the generality of our detection algorithms.

3.1 Data Capture for Optimal I/O: Mantaflow Experiments

In our initial round of experiments, our focus is on testing the utility of our framework, and quantifying whether it could be used for meaningful reductions in I/O. We begin by creating a reference implementation using Python (2022), Numpy (Walt et al. 2011), Scipy (Jones et al. 2001) and Scikit-Learn (Pedregosa et al. 2011). To simplify development and support rapid iteration, these experiments use Mantaflow (Thuerey and Pfaff 2018) – an open source library targeting fluid simulation research in computer graphics and machine learning – for the simulation. Despite being a serial code, Mantaflow's Python scene definition interface makes it ideal for integration and rapid testing with our algorithms. All of our Mantaflow experiments are conducted using two-dimensional (2D) simulations for speed and ease of visualization.

Fig. 4 Density field visualization from the *small plumes* Mantaflow simulation at one timestep. Darker colors signify higher density

To run the simulations, we created a driver script that loads an experiment definition file specifying the simulation setup, analysis partitions, simulation features to use for signature generation, as well as the signature, measure and decision functions to use for the analysis. Because the driver script also provides the simulation outer loop, it is trivial to run our analysis code alongside the simulation in situ.

We designed several Mantaflow simulations to test our event detection approach at different scales; for this chapter, we focus on our *small plumes* simulation, which has four state variables (density, pressure, x-velocity and y-velocity) and features three steady turbulent plumes of buoyant fluid using a 64×256 grid and running for 300 timesteps (Fig. 4).

Since the goal for our I/O use case is to minimize the amount of data saved to disk while simultaneously maximizing the number of events captured, a fundamental challenge is defining a sensible ground truth: for any given simulation, there is no well-defined way to specify which parts of the simulation should be considered events of interest (and thus flagged by our framework for subsequent storage to disk). To address this, we opted to create our own explicit ground truth by injecting random "depth charge" anomalies into the simulation. To do so, we generate a random value for each simulation cell at each timestep. At any cell where the random value exceeds a threshold, the simulation density is increased by a substantial amount, and the cell is marked as anomalous using an additional simulation state variable. Thus, the depth charge anomalies occur at random timesteps and locations within the simulation domain, and the anomalies state variable keeps track of where they occur (Fig. 5). The

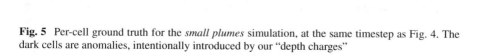

Fig. 5 Per-cell ground truth for the *small plumes* simulation, at the same timestep as Fig. 4. The dark cells are anomalies, intentionally introduced by our "depth charges"

overall impact is to introduce physically-implausible aberrations into the simulation which surely qualify as events worthy of detection. Having created the anomalies ourselves, we can then evaluate the algorithm's ability to flag them as events of interest. Note that, even with our explicitly injected anomalies, there is still ambiguity surrounding the question of which cells/partitions should be flagged as events: while the sudden onset of an anomaly is obviously an event worth noting, the threshold at which it should cease to be anomalous as it disperses is still arbitrary. Despite these shortcomings, our "depth charges" provide a quantitative way to compare performance among different algorithms tested using the framework.

The behavior of our driver script is as follows. First, at each timestep, we use the Mantaflow API to run the solver for that step. Next, we extract the simulation state (density, pressure, velocities and anomaly ground truth) and save the raw data to disk. We then divide the simulation grid into 8×8 analysis partitions, since our framework requires multiple analysis partitions even when there is a single processor, as is the case for the serial Mantaflow simulations. Next, we compute the per-partition signatures. To support computing temporal measures and because the Mantaflow simulations are so small, we store every signature computed at every timestep, though we assume in practice that an HPC simulation would retain a smaller number of the most recent signatures. The set of per-analysis-partition signatures are then passed to the measure function to generate per-partition measures. Since the measure function has access to the signatures for every partition and every timestep, it can calculate a measure based on a comparison of signatures across every analysis partition (a spatial measure), a comparison of signatures across time for a single partition (a temporal measure), or a hybrid of the two. Because our Mantaflow experiments run on a single process, no communication is necessary, unlike the HPC experiments described in Sect. 3.4. We save the measures computed for each partition to disk for subsequent visualization. Finally, the measure values are passed to the decision function to be flagged as *events* or not, and those decisions are written to disk.

Once the simulation is complete, we convert the simulation features, anomalies, measures and decisions stored on disk to color-mapped images, generating movies using the open source Imagecat (2022) library for compositing and Ffmpeg (2019) for encoding. The simulation movies provide a qualitative way to evaluate algorithm behaviors (Fig. 6).

For quantitative comparisons, we used the decision data to generate several metrics, including: (1) the percentage of simulation domain cells that are flagged as events by our framework, both per-timestep and for the simulation as a whole, and (2) the percentage of ground truth anomalous cells that are contained within partitions flagged as events, per-timestep and for the simulation as a whole. We refer to this latter metric as "recall".

Our early experiments were focused on identifying useful combinations of signature-measure-decision building blocks and developing intuition around their strengths and weaknesses. In this preliminary exploration, the percentage of simulation cells flagged as events ranges from 4.3% (excellent, a twenty-fold decrease in storage requirements) to 75% (likely not worth the effort), while our recall metric ranges from 35.4% (good) to 99.8% (excellent). One combination that produces con-

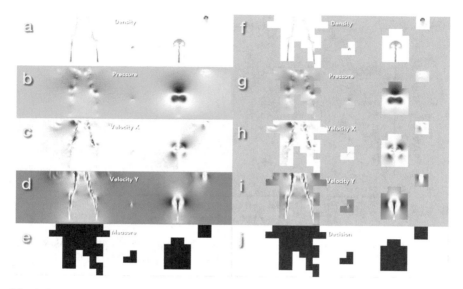

Fig. 6 Sample frame from a Mantaflow experiment movie: simulation state (**a**)–(**d**), per-analysis-partition measure (**e**) and decisions (**j**), simulation state masked by decisions (**f**)–(**i**)

sistently good results for a wide range of parameters used is the *quartile* signature, *dbscan* spatial measure with Euclidean distance, and *threshold* decision function. Figure 7 plots the total percentage of flagged analysis partitions (lower is better) versus the anomaly recall (higher is better) for a set of experiments using this combination. The result is intentionally evocative of a receiver operating characteristic curve, emphasizing the trade-offs inherent in our desire to maximize the number of detected events while minimizing the total number of analysis partitions flagged for storage to disk.

The *dbscan* measure used in these experiments has two main parameters: ε, the threshold distance below which two signatures are considered "neighbors"; and N_p, the minimum number of neighboring signatures required to form a "neighborhood." Once all of the neighborhoods in a collection of signatures are identified, any signatures not in a neighborhood are, by definition, flagged as interesting events.

We tested combinations of ε and N_p using grid search, varying ε values between 0.1 and 1.0 and N_p values between 1% and 50% of the total number of analysis partitions. At very low values of ε, we rapidly achieved high recall, approaching 100%. Values over 0.3 led to a rapid reduction in recall, dropping to around 8% for an ε of 1.0. Varying N_p had much less effect, with most values below 40% having little effect on recall. We are encouraged that many parameter combinations produce results near the knee of the curve in Fig. 7, indicating that the algorithm is robust for a wide range of reasonable DBSCAN parameters. We chose $\varepsilon = 0.2$ and $N_p = 2\%$ as the best parameters for this data, with results shown in Fig. 8.

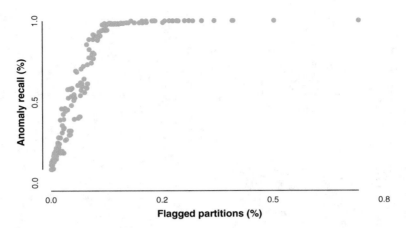

Fig. 7 Flagged analysis partitions versus anomaly recall for the *quartile-dbscan* Mantaflow experiments

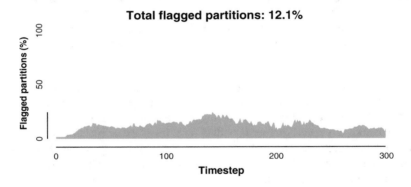

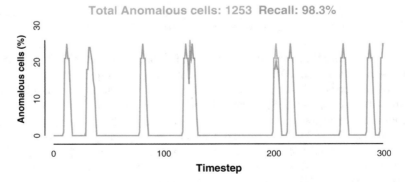

Fig. 8 Flagged analysis partitions (top) versus recall (bottom) for *quartile-dbscan* Mantaflow experiment with $\varepsilon = 0.2$ and $N_p = 2\%$

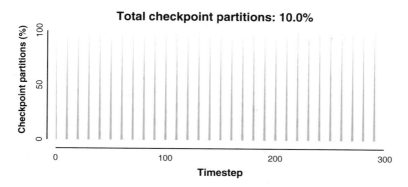

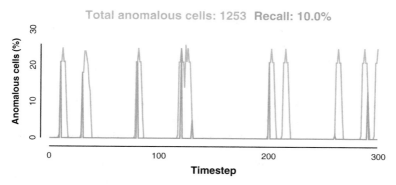

Fig. 9 Saved analysis partitions (top) versus recall (bottom) for a simulation saving a checkpoint at every tenth timestep for the Mantaflow experiment

In this case, an experimenter using the *quartile-dbscan* algorithm to decide which analysis partitions should be saved to disk would end up capturing 98.3% of the anomalies, while storing just 12.1% of the data. This is especially striking when we compare it to typical uniform temporal check-pointing of HPC simulation data: the experimenter who simply saves the entire simulation state at every tenth timestep as in Fig. 9 would use roughly the same amount of disk space (10% vs. 12.1%), while only capturing 10% of the interesting events!

We performed temporal anomaly detection experiments using similar techniques. One comparable result used the *minimax* signature, the *maxchange* measure, and the *threshold* decision function, producing a recall of 96.3% while flagging only 24% of the data.

3.2 Detecting Physical Phenomena: Marine Ice Sheet Instability (MISI)

While the in situ event detection framework described herein was originally developed for the purpose of optimizing HPC simulation output, the proposed approach can also be used to detect physical phenomena present in HPC simulation data to further our understanding of the underlying physical processes. Here, we describe a specific instance of this use case, in which our framework facilitates the study of the hypothesized Marine Ice Sheet Instability (MISI) using simulation data from the MPAS-Albany Land Ice (MALI) model (Hoffman et al. 2018), the land ice component of the U.S. Department of Energy's Energy Exascale Earth System Model (E3SM) (Leung et al. 2020).

The Marine Ice Sheet Instability, first introduced in the 1970s (Weertman et al. 1974; Thomas and Bentley 1978), hypothesizes that ice sheets grounded below sea-level may destabilize in a runaway fashion once the grounding line, the boundary between where the ice sheet is grounded and floating, reaches a point where the bedrock has a reverse slope gradient (Fig. 10) (Bamber et al. 2009). Once the bedrock beneath the grounding line is reverse sloping (i.e., it becomes deeper moving inland), ice thickness at the grounding line increases, leading to faster ice flow and greater ice flux divergence. As the flux at the grounding line increases, thinning at and upstream of the grounding line increases, causing the boundary between floating and grounded ice to move further inland. The result is a self-reinforcing mechanism that can cause rapid and irreversible ice sheet retreat and rapid sea level rise (Robel et al. 2019; Joughin and Alley 2019). Since the grounding line is often stabilized by the presence of an ice shelf (an extended region of floating ice that is dynamically connected to the grounded ice upstream of it), which has the effect of buttressing the ice and limiting ice flux at the grounding line, MISI is often triggered by the thinning or loss of ice shelves (Pattyn and Morlighem 2020). Satellite and modeling evidence suggests that MISI is underway in parts of the West (e.g., the Thwaites and Pine Island glacier) and East (e.g., the Totten glacier) Antarctic Ice Sheet (Robel et al. 2019; Joughin

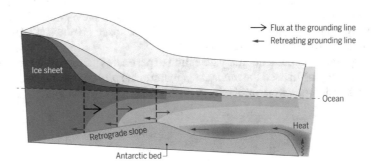

Fig. 10 Marine Ice Sheet Instability triggered by an unstable grounding line retreat on retrograde bedrock slope. Figure adapted from Pattyn and Morlighem (2020)

and Alley 2019; Gardner et al. 2018; Young et al. 2011). While it is theoretically possible to identify locations prone to MISI by combining bedrock elevation data with information on retrograde bedrock slopes, this approach is not feasible since bedrock elevation data are limited. Moreover, the retrograde bed slope alone is likely not a sufficient proxy for MISI, as it does not take into account important features relevant to MISI, e.g., ice flow speed and ice flux.

Our approach herein is to investigate the utility of our event detection framework in identifying the onset of MISI. Accordingly, we applied our event detection algorithms to two simulations datasets: (1) an idealized Antarctic BUttressing Model Intercomparison Project simulation (ABUMIP) (Sun et al. 2020), and (2) a predictive simulation of the Antarctic Ice Sheet with realistic climate forcing (Seroussi et al. 2020). Following the naming convention introduced in Sun et al. (2020) and Seroussi et al. (2020), respectively, we refer to these datasets as abuk and exp05, respectively. Both simulations start with a realistic present-day initial condition obtained by performing an adjoint-based optimization using the MALI model (Perego et al. 2014). They then simulate ice flow over Antarctica on a variable-resolution three-dimensional (3D) tetrahedral grid. The output from these simulations is subsequently mapped onto a 2D structured quadrilateral grid having a uniform resolution of 8 km (Fig. 11), for the purposes of analysis and comparison to other land ice models (Seroussi et al. 2020). In the abuk experiment, Antarctica's ice shelves are removed instantaneously, and we perform a simulation in which the formation of new floating ice is prevented and no change in external atmospheric or oceanic forcing is applied. Although unrealistic, this scenario provides an extreme upper bound on sea-level contributions from Antarctica, and exhibits the full potential of MISI (Sun et al. 2020). As such, the abuk dataset is ideal for "calibrating" (i.e., determining a reasonable set of features and analysis partition sizes) and "validating" (i.e., ensuring that reasonable analysis partitions are flagged as interesting) our event detection framework before applying it to the more realistic exp05 scenario. The second experiment, exp05, is a standard test case in the ISMIP6 (Ice Sheet Model Intercomparison Project 6) experiments (Seroussi et al. 2020), and is meant to be a realistic predictive simulation of the Antarctic Ice Sheet state with atmospheric and oceanic forcing[1] under the RCP8.5 (Representative Concentration Pathway 8.5) (IPCC 2021) radiative forcing emissions scenario, which corresponds to the likely outcome if society does not make concerted efforts to cut greenhouse emissions during the remainder of the twenty-first century (Edwards et al. 2021). For initial prototyping, our event detection algorithms are applied to the datasets a posteriori; integration of these algorithms into the MALI code for true in situ analyses will be the subject of future work. For the abuk dataset, there are 51 solution snapshots, corresponding to a 500 year simulation, with data saved every 10 years; for exp05, there are 86 solution snapshots, corresponding to an 85 year simulation, with data saved every year.

Prior to presenting our main results, we discuss some nuances pertaining to the generation of analysis partitions for the land ice datasets considered herein. For both the abuk and exp05 datasets, the underlying computational domain onto which the

---

[1] For details regarding these forcings, the reader is referred to Table 2 of Seroussi et al. (2020).

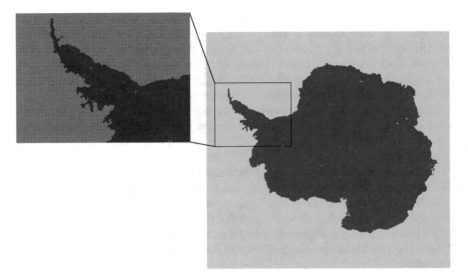

Fig. 11 "Full" 6088 km × 6088 km domain for the exp05 dataset, with active cells shown in blue. Left panel shows a close-up of the Antarctic peninsula and the structured 8 km quadrilateral mesh with which the problem is discretized

MALI output is mapped is a 6088 km × 6088 km square grid, discretized using 761 quadrilateral elements in each coordinate direction (Fig. 11). To determine which cells within this computational domain are "active" (ice-covered), a time-dependent mask derived from the ice thickness was computed at each timestep based on an ice thickness criterion: only cells in which the ice thickness is greater than 10 m are deemed "active" in each timestep. An important feature of masks derived in this way is that the masks, and hence the geometries on which the simulation proceeds, change in time: before solving for the ice sheet state at each time-step, inactive cells are removed from the mesh on which the simulation proceeds. While it would be possible to uniformly partition the "full" 761×761 element grid into P analysis partitions to use for our event detection workflow, such an approach would lead to an imbalanced set of partitions, in which many partitions would have few (or even zero) elements. Using an analysis partition set of this type could bias the event detection, especially when statistics-based signatures are employed. One approach to avoid this problem is to partition only the active grid, but this second approach also has several downsides: (1) its computational cost would likely preclude in situ analyses, and (2) with analysis partitions that change in time, it is not clear how to track temporal events using this methodology. To avoid these issues, we adopted a third approach, in which we created a mask (termed the "analysis partitioning mask") that was only slightly larger than the maximum ice extent across all simulation times for a given dataset, and created a single partition of the geometry defined by this mask prior to performing event detection. In the present study, we consider two types of analysis partitioning masks:

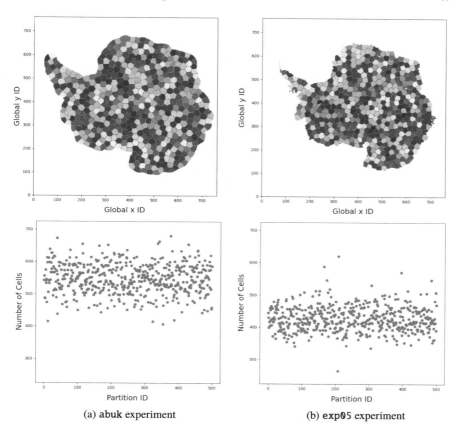

Fig. 12 Illustration of 500 analysis partitions (top panel) obtained using k-means clustering, and cell-counts for each analysis partition (bottom panel) for an active mesh with buffer (**a**) and the union of active meshes (**b**) analysis partitioning mask. The latter analysis partitioning mask was used for the `abuk` experiment, and the former was used for the `exp05` experiment. Different colors in the top panel represent distinct partitions

- *Active mesh with buffer:* a mask in which a buffer region is included around the maximum footprint of the underlying Antarctic geometry (Fig. 12a) is created;
- *Union of active meshes:* a mask is created by performing a union of the active cells across all simulation times (Fig. 12b).

Each approach to analysis partitioning mask creation has its pros and cons. The former approach is amenable to in situ analyses, but is likely to give rise to some analysis partitions with little to no elements. The latter approach minimizes the likelihood of empty/imbalanced analysis partitions, but would not be possible to generate in situ. Our preliminary numerical results, described below, suggest that both approaches to creating the analysis partitioning mask produce reasonable results for the datasets considered.

Having settled on an approach for dealing with the temporal variability of the active mesh in our land ice datasets, we now discuss the choice of partitioning scheme for generating the analysis partitions required by our event detection algorithm. We explored the use of several partitioning algorithms, including space-filling curve partitioning (e.g., Hilbert, Morton) (Sasidharan et al. 2015), quad-tree partitioning (Ansar et al. 2019), and k-means clustering (Hartigan and Wong 1979). Of these three approaches, k-means clustering produced the most balanced analysis partitions, shown in Fig. 12. These partitions are balanced in the sense that each partition has roughly the same number of cells, with the partition size appearing to be normally distributed around the target number of cells per partition. Our results below utilize the k-means partitioning algorithm implemented within Scikit-Learn (Pedregosa et al. 2011), seeded with a random initialization. The reader can observe by examining the bottom panel of Fig. 12 that this partitioning scheme produces a partitioning with fairly balanced cell counts per analysis partition. Applying the space-filling curve and quad-tree partitioning approaches to our datasets in contrast gives rise to partition sizes ranging from a single cell to the maximum number of cells/partition requested (partitions not shown). As mentioned earlier, having analysis partitions of widely disparate sizes is particularly problematic for statistics-based signatures within our framework, since these signatures are highly dependent on the number of cells per partition.

As discussed in Sun et al. (2020) and Seroussi et al. (2020), the abuk and exp05 datasets contain a number of fields that can be used as features in our event detection workflow. In the preliminary study presented here, we considered the following four solution fields as features, denoted by $\mathcal{F}_i$ for $i = 1, \ldots, 4$:

- $\mathcal{F}_1$: the ice sheet thickness,
- $\mathcal{F}_2$: the norm of the ice velocity at the ice surface,
- $\mathcal{F}_3$: the norm of the ice velocity at the ice base,
- $\mathcal{F}_4$: the norm of the ice velocity averaged over the vertical extent of the ice.

The ice sheet thickness is selected as a feature because it is a function of the bedrock geometry/topography; the ice velocity fields are used as features as fast-moving ice may correlate with the presence of MISI. In addition to employing the raw solution fields $\mathcal{F}_i$ in our analysis, we also considered logarithms of these fields, denoted by $\log(\mathcal{F}_i)$. We employed the *quartile* signature (Table 1), the *dbscan* measure with parameters $\varepsilon = 0.3$ and $N_p = 5\%$ (Ester et al. 1996) (see Table 2 and Sect. 3.1 for a discussion of this measure and parameters) and the *threshold* decision (Table 4). In this initial proof-of-concept study, only spatial events of interest were considered. The threshold decision flagged partitions with a measure less than zero. The k-means clustering algorithm was used to generate 14,000 partitions, each having approximately 16 cells, for both the abuk and exp05 experiments. For the abuk dataset, we partitioned the active mesh with a buffer region around it (Fig. 12a), whereas for the exp05 dataset, we partitioned an active mesh consisting of the union of all active meshes during the simulation (Fig. 12b).

Our main results are shown below, in Figs. 13, 14, 15 and 16, which plot the interesting analysis partitions in green, overlaying the ice thickness field feature used

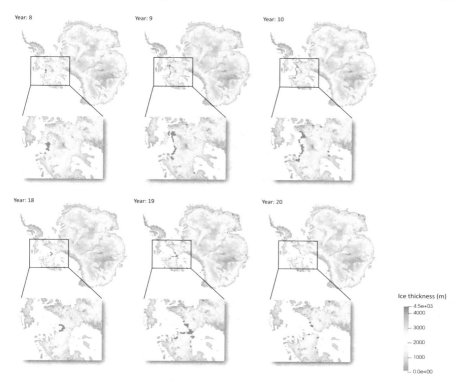

Fig. 13 Event detection results for abuk experiment with the four raw fields $\mathcal{F}_i$ for $i = 1, \ldots, 4$ as features. Analysis partitions identified as interesting are plotted in green, overlaying the ice thickness field for several years. Our results show that ice contained within analysis partitions identified as interesting in one timestep will in general melt (become inactive) in the following timestep

in the analysis. We emphasize that these results are preliminary and intended only to demonstrate the potential usefulness of the proposed framework in data-driven studies of land ice; scientific studies using our event detection framework will be the subject of future research.

3.2.1 Results for the abuk Experiment

We first apply our event detection framework to the abuk dataset, as this dataset is most likely to contain evidence of MISI. Figure 13 shows snapshots of the solution for the abuk dataset at several times, with a close-up in the vicinity of the Pine Island and Thwaites glaciers. Analysis partitions identified as interesting using our algorithm when employing the full set of fields $\{\mathcal{F}_i\}$ for $i = 1, \ldots, 4$ as features are plotted in green, overlaying the ice thickness field for several years. The reader can observe by inspecting this figure that cells comprising the analysis partitions

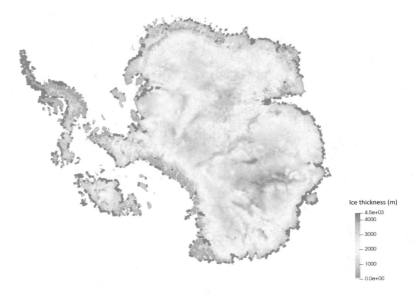

Fig. 14 Event detection results for the abuk experiment with $\mathcal{F}_1$ and $\log(\mathcal{F}_4)$ as features. Analysis partitions identified as interesting are plotted in green, overlaying the ice thickness field for year 33

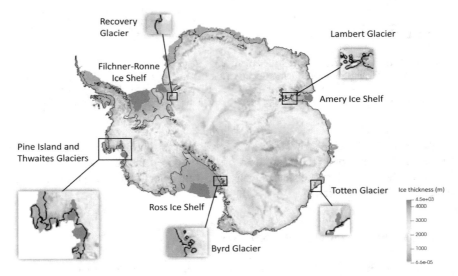

Fig. 15 Event detection results for exp05 experiment with the four raw fields $\mathcal{F}_i$ for $i = 1, \ldots, 4$ as features. Analysis partitions identified as interesting are plotted in green, overlaying the ice thickness field for year 33. The grounding line is shown with a black contour. Our event detection framework identifies the fastest moving areas along Antarctica's coast (ice shelves, outlet glaciers), where MISI is more likely to initiate

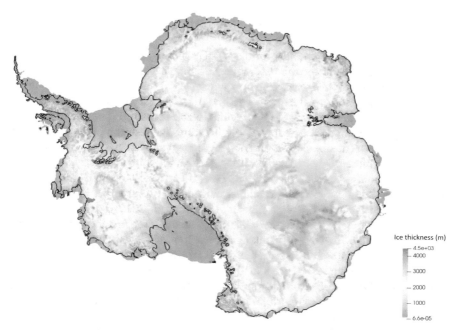

Fig. 16 Event detection results for the exp05 experiment with $\mathcal{F}_1$ and $\log(\mathcal{F}_4)$ as features. Analysis partitions identified as interesting are plotted in green, overlaying the ice thickness field for year 33. The grounding line is shown with a black contour

identified as interesting in one timestep subsequently become inactive (based on previously-described active mask criterion) in the following timestep. In other words, the ice that is flagged by our algorithm melts shortly after it is flagged, a behavior consistent with MISI.

Next, we perform event detection using a reduced set of features, namely $\mathcal{F}_1$ and $\log(\mathcal{F}_4)$. Figure 14 plots the anomalies identified by our framework in year 33 of the simulation, again in green and overlaying the norm of the ice thickness field. It is interesting to observe that significantly more interesting partitions are identified with the new set of features. This is not surprising, as applying a logarithm transform of an analysis feature when using the *dbscan* measure has the effect of emphasizing small differences in small-magnitude values. An additional noteworthy observation is that, with the new set of features, not all of the interesting analysis partitions identified by our algorithm are at or near the grounding line. In particular, several of the flagged locations are located a large distance inland. These locations appear to be regions where the ice retreats the fastest, and should be inspected further in search of MISI.

3.2.2 Results for the `exp05` Experiment

Having obtained plausible results for the `abuk` experiment, we now turn our attention to the more realistic `exp05` case. Figure 15 plots results for the `exp05` dataset corresponding to year 33 in the simulation, with analysis partitions identified as interesting plotted in green and the grounding line (the boundary between where the ice sheet is grounded and floating) plotted with a black contour. From this figure, one can see that our event detection framework identifies the fastest moving areas along Antarctica's coast (the ice shelves and outlet glaciers) as events. These are locations where MISI is more likely to originate. In particular, the following glaciers are identified as containing events of interest: Pine Island, Thwaites, Totten, Byrd, Recovery and Lambert (see Fig. 15). Observational evidence suggests MISI is underway at Thwaites, Pine Island and Totten glaciers (Robel et al. 2019; Joughin and Alley 2019; Gardner et al. 2018; Young et al. 2011). The other regions identified as interesting by our framework are worth taking a closer look at – in both model simulations and observational datasets – in search of MISI (Hoffman et al. 2022).

The most intriguing result is shown in Fig. 16, which plots the interesting analysis partitions for the `exp05` dataset with the new set of features, again in green. The reader can observe that our algorithm flags several regions located inland relative to the grounding line (shown by a black contour). Additionally, the analysis partitions identified as interesting on and near Antarctica's ice shelves closely match the locations that have a significant impact on grounding line flux identified by Reese et al. (2018). While a more rigorous study is required for validating this result, the fact that there is corroboration with previously published results appears promising. A more rigorous investigation, towards understanding the physical mechanisms driving the events identified by our framework, will be the subject of future work. Future work will also explore the use of alternate features in the event detection workflow (including lateral buttressing in shear zones, basal friction, and flux fields, such as the ice velocity flux divergence), as well as alternate signatures and measures, including temporal measures (Table 3). We additionally plan to apply our methodology to higher-resolution datasets (e.g., 3D unstructured datasets produced by running the MALI model/code (Hoffman et al. 2018)) and to land ice datasets expected to exhibit stochastic behavior, e.g., simulations that include parameterizations of physical processes for ice calving and subglacial hydrology.

Finally, it is worth remarking that interesting events or anomalous behaviors identified in land ice simulations using the proposed framework could be relevant for scientists even if they are not an indication of MISI. In this context, an analysis partition flagged by our framework could be indicative of something incorrect in the data or underlying land ice model (e.g., a software flaw or missing physics), or of interesting physical phenomena other than MISI.

3.3 Reduced Order Modeling: Sample Mesh Generation for Hyper-Reduction

To highlight the breadth of application spaces that can benefit from the proposed event detection algorithms, we discuss a fundamentally new use case for our framework within the field of projection-based model reduction.

Projection-based reduced order modeling is a promising strategy for reducing the computational cost of high-fidelity HPC simulations, which are often too expensive for use in a design or analysis setting (e.g., optimization, UQ). Reduced order models (ROMs) have two key features: they are constructed to retain the essential physics and dynamics of their corresponding full order models (FOMs) and they incur a substantially lower (in some cases by orders of magnitude) computational cost. In projection-based model reduction, the state variables are approximated within a low-dimensional subspace, which is typically obtained offline by first applying data compression on a set of snapshots collected from a high-fidelity simulation or physical experiment. A typical projection-based ROM workflow consists of three steps, depicted in Fig. 17 and described succinctly below. In this figure, and the discussion that follows, it is assumed that the FOM is given by the following nonlinear ordinary differential equation (ODE):

$$\frac{d\mathbf{w}}{dt} = \mathbf{f}(\mathbf{w}; t, \boldsymbol{\mu}), \tag{1}$$

where $\mathbf{w}$ denotes the solution vector t denotes time, $\boldsymbol{\mu}$ is a vector of parameters Note that (1) is very generic: an ODE of the form (1) is obtained, for example, by semi-discretizing the partial differential equations (PDEs) defining the FOM in space using a numerical method, such as the finite element or the finite volume method.

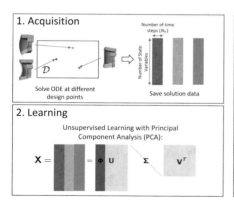
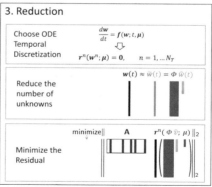

Fig. 17 Illustration of a projection-based model reduction workflow using the POD/LSPG method with hyper-reduction of a full-order model given by the ODE $\frac{d\mathbf{w}}{dt} = \mathbf{f}(\mathbf{w}; t, \boldsymbol{\mu})$. In this figure, $(\cdot)$ denotes "function of" rather than multiplication. The matrices and vectors appearing in this figure have the following dimensions: $\mathbf{X} \in \mathbb{R}^{N \times K}$; $\boldsymbol{\Phi} \in \mathbb{R}^{N \times M}$; $\mathbf{w}, \tilde{\mathbf{w}}, \mathbf{f}, \mathbf{r}^n \in \mathbb{R}^N$; $\hat{\mathbf{w}}\hat{\mathbf{v}} \in \mathbb{R}^M$; $\mathbf{A} \in \mathbb{R}^{q \times N}$; $\boldsymbol{\mu} \in \mathbb{R}^L$, where $L \in \mathbb{N}$ is the number of parameters

Step 1. Acquisition of high-fidelity snapshot data. The first step in a typical projection-based model reduction workflow is the acquisition of a set of K instantaneous snapshots of a numerical solution field. Typically snapshots are collected for K values of a parameter of interest (see Fig. 17), at K different times, or both.

Step 2. Learning a reduced basis. Given an ensemble of high-fidelity snapshots denoted by $\{w^n\}_{n=1}^K$, the next step is the calculation of a basis of reduced dimension $M \ll N$, where N denotes the number of degrees of freedom (dofs) in the FOM. There are numerous approaches in the literature for computing a low-dimensional subspace, but we restrict the discussion here to the Proper Orthogonal Decomposition (POD) method (Sirovich 1987; Holmes et al. 1996) for calculating reduced bases, due to its simplicity and prevalence in practice. Mathematically, POD is closely related to Principal Component Analysis (PCA), and seeks an M-dimensional subspace (with $M \ll K$) spanned by a set of modes $\{\boldsymbol{\phi}_i\}_{i=1}^M$ such that the difference between the snapshot ensemble $\{w^n\}_{n=1}^K$ and the projection of this ensemble onto the reduced subspace is minimized on average. It is a well-known result that the solution to the POD optimization problem reduces to a singular value decomposition problem involving the snapshot matrix $\mathbf{X}$, as shown in Fig. 17; specifically, the modes $\{\boldsymbol{\phi}_i\}_{i=1}^M$ are the M left singular vectors corresponding to the M largest singular values of $\mathbf{X}$. The interested reader is referred to Holmes et al. (1996), Kunisch and Volkwein (2002), Rathinam and Petzold (2003) for details.

Step 3. Projection-based reduction. The final step is the actual reduction, obtained by projecting the equations defining the FOM onto the reduced basis, denoted by $\boldsymbol{\Phi} := [\boldsymbol{\phi}_1, \ldots, \boldsymbol{\phi}_M] \in \mathbb{R}^{N \times M}$. Common projection methods are Galerkin projection and Least-Squares Petrov-Galerkin (LSPG) projection; herein, we focus on the latter approach, as it has been shown to exhibit better stability properties, especially for fluid systems (Carlberg et al. 2017). This approach operates on a FOM that has been fully discretized in both space and time, which can be written as:

$$r^n(w^n; \boldsymbol{\mu}) = \mathbf{0}, \tag{2}$$

where r denotes the residual, and the super-script n denotes the time index, with $n = 1, \ldots, N_T$, so that $w^n := w(t^n)$, where t^n is the nth timestep within a simulation based on (2). The high-fidelity solution $w(t)$ is approximated as a linear combination of the reduced basis modes:

$$w(t) \approx w_M(t) = \boldsymbol{\Phi}\hat{w}(t), \tag{3}$$

where $\hat{w}(t) \in \mathbb{R}^M$, with $M \ll N$. Given this definition, in the LSPG approach, solving for the ROM solution amounts to solving the following least-squares optimization problem:

$$\hat{w}^n = \arg \min_{y \in \mathbb{R}^M} ||r^n(\boldsymbol{\Phi}y; \boldsymbol{\mu})||_2^2, \tag{4}$$

for $n = 1, \ldots, N_T$ and $\hat{\boldsymbol{w}}^n := \hat{\boldsymbol{w}}(t^n)$. Equation (4) can be solved using the Gauss-Newton approach following the method of Carlberg et al. (2013). Unfortunately, the approach described thus far is inefficient for nonlinear problems, as the solution of the ROM problem (4) requires algebraic operations that scale with N, the dimension of the original FOM. This problem can be circumvented through the use of hyper-reduction, the basic idea of which is to compute the residual at some small number of points q with $q \ll N$, encapsulated in a "sampling matrix" $\mathbf{A}$ computed as a pre-processing step of the model reduction procedure using available snapshot data. The set of q points is typically referred to as the "sample mesh", and a variety of quasi-optimal approaches aimed to minimize the representation error of a given nonlinear function appearing in the FOM residual exist—examples include the (discrete) empirical interpolation method (D)EIM (Barrault et al. 2004; Chaturantabut and Sorensen 2010), "best points" interpolation (Nguyen et al. 2008; Nguyen and Peraire 2008), collocation (LeGresley 2006), gappy POD (Everson and Sirovich 1995), and p–sampling (Drmac and Gugercin 2016). These approaches approximate the solution to the NP-hard optimization problem of minimizing the representation of a nonlinear residual using different greedy approaches. Typically, as one may expect based on intuition, the sample mesh points returned by these algorithms are clustered in regions where the simulated solution exhibits "interesting" behavior/features, e.g., shocks, vortices, etc. (see e.g., Fig. 18). With the introduction of hyper-reduction, the LSPG optimization problem takes the form

$$\hat{\boldsymbol{w}}^n = \arg \min_{\boldsymbol{v} \in \mathbb{R}^M} ||\mathbf{A}\boldsymbol{r}^n(\boldsymbol{\Phi}\boldsymbol{y}; \boldsymbol{\mu})||_2^2. \tag{5}$$

As illustrated in Fig. 17, the matrix $\mathbf{A} \in \mathbb{R}^{q \times N}$ is sparse, and has the effect of "sub-selecting" the residual $\boldsymbol{r}$ at some small number of points q, corresponding to the non-zero columns of $\mathbf{A}$.

Current state-of-the-art methods employ a *single static* sample mesh computed offline, and use the *same* sample mesh for hyper-reduction for all the timesteps at which the ROM solution is computed. It has been observed that, for certain applications, sample meshes computed using standard hyper-reduction methods (gappy POD (Everson and Sirovich 1995), p–sampling (Drmac and Gugercin 2016)) are inadequate; in particular, they yield ROMs that are less accurate than ROMs constructed with a *random* sample mesh that knows nothing about the problem dynamics (Blonigan et al. 2021).

We hypothesize herein that it may be possible to improve the accuracy of hyper-reduced ROMs through the creation of a set of *evolving* sample meshes, calculated using the unique features present in the solution at each time, or within time windows. The parallel to AMR (Berger and Oliger 1984) should be clear. To explore this idea, we perform a preliminary study in which we use our event detection framework to calculate dynamically-changing sample meshes, with readily-available snapshots of the FOM solution and the solution residual as features. In this approach, we use the analysis partitions flagged as anomalous to define the sample mesh points.

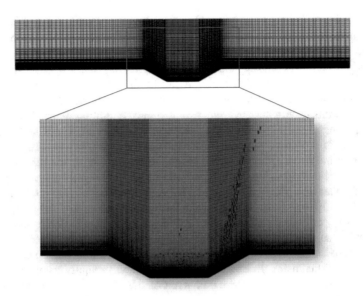

Fig. 18 Computational domain (top) and representative sample mesh points shown in red (bottom) for the 2D open cavity geometry. The sample mesh was obtained using the p–sampling approach (Drmac and Gugercin 2016)

Below, we present and describe some preliminary results exploring the viability of our proposed approach to dynamic sample mesh generation in the context of a problem involving a 2D viscous compressible flow with a Reynolds number of 10,000 over an open cavity geometry, pictured in Fig. 18. To generate a FOM of the form (2), the governing compressible Navier-Stokes equations are discretized in space using a third order Discontinuous Galerkin (DG) method with 600×240 elements in the streamwise and wall-normal direction, respectively, and in time with a Crank-Nicolson time-stepper having a timestep of 5×10^{-3}. The mesh for this geometry is obtained by discretizing a rectangular region with a uniform 600×240 mesh, and transforming it to fit the cavity geometry of interest. More details pertaining to the high-fidelity discretization can be found in Parish and Carlberg (2021) and are not repeated here for the sake of brevity. The free-stream Mach number is unity, which causes a shock to form in the problem solution (see Fig. 19, top row). A POD basis is constructed from 1000 snapshots of the high-fidelity solution. These same snapshots are employed to calculate a sample mesh having 1000 points using the p–sampling approach. This sample mesh is shown in Fig. 18.

The objective of the present section is to explore the viability of constructing *dynamic* sample meshes using our event detection framework. The natural choice of features to use for this task are the solution (Fig. 19, top row) and the solution residual (Fig. 19, second row). The former is a vector of the four primary conserved variables, ρ, ρu, ρv and ρe, where ρ is the fluid density, u and v are the fluid velocities, and e

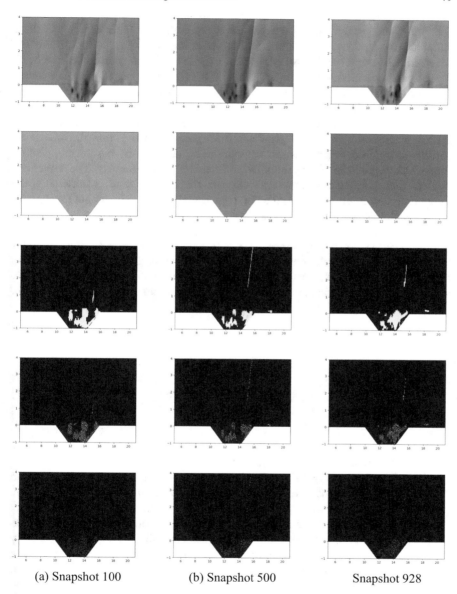

(a) Snapshot 100 (b) Snapshot 500 Snapshot 928

Fig. 19 Plots of the density solution (top row), the density residual (second row) and dynamic sample meshes calculated using our event detection framework (rows 3–5) for the 2D compressible cavity flow problem at the times of snapshots 100 (**a**), 500 (**b**) and 928 (**c**). In rows 3–5, sample mesh points are shown in yellow. The sample meshes in rows 4 and 5 are obtained by randomly selecting one-fourth and one-sixteenth of the points, respectively, within each interesting analysis partition shown in the third row

is the fluid energy; the latter is the residual of the governing PDEs for each of these variables, which contains the nonlinear terms in the governing partial differential equations, the compressible Navier-Stokes equations. For the purpose of the event detection, we partition our geometry into 150×60 analysis partitions, each having 4×4 cells. In this preliminary study, we consider the *quartile* signature (Table 1), the *dbscan* measure (Table 2) with $\varepsilon = 0.3$ and $N_p = 1\%$ (Table 2) and the *threshold* decision with a threshold of 0.5 (Table 4). The sample meshes returned by this approach are plotted in Fig. 19. Row 3 of this figure shows in yellow the interesting partitions, which define a dynamic sample mesh, identified by our event detection framework at the time of snapshots 100, 500, and 928, respectively. The reader can observe that the dynamic sample meshes are changing in time. Additionally, the sample mesh points are in general concentrated within the cavity and in the vicinity of the shock that is seen in the density solutions (Fig. 19, top row).

The reader can observe by comparing the third row of Fig. 19 with Fig. 18 that the sample meshes identified by our event detection framework are qualitatively similar to the static sample mesh obtained using the p–sampling algorithm. In an effort to measure the quality of the dynamic sample meshes calculated using our framework, we calculate the following quantity given a sample mesh represented by the matrix $\mathbf{A}$:

$$\epsilon := \frac{||\mathbf{w} - \mathbf{w}_s||_2}{||\mathbf{w}||_2}, \tag{6}$$

where $\mathbf{w}_s := \mathbf{\Phi}\hat{\mathbf{w}}_s$ and

$$\hat{\mathbf{w}}_s = \arg \min_{\hat{w} \in \mathbb{R}^M} ||\mathbf{A}\mathbf{X} - \mathbf{A}\mathbf{\Phi}\hat{\mathbf{x}}||_2^2. \tag{7}$$

In this context, $\mathbf{x}_s$ is the optimal state one can reconstruct given knowledge of only the FOM state and the sample mesh. The quantity (6) has the advantage that it is computable offline (without running the full model reduction workflow).

Figure 20a plots the quantity ϵ from (6) for the fluid density solution as a function of time for the dynamic sample meshes obtained using our approach and for the static sample mesh obtained using p–sampling. As noted earlier, this comparison is not entirely consistent, since our dynamic sample meshes contain far more points than the static sample mesh we are comparing to (see Fig. 20b). A very simple strategy for reducing the sizes of our dynamic samples is to randomly drop a fixed fraction of the sample mesh points within each analysis partition flagged by our approach. Figure 19 shows the resulting sample meshes when one-quarter (fourth row) and one-sixteenth (fifth row) of the sample mesh points are kept within each interesting analysis partition. By randomly selecting just one sample mesh point within each interesting analysis partition (which corresponds to the one-sixteenth sub-sampling shown in Fig. 19, the fifth row), it is possible to reduce the sizes of our dynamic sample meshes to be on the order of the static sample mesh obtained through p–sampling (Fig. 20b). Remarkably, as the reader can see from examining Fig. 20a, reducing the number of sample mesh points in this way does not increase the error (6). While the

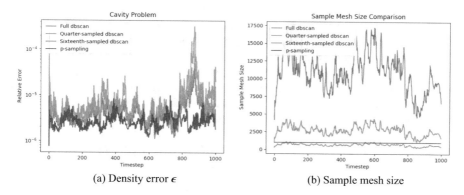

(a) Density error ϵ (b) Sample mesh size

Fig. 20 Comparison of errors ϵ in the density solution and the sample mesh size as a function of time for the cavity flow problem for sample meshes calculated using our event detection framework versus p–sampling

fact that the error (6) for the dynamic sample meshes obtained using our approach are roughly comparable to the errors of the p–sampling sample mesh may seem negative, it is actually encouraging, given that our approach is unsupervised and not based on an underlying optimization problem. Future work will focus on improving the sample meshes calculated using our approach, e.g., by bringing in ideas from traditional sample mesh approaches, which are based on minimizing the approximation error on a given sample mesh. Additionally, we plan to deploy our approach on test cases with more sophisticated dynamics, for which a dynamic sample mesh procedure will likely yield a greater benefit (e.g., problems with moving shocks). Future work will also include the design of signature-measure pairs that can guarantee that a given number of analysis partitions are selected at any given timestep; in order to achieve this, it is necessary to use a non-boolean measure.

3.4 HPC Experiments

As discussed in Sect. 1, an important requirement for an in situ event detection framework is that it be scalable and communication-minimizing. In this section, we verify the scalability of our framework in an HPC application utilizing MPI (Message Passing Interface Forum 1994) for coordinating the parallel communication and computation. In order to perform this study, we embedded a Python interpreter in the S3D combustion simulation code (Chen et al. 2009) which is written in Fortran 90. References to the raw data from the Fortran side were passed to the Python framework at each timestep, without duplication. The mpi4py package (Dalcín et al. 2005) was used to access the MPI environment from Python and perform collective communication between processors.

We ran our experiment using the Cori Cray XC40 machine at NERSC. The simulation represented conditions of a homogeneous charge compression ignition (HCCI) combustion of ethanol-air mixture at conditions typical of internal combustion engines. The mixture undergoes compression heating and auto-ignition kernels appear locally in small pockets, as shown in Fig. 21, that lead to the eventual combustion of the entire mixture. The goal for an event detection algorithm in this case is to identify the partitions where the auto-ignition kernels appear.

We decomposed the 2D simulation domain into 1024 partitions, with one partition per MPI rank, and processed 626 snapshots with 3136 grid points per partition and 33 features at each grid point. The event detection involved the following steps:

- global *min-max* pre-processing utilizing two MPI all-reduces, one each for the per-feature global *min* and *max*, over a vector of a size equal to the number of features.
- *mean* signature on the data locally on each partition (no MPI communication involved).
- *msd* measure, which involves computing a global mean of signatures and requires an MPI all-reduce of a vector of size equal to the number of features.

In a previous work (Konduri et al. 2018), we used this simulation as a motivation for designing a new signature – *feature moment metric* (fmm) – which represents the distribution of a given joint statistical moment (e.g., Kurtosis) across all the features. Here our focus is only on demonstrating the parallel performance of the framework and hence we use the simpler *mean* signature.

The execution times for the solver and the event detection components were recorded for the simulation. The solver execution time was 0.126 s for every simulation timestep. The event detection execution time ranged from a minimum of 0.012 s per timestep to a maximum of 2.28 s, with an average of 0.2 s. Because the

Fig. 21 Contour plot of heat release rate ($J/m^3/s$) at an early instant of the HCCI combustion simulation. A 12×12 partitioning of the domain is shown with white lines, and auto-ignition kernels are denoted by regions of high (red) heat release rate

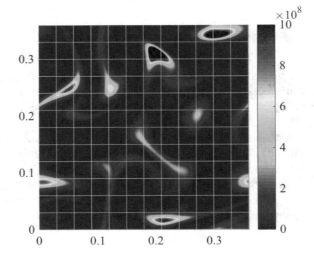

workflow was identical from one timestep to the next, the large variation in the times can be attributed to system noise. While not negligible, the average event detection time was on the same order of magnitude as the solver, and thus within the realm of practicality, depending on the application. Encouragingly, the minimum time was an order of magnitude smaller than simulation time, suggesting that – under conditions free of system noise – the event detection could be performed in a fraction of the simulation time.

Note too that we used Python in situ to run this experiment for expediency, and that the analysis time could be drastically reduced by porting our framework to compiled code. Finally, analysis overhead could be further reduced for large-scale applications by reducing the number of event detection checks. Performing the event detection at, for instance, every Nth timestep would be an effective compromise between traditional check-pointing and fine-grained event detection, reducing the event detection load to a negligible portion of the runtime.

4 Conclusion

This work represents a first step in the development of event detection algorithms that can automatically identify events of interest in situ. Specifically, we presented a signatures-measures-decisions framework for the development of in situ HPC event detection algorithms. This framework is a useful decomposition that supports generalizability, unsupervised detection, low communication requirements and online processing. We have developed components under this framework which enable the use of standard event detection algorithms under the aforementioned constraints, in addition to entirely new combinations. We illustrated how example algorithms made from these components can optimize I/O while running an HPC simulation, leading to the capture of many more interesting events than typical uniform check-pointing. We highlighted two additional use cases for the proposed framework: detecting interesting events in HPC simulations (the Marine Ice Sheet Instability in land ice data), and identifying optimal space-time subregions for the hyper-reduction step of a typical projection-based model reduction workflow. Finally, we demonstrated, in a study using HPC and MPI, that in situ event detection overhead can be on the order of magnitude of the simulation, and performance can be improved further with minor adjustments.

This work enables future research in several areas, such as the question of what should constitute an "interesting" event for a given simulation, or, ideally, how to define "interesting" for *any* given simulation. Apart from detecting events the proposed approaches can also identify numerical anomalies, which can help with debugging and interpretation of simulation results. In addition, it is possible that this framework can be used to classify events either in situ or as a post-processing technique by analyzing the signatures themselves; the signatures distill information from a large number of samples and are less expensive to analyze. Finally, we hope that experiments done using this framework will inspire HPC simulation code developers to incorporate these capabilities into native code, allowing for even more efficient in situ event detection.

Acknowledgements This work was funded through U.S. Department of Energy Advanced Scientific Computing Research (ASCR) grant FWP #18019471. Sandia National Laboratories is a multi-mission laboratory managed and operated by National Technology and Engineering Solutions of Sandia, LLC., a wholly owned subsidiary of Honeywell International, Inc., for the U.S. Department of Energy's National Nuclear Security Administration under contract DE-NA-0003525. The views expressed in the article do not necessarily represent the views of the U.S. Department of Energy or the United States Government.

The authors gratefully acknowledge Dr. Stephen Price, Dr. Matt Hoffman and Dr. Mauro Perego for providing the MALI simulation data analyzed in Sect. 3.2, for engaging in many fruitful discussions regarding the physics of the Marine Ice Sheet Instability (MISI), and for assisting with the interpretation of our results within the context of MISI.

References

Ahmed T (2009) Online anomaly detection using KDE. In: IEEE global telecommunications conference, pp 1–8

Ansar S, Hussain M, Mazhar S, Manzoor T, Siddiqui K, Abid M, Jamal H (2019) Mesh partitioning and efficient equation solving techniques by distributed finite element methods: a survey. Arch Comput Meth Eng 26:1–16

Bamber J, Riva R, Vermeersen B, LeBrocq A (2009) Reassessment of the potential sea-level rise from a collapse of the West Antarctic Ice Sheet. Science 324(5929):901–903

Bandaragoda TR, Ting KM, Albrecht D, Liu FT, Wells JR (2014) Efficient anomaly detection by isolation using nearest neighbour ensemble. In: IEEE international conference on data mining, pp 698–705

Barrault M, Maday Y, Nguyen NC, Patera AT (2004) An 'empirical interpolation' method: application to efficient reduced-basis discretization of partial differential equations. Comptes Rendus Mathematique 339(9):667–672

Bennett JC, Bhagatwala A, Chen JH, Pinar A, Salloum M, Seshadhri C (2016) Trigger detection for adaptive scientific workflows using percentile sampling. SIAM J Sci Comp 38:240–260

Berger MJ, Oliger J (1984) Adaptive mesh refinement for hyperbolic partial differential equations. J Comput Phys 53(3):484–512

Blonigan PJ, Rizzi F, Howard M, Fike JA, Carlberg KT (2021) Model reduction for steady hypersonic aerodynamics via conservative manifold least-squares petrov-galerkin projection. AIAA J 59(4):1296–1312

Carlberg K, Farhat C, Cortial J, Amsallem D (2013) The gnat method for nonlinear model reduction: effective implementation and application to computational fluid dynamics and turbulent flows. J Comput Phys 242:623–647

Carlberg K, Barone M, Antil H (2017) Galerkin v. least-squares petrov–galerkin projection in nonlinear model reduction. J Comput Phys 330:693–734

Chaturantabut S, Sorensen DC (2010) Nonlinear model reduction via discrete empirical interpolation. SIAM J Sci Comp 32(5):2737–2764

Chen JH, Choudhary A, Supinski BD, DeVries M, Hawkes ER, Klasky S, Liao W-K, Ma K-L, Crummey JM, Podhorszki N et al (2009) Terascale direct numerical simulations of turbulent combustion using S3D. Comp Sci Discov 2(1):015001

Dalcín L, Paz R, Storti M (2005) Mpi for python. J Par Distri Comp 65(9):1108–1115

Drmac Zl, Gugercin S (2016) A new selection operator for the discrete empirical interpolation method—improved a priori error bound and extensions. SIAM J Sci Comp 38(2):A631–A648

Edwards TL, Nowicki S, Marzeion B et al (2021) Projected land ice contributions to twenty-first-century sea level rise. Nature 593:74–82

Ester M, Kriegel H-P, Sander J, Xu X (1996) A density-based algorithm for discovering clusters in large spatial databases with noise. In: Proceedings of the second international conference on knowledge discovery and data mining. AAAI Press, pp 226–231

Everson R, Sirovich L (1995) Karhunen-loève procedure for gappy data. J Opt Soc Am A 12(8):1657–1664

Ffmpeg (2019) Online; accessed 2019-09-18. https://ffmpeg.org

Gardner AS, Moholdt G, Scambos T, Fahnstock M, Ligtenberg S, van den Broeke M, Nilsson J (2018) Increased West Antarctic and unchanged East Antarctic ice discharge over the last 7 years. The Cryosphere 12(2):521–547

Hartigan JA, Wong MA (1979) Algorithm AS 136: a K-Means clustering algorithm. App Stat 28(1):100–108

Hoffman M (2022) Personal correspondance

Hoffman MJ, Perego M, Price SF, Lipscomb WH, Zhang T, Jacobsen D, Tezaur I, Salinger AG, Tuminaro R, Bertagna L (2018) Mpas-albany land ice (mali): a variable-resolution ice sheet model for earth system modeling using voronoi grids. Geosci Model Develop 11(9):3747–3780

Holmes P, Lumley JL, Berkooz G (1996) Turbulence, coherent structures, dynamical systems and symmetry. Cambridge University Press

Imagecat (2022) Online; accessed 2022-01-11. https://imagecat.readthedocs.io

IPCC (2021) Representative Concentration Pathways (RCPs). https://sedac.ciesin.columbia.edu/ddc/ar5_scenario_process/RCPs.html

Jones E, Oliphant T, Peterson P et al (2001) SciPy: Open source scientific tools for Python –. Online; accessed 2019-09-18. http://www.scipy.org

Joughin I, Alley R (2019) Stability of the West Antarctic ice sheet in a warming world. Nature 4:506–513

Konduri A, Kolla H, Kegelmeyer WP, Shead TM, Ling J, Davis WL (2018) Anomaly detection in scientific data using joint statistical moments. J Comput Phys 387:522–538

Kunisch K, Volkwein S (2002) Galerkin proper orthogonal decomposition for a general equation in fluid dynamics. SIAM J Num Anal 40(2):492–515

LeGresley P (2006) Application of proper orthogonal decomposition (POD) to design decomposition methods. PhD thesis, Stanford University

Leung LR, Bader DC, Taylor MA, McCoy RB (2020) An introduction to the e3sm special collection: goals, science drivers, development, and analysis. J Adv Model Earth Sys 12(11):e2019MS001821

Ling J, Kegelmeyer WP, Aditya K, Kolla H, Reed KA, Shead TM, Davis WL (2017) Using feature importance metrics to detect events of interest in scientific computing applications. In: 2017 IEEE 7th symposium on large data analysis and visualization (LDAV). IEEE, pp 55–63

Liu FT, Ting KM, Zhou ZH (2012) Isolation-based anomaly detection. ACM Trans Knowl Disc Data 6:1–39

Message Passing Interface Forum (1994) Mpi: a message-passing interface standard. Technical report, University of Tennessee, USA

Nguyen NC, Peraire J (2008) An efficient reduced-order modeling approach for non-linear parametrized partial differential equations. Int J Num Meth Eng 76(1):27–55

Nguyen N, Patera A, Peraire J (2008) A 'best points' interpolation method for efficient approximation of parametrized functions. Int J Num Meth Eng 73:521–543

Parish EJ, Carlberg KT (2021) Windowed least-squares model reduction for dynamical systems. J Comput Phys 426:109939

Pattyn F, Morlighem M (2020) The uncertain future of the antarctic ice sheet. Science 367(6484):1331–1335

Pedregosa F, Varoquaux G, Gramfort A, Michel V, Thirion B, Grisel O, Blondel M, Prettenhofer P, Weiss R, Dubourg V, Vanderplas J, Passos A, Cournapeau D, Brucher M, Perrot M, Duchesnay E (2011) Scikit-learn: machine learning in Python. J Mach Learn Res 12:2825–2830

Perego M, Price S, Stadler G (2014) Optimal initial conditions for coupling ice sheet models to earth system models. J Geophys Res, Earth Surface 119(9):1894–1917

Python. Online; accessed 2019-09-18. https://www.python.org

Rathinam M, Petzold L (2003) A new look at proper orthogonal decomposition. SIAM J Num Anal 41(5):1893–1925

Reese R, Gudmundsson G, Levermann A, Winkelmann R (2018) The far reach of ice-shelf thinning in Antarctica. Nat Clim Change 8:53–57

Robel AA, Seroussi H, Roe GH (2019) Marine ice sheet instability amplifies and skews uncertainty in projections of future sea-level rise. Proc Natl Acad Sci 116(30):14887–14892

Sasidharan A, Dennis JM, Snir M (2015) A general space-filling curve algorithm for partitioning 2d meshes. In: 2015 IEEE 17th international conference on high performance computing and communications, 2015 ieee 7th international symposium on cyberspace safety and security, and 2015 ieee 12th international conference on embedded software and systems, pp 875–879

Seroussi H, Nowicki S, Payne AJ, Goelzer H, Lipscomb WH, Abe-Ouchi A, Agosta C, Albrecht T, Asay-Davis X, Barthel A, Calov R, Cullather R, Dumas C, Galton-Fenzi BK, Gladstone R, Golledge NR, Gregory JM, Greve R, Hattermann T, Hoffman MJ, Humbert A, Huybrechts P, Jourdain NC, Kleiner T, Larour E, Leguy GR, Lowry DP, Little CM, Morlighem M, Pattyn F, Pelle T, Price SF, Quiquet A, Reese R, Schlegel N-J, Shepherd A, Simon E, Smith RS, Straneo F, Sun S, Trusel LD, Van Breedam J, van de Wal RSW, Winkelmann R, Zhao C, Zhang T, Zwinger T (2020) ISMIP6 Antarctica: a multi-model ensemble of the Antarctic ice sheet evolution over the 21st century. The Cryosphere 14:3033–3070

Sirovich L (1987) Turbulence and the dynamics of coherent structures, Part III: dynamics and scaling. Q Appl Math 45(3):583–590

Sun S, Pattyn F, Simon EG, Albrecht T, Cornford S, Calov R, Dumas C, Gillet-Chaulet F, Goelzer H, Golledge NR et al (2020) Antarctic ice sheet response to sudden and sustained ice-shelf collapse (ABUMIP). J Glaciol 66(260):891–904

Thomas RH, Bentley CR (1978) A model for holocene retreat of the west antarctic ice sheet. Quat Res 10(2):150–170

Thuerey N, Pfaff T (2018) MantaFlow. Online; accessed 2019-09-18. http://mantaflow.com

Ullrich PA, Zarzycki CM (2016) Tempestextremes v1.0: a framework for scale-insensitive pointwise feature tracking on unstructured grids. In: Geoscientific model development discussion

Walt SVD, Colbert SC, Varoquaux G (2011) The numpy array: a structure for efficient numerical computation. Comp Sci Eng 13(2):22

Weertman J (1974) Stability of the junction of an ice sheet and an ice shelf. J Glaciol 13(67):3–11

Welch P (1967) The use of fast fourier transform for the estimation of power spectra: a method based on time averaging over short, modified periodograms. IEEE Trans Audio Electroacoust 15(2):70–73

Wu K, Zhang K, Fan W, Edwards A, Philip SY (2014) Rs-forest: a rapid density estimator for streaming anomaly detection. In: IEEE international conference on data mining, pp 600–609

Young D, Wright A, Roberts J, Warner R, Young N, Greenbaum J, Schroeder D, Holt J, Sugden D, Blankenship D, vanOmmen T, Siegert M (2011) A dynamic early East Antarctic Ice Sheet suggested by ice-covered fjord landscapes. Nature 474:72–75

Zhao M, Held IM, Lin SJ, Vecchi GA (2009) Simulations of global hurricane climatology, interannual variability, and response to global warming using a 50-km resolution GCM. J Clim 22:6653–6678

Machine-Learning for Stress Tensor Modelling in Large Eddy Simulation

Z. M. Nikolaou, Y. Minamoto, C. Chrysostomou, and L. Vervisch

Abstract The accurate modelling of the unresolved stress tensor is particularly important for Large Eddy Simulations (LES) of turbulent flows. This term affects the transfer of energy from the largest to the smallest scales and *vice versa*, thus controlling the evolution of the flow field-in reacting flows, the flow field transports scalar fields such as mass fractions and temperature both of which control the species production and destruction rates. A large number of models have been developed in past years for the stress tensor in incompressible and non-reacting flows. A common characteristic of the majority of the classical models is that simplifying assumptions are typically involved in their derivation which limits their predictive ability. At the same time, various tunable parameters appear in the relevant closures whose value depends on the flow geometry/configuration/spatial location, and which require careful regularisation. Data-driven methods for the stress tensor is an emerging alternative modelling approach which may help to circumvent the above issues, and in recent studies several such models were developed and evaluated. This chapter discusses the modelling problem, presents some of the most popular algebraic models, and reviews some recent advances on data-driven methods.

Z. M. Nikolaou (✉) · L. Vervisch
CORIA-CNRS, Normandie Université, INSA de Rouen, Normandy, France
e-mail: znikolaou@insa-rouen.fr

L. Vervisch
e-mail: luc.vervisch@insa-rouen.fr

Y. Minamoto
Department of Mechanical Engineering, Tokyo Institute of Technology, 2-12-1 Ookayama, Meguro, Tokyo 152-8550, Japan
e-mail: minamoto.y.aa@m.titech.ac.jp

C. Chrysostomou
The Cyprus Institute, 20 Konstantinou Kavafi Street 2121, Nicosia, Cyprus
e-mail: c.cchrysostomou@cyi.ac.cy

© The Author(s) 2023
N. Swaminathan and A. Parente (eds.), *Machine Learning and Its Application to Reacting Flows*, Lecture Notes in Energy 44, https://doi.org/10.1007/978-3-031-16248-0_4

1   Introduction

LES is a powerful tool for simulating a wide range of flows including turbulent and reacting flows. Although LES is more expensive than Reynolds Averaged Navier Stokes (RANS) simulations, with the rapid advances of fast and efficient computer hardware and scalable but also readily available software, LES is increasingly being used in a wide range of industries (aerospace, automotive, energy, chemical) for modelling fluid flows in complex and often realistic-size geometries (Gicquel et al. 2012; Pitsch 2006). In comparison to Direct Numerical Simulations (DNS) where all length and time-scales are resolved, LES reduces the computational load substantially by resolving only the largest scales.

LES comes in two main flavours: implicit and explicit (Gicquel et al. 2012; Sagaut 2001). In implicit LES, the filtering is essentially done through the numerical scheme whereby the goal is to obtain steady or at least bounded solutions for a given mesh size/time-step. In explicit LES, a spatial filter having a width Δ is applied to the governing equations, and unresolved terms appearing in the resulting equation set are modelled explicitly. This is done either by developing suitable algebraic functions involving the resolved variables on the mesh, and/or by developing and solving suitable transport equations. In the majority of classic approaches the mesh spacing h to filter ratio $h/\Delta = 1$ but this need not necessarily be the case as we discuss later on. Each of these two approaches has its merits and drawbacks and in this chapter we focus on explicit LES which solves the filtered equations. The filtered compressible momentum equation reads,

$$\frac{\partial \overline{\rho} \tilde{u}_i}{\partial t} + \frac{\partial \overline{\rho} \tilde{u}_i \tilde{u}_j}{\partial x_j} = -\frac{\partial \overline{p}}{\partial x_i} + \frac{\partial \tau_{ij}^r}{\partial x_j} - \frac{\partial \tau_{ij}}{\partial x_j}, \tag{1}$$

where the overbar denotes spatial filtering using a suitable filter i.e.

$$\overline{\phi}(\underline{x}, t) = \int_{-\infty}^{\infty} G(\underline{x} - \underline{x}'; \Delta) \phi(\underline{x}') d\underline{x}', \tag{2}$$

where G is the LES filter and ϕ the quantity being filtered. Note that $\tilde{\ }$ denotes Favre-filtering i.e. $\tilde{\phi} = \overline{\rho \phi}/\overline{\rho}$. The resolved and unresolved stress tensors τ_{ij}^r and τ_{ij} are given by,

$$\tau_{ij}^r = \overline{\mu \left(\frac{\partial u_i}{\partial x_j} + \frac{\partial u_j}{\partial x_i} \right)} - \frac{2}{3} \delta_{ij} \overline{\mu \frac{\partial u_k}{\partial x_k}}, \tag{3}$$

and

$$\tau_{ij} = \overline{\rho} (\widetilde{u_i u_j} - \tilde{u}_i \tilde{u}_j), \tag{4}$$

respectively. The resolved stress tensor is typically closed using the gradients of the filtered velocity components (hence called resolved, not because it is actually resolved but because the approximation below is such a good one),

$$\tau_{ij}^r \simeq \bar{\mu} \left(\frac{\partial \tilde{u}_i}{\partial x_j} + \frac{\partial \tilde{u}_j}{\partial x_i} - \frac{2}{3} \delta_{ij} \frac{\partial \tilde{u}_k}{\partial x_k} \right) = 2\bar{\mu} \left(\tilde{S}_{ij} - \frac{1}{3} \delta_{ij} \tilde{S}_{kk} \right), \tag{5}$$

where

$$\tilde{S}_{ij} = \frac{1}{2} \left(\frac{\partial \tilde{u}_i}{\partial x_j} + \frac{\partial \tilde{u}_j}{\partial x_i} \right), \tag{6}$$

is the (resolved) rate of strain tensor. Clearly τ_{ij} is an unclosed term and requires modelling in order to produce a closed equation set. This term is very important since it determines the dissipation/back-scatter of kinetic energy (Sagaut 2001)-multiplying Eq. 1 with $\tilde{u}_i$ and summing it is straightforward to show that the contribution of the unresolved stress tensor to the resolved total kinetic energy $e_r = 1/2\tilde{u}_i\tilde{u}_i$ is $-\tilde{u}_i \partial \tau_{ij}/\partial x_j$.

A large number of different models have been developed in the literature throughout the years for τ_{ij} aimed mainly at incompressible and non-reacting flows (Meneveau and Katz 2000). In the classic modelling approach, the stress tensor is modelled by developing suitable algebraic functions of the resolved quantities. In incompressible flows for instance, these include the filtered velocity components $\bar{u}_i$ as well as any other derived quantities such as their gradients and/or functions of their gradients, higher-order filtered values of the aforementioned quantities etc. The majority of these models are relatively straightforward to implement while the computational cost depends on the formulation: the dynamic evaluation of model parameters can be substantially more expensive than the static approach (where a constant value for a certain parameter is assumed). A common characteristic of all of the aforementioned models is that they usually involve some simplifying assumption in their development which may or may not be valid for conditions other than those originally developed for. For example, the Boussinesq assumption is a rather strong one (Schmitt 2007). Previous theoretical as well as experimental work showed that this assumption is invalid both for non-reacting (Tao et al. 2000, 2002) and reacting flows (Klein et al. 2015; Pfandler et al. 2010). Another issue with classic algebraic models is that they involve tunable parameters whose spatio-temporal variation depends on the flow regime and/or reaction mode. As a result, a single universal method for accurate parameterisation/regularisation of the models' constants is difficult to obtain.

Despite the aforementioned issues, the standard approach in reacting LES is to employ models originally developed and validated for incompressible and non-reacting flows. Reacting flows, however, bring additional challenges. The heat release causes large variations in density, temperature, velocity, and viscosity across the flame-front. All of these quantities affect the modelling of the stress tensor. Models developed for non-reacting and incompressible flows do not account for such effects. For instance, it was shown in Klein et al. (2015) as well as in previous theoretical and experimental studies (Bray et al. 1981; Chomiak and Nisbet 1995) that even for simple flow configurations such as freely-propagating premixed flames classic models are inadequate. In particular, it was shown (Klein et al. 2015) that counter-gradient transport also occurs for the components of the stress tensor, and as a result classic static gradient-type models cannot capture counter-gradient transport. Even dynamic models where the sign of the dynamic parameter can in principle change, fail to capture counter-gradient transport (Klein et al. 2015). In addition, it was shown in Klein et al. (2015) that the standard averaging procedure for regularising the dynamic parameters e.g. C_D in the Smagorinsky model is not suitable for reacting flows. The behaviour and performance of these models for more demanding configurations such as shear-induced flows with a larger spatial in-homogeneity is unclear, and the deficiencies of such models can only be unveiled through further investigation using both a priori as well as a posteriori studies. All of these issues essentially limit the predictive ability of LES to conditions where the models for the unresolved terms are known to perform well.

In light of the aforementioned long-standing issues, in the past few years a wide range of alternative non-classic modelling strategies have been proposed and evaluated (Domingo et al. 2020) including machine-learning which has the potential to circumvent such issues. Data-driven methods which include a wide range of network architectures have been widely used to solve classification and regression problems in image recognition (Krizhevsky et al. 2012), text translation (Sutskever et al. 2014), decision making (Mnih et al. 2015; Silver et al. 2016), gene profiling (Khan et al. 2001) etc. by directly exploiting the abundance of information contained within very large data sets. In the field of fluid mechanics databases are also quite substantial-DNS databases of non-reacting flows for instance are of the order of petabytes (Kanov et al. 2015). In reacting flows, simulations using DNS with detailed chemistry and multi-step reduced chemistry are slowly yet steadily becoming more common (Aspden et al. 2016; Minamoto et al. 2011; Nikolaou and Swaminathan 2014, 2015; Wang et al. 2017) while numerical solvers are being developed for DNS aimed at the exascale (Treichler et al. 2017) and exploiting hybrid architectures (Perez et al. 2018). As a result, the application of machine-learning techniques using data from such high-fidelity simulations for modelling purposes in LES appears to be a timely one.

In the text which follows we present in Sect. 2 some fundamental/popular models in the literature which have been the subject of recent and extensive testing in reacting flows (Nikolaou et al. 2019, 2021). In Sect. 3 another emerging approach namely deconvolution is discussed, and in Sect. 4 a review of the main approaches used for machine-learning is given. The main challenges and caveats associated with machine-learning methods are summarised in Sect. 6.

2   Classic Stress Tensor Models

2.1   *Smagorinsky*

The Smagorinsky model is an eddy-diffusivity type of model originally developed for application to atmospheric flows (Moin et al. 1991; Smagorinsky 1963). The stress tensor closure reads,

$$\tau_{ij} - \frac{1}{3}\delta_{ij}\tau_{kk} = -2\bar{\rho}\nu_t \left(\tilde{S}_{ij} - \frac{1}{3}\delta_{ij}\tilde{S}_{kk} \right), \tag{7}$$

where the turbulent viscosity ν_t is modelled using $\nu_t = (C_D\Delta)^2|\tilde{S}|$ with $|\tilde{S}| = \sqrt{2\tilde{S}_{ij}\tilde{S}_{ij}}$. In the original (static) version C_D is replaced by C_S^2 with $C_S \simeq 0.2$. It is a very popular model as it is relatively straightforward to implement and computationally efficient. However from a theoretical point of view there are some key issues to highlight. Firstly, it is a purely dissipative model whereas a reverse flow of energy (backscatter) is known to exist from the smaller scales to the larger scales both in 2D flows as shown by Fjortof (1953) and in 3D flows (Domaradzki et al. 1993; Kerr et al. 1996; Piomelli et al. 1991). In addition, the assumption of the unresolved stress tensor being aligned to the resolved rate of strain tensor is a rather strong one as shown by previous experimental and numerical studies (Tao et al. 2000, 2002). Another issue, is that the model predictions are sensitive to the value of C_S (Smagorinsky constant) which depends on the flow regime (Deardoff 1970; Lilly 1966), but also on the filter width and mesh spacing (Mason and Callen 1986).

These limitations soon became apparent with the static Smagorinsky model performing relatively well for homogeneous and isotropic decaying turbulence but poorly for shear-dominated flows such as turbulent channel flow. In such configurations the value $C_S \simeq 0.2$ in the near-wall region was found to be excessive and a reduction was required to obtain the correct (lower) dissipation. This led to the development of a dynamic version by Germano et al. (1991) where C_D was no longer constant but calculated dynamically (during the simulation) from the resolved flow variables. The dynamic Smagorinsky model showed considerable improvement over its static version, particularly in shear flows (Germano et al. 1991), and was later adapted to compressible flows by Moin et al. (1991) whereby C_D is typically calculated using the least-squares approach (Lilly 1992; Salvetti 1994),

$$C_D = \frac{\langle -(L_{ij} - \frac{1}{3}\delta_{ij}L_{kk})M_{ij}\rangle}{\langle 2\Delta^2 M_{ij}M_{ij}\rangle}, \tag{8}$$

where $<>$ indicates a suitable averaging (regularisation) procedure, and $\hat{}$ indicates test-filtering with a filter $\hat{\Delta}$. The ratio $\gamma = \hat{\Delta}/\Delta$ is typically taken to equal 2. The Leonard term L_{ij} is given by,

$$L_{ij} = \widehat{\overline{\rho \tilde{u}_i \tilde{u}_j}} - (\widehat{\overline{\rho \tilde{u}_i}})(\widehat{\overline{\rho \tilde{u}_j}})/\hat{\bar{\rho}}, \tag{9}$$

and

$$M_{ij} = \alpha^2 \hat{\bar{\rho}} |\hat{\tilde{S}}| \left(\hat{\tilde{S}}_{ij} - \frac{1}{3}\delta_{ij}\hat{\tilde{S}}_{kk} \right) - \left(\widehat{\bar{\rho} |\tilde{S}| \tilde{S}_{ij}} - \frac{1}{3}\delta_{ij}\widehat{\bar{\rho} |\tilde{S}| \tilde{S}_{kk}} \right), \tag{10}$$

An important point to note is that the Smagorinsky model does not apply for the normal (isotropic) components of the stress tensor. Typically, the static Yoshizawa approximation is used to explicitly model τ_{kk} (Yoshizawa 1986) as follows,

$$\tau_{kk} = 2\bar{\rho}C_I \Delta^2 |\tilde{S}|^2, \tag{11}$$

where in the static version the model parameter C_I is a constant. Yoshizawa suggested a value of $\simeq 0.089$ (Yoshizawa 1986), however values ranging from 0.0025–0.009 were reported while dynamically evaluating C_I in the study of Moin et al. (1991). In the dynamic version, C_I is calculated using (Moin et al. 1991),

$$C_I = \frac{< L_{kk} >}{< P >} \tag{12}$$

where L_{kk} is the trace of the Leonard term, and the term P is given by,

$$P = 2 \left(\hat{\bar{\rho}}\hat{\Delta}^2 |\hat{\tilde{S}}|^2 - \Delta^2 \widehat{\bar{\rho} |\tilde{S}|}^2 \right)$$

From the equations just presented above it becomes apparent that even for a simple model like Smagorinsky the evaluation can be rather complicated: it involves the calculation of tensor variables which include gradients, and filtering as well as test-filtering operations, a process which introduces an additional ad-hoc parameter (test-filter to filter-width ratio) etc. It is also important to note that a regularization procedure for the evaluation of dynamic parameters is almost always required to render them spatially smooth, thus avoiding numerical instabilities. This process is not always unique or justifiable, and typically involves averaging in homogeneous directions (if any), thresholding, smoothing, or otherwise if no homogeneous directions exist. Other more practical issues pertain to the division by near-zero numbers as in the equations for C_D, C_I and so on.

2.2 Scale Similarity

Consider an incompressible flow in which case the unresolved stress tensor is now simply $\tau_{ij} = \overline{u_i u_j} - \bar{u}_i \bar{u}_j$. The closure problem reduces to finding a suitable approximation for $\overline{u_i u_j}$. Consider $u_i' = u_i - \bar{u}_i$ i.e. the difference between the unfiltered and filtered fields. Then we have upon expansion of the filtered product,

$$\overline{u_i u_j} = \overline{(\bar{u}_i + u'_i)(\bar{u}_j + u'_j)}$$
$$= \overline{\bar{u}_i \bar{u}_j} + \overline{\bar{u}_i u'_j} + \overline{\bar{u}_j u'_i} + \overline{u'_i u'_j}$$
$$= \overline{\bar{u}_i \bar{u}_j} + \overline{\bar{u}_i (u_j - \bar{u}_j)} + \overline{\bar{u}_j (u_i - \bar{u}_i)} + \overline{(u_i - \bar{u}_i)(u_j - \bar{u}_j)} \tag{13}$$

Up to this point the expansion is perfectly fine however the problem has not disappeared since we are left with further unclosed terms namely the last three terms in the equation above. The main step which follows in scale-similarity models to solve this problem is to assume that (Bardina et al. 1983),

$$\overline{\bar{u}_i (u_j - \bar{u}_j)} \simeq \bar{\bar{u}}_i \overline{(u_j - \bar{u}_j)} = \bar{\bar{u}}_i (\bar{u}_j - \bar{\bar{u}}_j) \tag{14}$$

and that,

$$\overline{(u_i - \bar{u}_i)(u_j - \bar{u}_j)} \simeq \overline{(u_i - \bar{u}_i)} \cdot \overline{(u_j - \bar{u}_j)} = (\bar{u}_i - \bar{\bar{u}}_i)(\bar{u}_j - \bar{\bar{u}}_j) \tag{15}$$

i.e. essentially that that filtering operations commute to the individual components of each product. The above assumptions eventually lead to,

$$\tau_{ij} = \overline{\bar{u}_i \bar{u}_j} - \bar{\bar{u}}_i \bar{\bar{u}}_j \tag{16}$$

which is the scale-similarity model (SIMB) of Bardina for incompressible flows (Bardina et al. 1983). The compressible version derived following analogous arguments reads,

$$\tau_{ij} = \bar{\rho}(\overline{\tilde{u}_i \tilde{u}_j} - \bar{\tilde{u}}_i \bar{\tilde{u}}_j), \tag{17}$$

Scale-similarity models are able to predict backscatter unlike the static Smagorinsky model however when applied to LES they have long been known to provide insufficient dissipation, clearly a result of the assumptions involving the filtering operations. In an attempt to improve the predictions of the scale-similarity model Andreson and Domaradzki proposed an improved version (Anderson and Domaradzki 2012). Based on the Inter-Scale Energy transfer model of Anderson and Domaradzki (2012) Klein et al. then (2015) suggested a modified version for application to reacting flows (SIMET). This model reads,

$$\tau_{ij} = \bar{\rho} \left(\widehat{\tilde{u}_i \tilde{u}_j} + \widehat{\hat{\tilde{u}}_j \tilde{u}_i} - \hat{\tilde{u}}_i \hat{\tilde{u}}_j - \widehat{\hat{\tilde{u}}_i \hat{\tilde{u}}_j} \right), \tag{18}$$

In fact, there exist a plethora of scale-similarity models in the literature and a common characteristic of the majority of them is insufficient dissipation. As a result, the most usual application of scale-similarity models is in mixed models. In such models as the name suggests different models are mixed together with the most usual approach being the addition of an eddy-diffusivity type of model (typically Smagorinsky) to a scale-similarity model in order to provide sufficient dissipation.

2.3 Gradient Model

The gradient model (GRAD) can be derived by expanding in Taylor series the filtered velocity product in the expression for τ_{ij} (Vreman et al. 1996) and retaining the leading term in the expansion (Clark 1979) leading to,

$$\tau_{ij} = \bar{\rho}\frac{\Delta^2}{12}\frac{\partial \tilde{u}_i}{\partial x_k}\frac{\partial \tilde{u}_j}{\partial x_k}, \tag{19}$$

Models of the above kind typically give very good results in a priori studies and provided the filter width is sufficiently small so that the contribution from the terms dropped in the Taylor series expansion is small. However, like the scale-similarity models gradient-type models were also found to provide insufficient dissipation in LES, and as a result they are mainly used in mixed models. An interesting point with the gradient model is that it is essentially a low-order deconvolution-based model (discussed later on).

2.4 Clark Model

Vreman et al. (1996) built upon the mixed model of Clark (1979) to produce the following dynamic mixed model,

$$\tau_{ij} = \bar{\rho}\frac{\Delta^2}{12}\frac{\partial \tilde{u}_i}{\partial x_k}\frac{\partial \tilde{u}_j}{\partial x_k} - C_C \bar{\rho}\Delta^2 |\tilde{S}'|\tilde{S}'_{ij}, \tag{20}$$

where

$$S'_{ij}(\tilde{\mathbf{u}}) = \frac{\partial \tilde{u}_i}{\partial x_j} + \frac{\partial \tilde{u}_j}{\partial x_i} - \frac{2}{3}\delta_{ij}\frac{\partial \tilde{u}_k}{\partial x_k} = 2\left(\tilde{S}_{ij} - \frac{1}{3}\delta_{ij}\tilde{S}_{kk}\right), \tag{21}$$

and $|S'| = (S'_{ij}S'_{ij}/2)^{1/2}$. In the static version $C_C = 0.172$ and in the dynamic version it is calculated using,

$$C_C = \frac{\langle M'_{ij}(L_{ij} - H_{ij})\rangle}{\langle M'_{ij}M'_{ij}\rangle}. \tag{22}$$

Denoting $v_i = \widehat{\bar{\rho}\tilde{u}_i}/\hat{\bar{\rho}}$, the tensors H_{ij} and M_{ij} are given by

$$H_{ij} = \hat{\bar{\rho}}\frac{\hat{\Delta}^2}{12}\frac{\partial v_i}{\partial x_k}\frac{\partial v_j}{\partial x_k} - \frac{\Delta^2}{12}\left(\bar{\rho}\frac{\partial \tilde{u}_i}{\partial x_k}\frac{\partial \tilde{u}_j}{\partial x_k}\right), \tag{23}$$

and

$$M'_{ij} = -\hat{\bar{\rho}}\hat{\bar{\Delta}}^2 |S'(\underline{v})| S'_{ij}(\underline{v}) + \Delta^2 \left(\bar{\rho}|S'(\tilde{\mathbf{u}})| S'_{ij}(\tilde{\mathbf{u}})\right), \tag{24}$$

The Clark model is a mixed model with the first part consisting of a gradient component and the second consisting of a Smagorinsky-type component to provide the necessary dissipation. This model gave good results for the temporal mixing layer in Vreman et al. (1996, 1997) and was also one of the models selected for testing in Nikolaou et al. (2021) in order to elucidate any difference with the gradient model and to shed light as to whether the eddy-diffusivity part improves the predictions or not.

2.5 Wall-Adapting Local Eddy-Viscosity (WALE)

This model was used to simulate a wall-impinging jet with overall good results in Lodato et al. (2009). It is a mixed model with a Smagorinsky-type component and a scale-similarity component,

$$\tau_{ij} - \frac{1}{3}\delta_{ij}\tau_{kk} = -2\bar{\rho}\nu_t \left(\tilde{S}_{ij} - \frac{1}{3}\delta_{ij}\tilde{S}_{kk}\right) + \bar{\rho}(\widetilde{\bar{u}_i \bar{u}_j} - \hat{\bar{u}}_i \hat{\bar{u}}_j), \tag{25}$$

The turbulent viscosity is calculated from the velocity gradient and shear rate tensors using,

$$\nu_t = (C_W \Delta^2) \frac{(\tilde{s}_{ij}^d \tilde{s}_{ij}^d)^{3/2}}{(\tilde{S}_{ij}\tilde{S}_{ij})^{5/2} + (\tilde{s}_{ij}^d \tilde{s}_{ij}^d)^{5/4}}, \tag{26}$$

The model constant $C_W = 0.5$, and $\tilde{s}_{ij}^d$ is the traceless symmetric part of the squared resolved velocity gradient tensor $\tilde{g}_{ij} = \partial \tilde{u}_i / \partial x_j$,

$$\tilde{s}_{ij}^d = \frac{1}{2}(\tilde{g}_{ij}^2 + \tilde{g}_{ji}^2) - \frac{1}{3}\delta_{ij}\tilde{g}_{kk}^2, \tag{27}$$

where $\tilde{g}_{ij}^2 = g_{ik}g_{kj}$. Note that in this case as well, the static Yoshizawa closure is used to model the trace of the stress tensor as discussed above.

3 Deconvolution-Based Modelling

Deconvolution methods were probably first introduced in fluid mechanics research in the works of Leonard and Clark (Clark 1979; Leonard 1974). Deconvolution aims to invert the filtering operation in LES in order to obtain an approximation of the unfiltered field ϕ^* from the filtered field $\bar{\phi}$ which is resolved by the LES. Then, the filtered non-linear functions of ϕ can be approximated using the deconvoluted fields i.e. $\overline{f(\phi)} \simeq f(\overline{\phi^*})$. In the case of the unresolved stress tensor τ_{ij} is a function of the three velocity components therefore the term is closed using $\tau_{ij} \simeq \bar{\rho}(\widetilde{u_i^* u_j^*} - \tilde{u}_i \tilde{u}_j)$. Since the deconvolution operation is a purely mathematical operation relating filtered and unfiltered fields such methods do not include any assumptions and/or any modelling parameters/constants. As a result, in principle, they can be used to model a wide range of unresolved terms in the governing equations for different flow configurations including both reacting and non-reacting flows. The deconvolution can be accomplished with (a) Approximate methods, (b) Iterative methods and (c) using machine-learning.

Approximate methods are based on truncated Taylor series expansions of the inverse filtering operation. This approach was used to derive explicit algebraic models for the Reynolds stresses in non-reacting flows (Domaradzki and Saiki 1997; Geurts 1997). In the works of Stolz and Adams (1999) an Approximate Deconvolution Method (ADM) based on a truncated expansion of the inverse filter operation was used, and the deconvoluted signal was then explicitly filtered to obtain closures for the Reynolds stresses. The method was later used by the same authors to model the Reynolds stress terms in wall-bounded flows as well (Stolz and Adams 2001) where classic models such as the static Smagorinsky model are otherwise too dissipative. Approximate deconvolution methods have also been applied to reacting flows (Domingo and Vervisch 2015, 2017; Mathew 2002; Mehl and Fiorina 2017) with overall good results.

Iterative deconvolution methods include the use of reconstruction algorithms such as van Cittert iterations (Nikolaou et al. 2019; Nikolaou and Vervisch 2018; Nikolaou et al. 2018) or otherwise (Wang and Ihme 2017). The classic van Cittert algorithm with a constant coefficient b reads,

$$\phi^{*n+1} = \phi^{*n} + b(\bar{\phi} - G * \phi^{*n}) \tag{28}$$

where $\phi^{*0} = \bar{\phi}$, and ϕ^{*n} is the approximation of the un-filtered field for a given iteration count. In the case $\phi = \rho u_i$ and $\phi = \rho$ with $b = 1$ (typical value), the first two iterations result in the following approximations for the unfiltered density and density-velocity product,

$$\rho^{*0} = \bar{\rho}$$
$$\rho^{*1} = 2\bar{\rho} - \bar{\bar{\rho}}$$
$$\{\rho u_i\}^{*0} = \overline{\rho u_i}$$
$$\{\rho u_i\}^{*1} = 2\overline{\rho u_i} - \overline{\overline{\rho u_i}}$$

The ^{n} approximation of $\rho u_i u_j$ is calculated using $\{\rho u_i u_j\}^{*n} = \{\rho u_i\}^{*n} \{\rho u_j\}^{*n} / \rho^{*n}$, and the corresponding approximation of the unresolved stress tensor is calculated using $\tau_{ij}^n = \bar{\rho}(\{\rho u_i u_j\}^{*n} / \bar{\rho} - \tilde{u}_i \tilde{u}_j)$. It is straightforward to show that the first two are,

$$\tau_{ij}^0 = \overline{\bar{\rho}\tilde{u}_i\tilde{u}_j} - \bar{\rho}\tilde{u}_i\tilde{u}_j$$

$$\tau_{ij}^1 = \left(\frac{\overline{4\overline{\rho u_i}\cdot\overline{\rho u_j} - 2\overline{\rho u_i}\cdot\overline{\overline{\rho u_j}} - 2\overline{\rho u_j}\cdot\overline{\overline{\rho u_i}} + \overline{\overline{\rho u_i}}\cdot\overline{\overline{\rho u_j}}}}{2\bar{\rho} - \bar{\bar{\rho}}}\right) - \bar{\rho}\tilde{u}_i\tilde{u}_j$$

Note that for $n = 0$, a Bardina-like scale-similarity model is recovered. For $n = 1$ an extended similarity-like model is obtained which involves double and triple-filtered quantities and so on for higher-order approximations. Successive iterations lead to higher-order approximations of the unfiltered fields and of the unresolved stress tensor as shown by Stolz and Adams (2001). For example, four iterations are sufficient to recover the gradient model supplemented by the next term in the series (Eq. B9 in Stolz and Adams 1999).

It is important to note that deconvolution methods only recover wavenumbers which are resolved by the LES mesh. As a result, deconvolution methods require $h/\Delta < 1$ so that wavenumbers below Δ can be recovered. As for the van Cittert algorithm it is a linear one, and for periodic signals it is straightforward to show that for a sufficiently large number of iterations, and provided $0 < b < 2$, the algorithm is stable and converges to the original value of the un-filtered field for all finite wavenumbers on the mesh (Nikolaou and Vervisch 2018). b is typically taken to equal 1 for non-oscillatory convergence as shown in Nikolaou and Vervisch (2018). The maximum number of iterations required for a sufficiently small reconstruction error, depends on the largest wavenumber resolved by the mesh i.e. on the h/Δ ratio with increasing resolution requiring a larger number of iterations.

4 Machine-Learning Based Models

The theoretical justification for using machine-learning methods and specifically artificial neural networks can be justified by the seminal work of Hornik (1991) where it was proven that a feed-forward neural network, even with a single hidden layer, acts as a universal function approximator (for functions with certain properties), in the limit of a sufficiently large number of nodes. As a result, algebraic closures of increased order of complexity can in principle be developed e.g. for the stress tensor by adjusting the number of layers and/or nodes. Machine-learning methods with regards to modelling the stress-tensor in the context of LES can (thus far) be roughly divided into three distinct categories:

(a) Optimization/tuning of existing model parameters and/or their evaluation procedures.
(b) Direct modelling of the stress tensor using as inputs variables which are resolved by the LES.
(c) Deconvolution-based approaches.

In comparison to non-reacting flows the use of machine-learning for modelling purposes in reacting flows is scarce and has been primarily used to model/accelerate the chemical kinetics (Chatzopoulos and Rigopoulos 2013; Ihme et al. 2009; Sen and Menon 2009; Sen et al. 2010). In terms of modelling, convolutional networks were successfully employed to model the Flame Surface Density (FSD) in Lapeyre et al. (2019) which is an important term in reacting LES (Nikolaou and Swaminathan 2018), and was shown to outperform classic state of the art algebraic models. In Nikolaou et al. (2018, 2019) convolutional networks were used in a deconvolution-based context to model the scalar variance, a key modelling parameter in flamelet methods while Seltz et al. (2019) employed convolutional neural networks to provide a unified modelling framework for both the source and scalar flux terms in the filtered scalar transport equation. With regards to modelling the stress tensor, categories (a)–(c) are discussed in the text which follows.

4.1 Type (a)

Probably the first application of machine-learning in LES with regards to the stress tensor dates to the work of Sarghini et al. (2003) in which a neural network was trained to predict the turbulent viscosity parameter in the Smagorinsky part of a mixed model (Smagorinsky+Bardina). The network was trained by first running LES at $Ret = 180$ with Bardina's model and the viscosity parameter calculated using the classic dynamic procedure. The data generated from the LES were then used to train the network to essentially replace the more expensive dynamic calculation of the viscosity parameter. The inputs consisted of the nine velocity gradients $\partial \bar{u}_i / \partial x_j$ and the six velocity fluctuation products $u_i' u_j'$. The network was four layers deep, 1(15)-2(12)-3(6)-4(1) with the numbers in parentheses indicating the number of neurons in each layer, and fully connected. The authors reported a 20% speedup in comparison to using the dynamic procedure and that the network performed well for a certain range of Ret close to the training Reynolds number. For larger Reynolds numbers at $Ret = 1050$ a novel training procedure was concluded to be required.

In a more recent study (Xie et al. 2019) a version of the Clark model presented in Sect. 2 was adopted having two tunable parameters instead of one: one for the gradient part and the other for the Smagorinsky part. DNS data of compressible decaying turbulence were then used to train a neural network to predict these two parameters using as inputs the filtered velocity divergence $\partial \tilde{u}_i / \partial x_i$, the filtered vorticity magnitude $|\epsilon_{ijk} \partial \tilde{u}_i / \partial x_j|$, the filtered velocity gradient magnitude $\sqrt{\partial \tilde{u}_i / \partial x_j \partial \tilde{u}_i / \partial x_j}$ and the filtered strain rate tensor magnitude $\sqrt{S_{ij} S_{ij}}$. The developed networks showed improved performance over the static/dynamic Smagorinsky and classic Clark models in the a posteriori testing which followed.

4.2 Type (b)

The first direct modelling approach dates to the work of Gamahara and Hattori (2017) where DNS data of turbulent channel flow at $Ret = 180$ were used for training the networks in the usual approach whereby the DNS data are filtered to simulate an LES. A range of possible inputs were tested: (a) $\{y, S_{ij}\}$, (b) $\{y, S_{ij}, \Omega_{ij}\}$, (c) $\{y, \partial \bar{u}_i/\partial x_j\}$ and (d) $\{\partial \bar{u}_i/\partial x_j\}$, where $\Omega_{ij} = (\partial \bar{u}_i/\partial x_j - \partial \bar{u}_j/\partial x_i)/2$ is the rotation-rate tensor, and y is the distance from the wall. In total six three-layer fully connected networks were trained i.e. one for each component of the stress tensor. Correlation coefficients were then extracted between the predicted and as-extracted from the DNS components of the stress tensor. For the largest and most dominant streamwise component τ_{11}, all four sets showed similar correlations in the region of 0.8 with group (c) having the highest. This group was then tested (a-priori) against DNS data of higher Reynolds number at $Ret = 400$ and $Ret = 800$ with overall good results. A posteriori tests at $Ret = 180$ and $Ret = 400$ were also conducted in the same study with overall good results in comparison to the classic Smagorinsky model even though no obvious advantage was reported by the authors.

In the same spirit of Gamahara and Hattori (2017), Wang et al. (2018) used DNS data to train a network to directly predict the stress tensor. The DNS data corresponded to homogeneous decaying turbulence at $Re_\lambda = 220$. Five different sets of inputs were tested using four-layer and five-layer networks: (a) $\bar{u}_i$: 1(3)-2(20)-3(10)-4(1), (b) $\partial \bar{u}_i/\partial x_j$: 1(9)-2(40)-3(20)-4(1), (c) $\partial^2 \bar{u}_i/\partial^2 x_j$: 1(9)-2(40)-3(20)-4(1), (d) $\partial^2 \bar{u}_i/\partial x_j \partial x_k$ 1(9)-2(40)-3(20)-4(1) and (e) all of the previous inputs: 1(30)-2(90)-3(60)-4(30)-5(1). As in Gamahara and Hattori one network for each component of the stress tensor was developed. Of all the inputs tested groups (b) and (e) produced the highest correlations in a priori testing, with group (e) however only improving marginally the correlations at the expense of having a more complex network. Therefore the importance of using the velocity gradients much like in the study of Gamahara and Hattori was confirmed albeit in a different configuration. Of course this is not surprising since the velocity gradients appear in many models for the stress tensor. Moving on, a further refined network based on group (b) was then developed and tested a posteriori in LES and compared against the static and dynamic Smagorinsky models. The ANN model showed improved agreement in comparison to the two classic models both in predicting the temporal evolution of the kinetic energy and its dissipation rate. In terms of computational cost, the ANN model was found to be 3.6 times slower than the static Smagorinsky model and 1.8 times slower than the dynamic Smagorinsky model, indicating that neural network models need to be as simple as possible to limit computational cost.

Following Wang et al. (2018) in Zhou et al. (2019) a similar procedure was applied to the same configuration i.e. decaying homogeneous turbulence in order to develop a network for the stress tensor. In contrast to the the previous works (Gamahara and Hattori 2017; Wang et al. 2018) a single network was trained for all six components of the stress tensor while additionally taking into account the filter width which along with the nine velocity gradients constituted the input set to the network. The evaluation was performed both a priori against the DNS data and a posteriori with LES with the ANN-based model showing an overall improved performance in comparison to the dynamic Smagorinsky model.

In a more recent study (Park and Choi 2021) the case of turbulent channel flow was revisited. As in the work of Gamahara and Hattori (2017) similar inputs were tested with a four-layer network and six outputs instead. The inputs tested included single-point but also multiple-point variables along the streamwise and spanwise directions. The inputs consisted of (a) S_{ij}-single point (b) $\partial \bar{u}_i/\partial x_j$-single point, (c) S_{ij}-multiple points, (d) $\partial \bar{u}_i/\partial x_j$-multiple points and (e) $\{\bar{u}_i, \partial \bar{u}_i/\partial x_j\}$-multiple points. In the a priori tests it was found that the groups (c) and (d) provided the highest correlations and reasonably predicted the backscatter. However, in a posteriori tests it was found that these inputs led to instabilities unless backscatter clipping was used. The single-point group (a) on the other hand showed very good agreement in the a posteriori tests despite the lower correlations observed in the a priori tests.

In reacting flows, an posteriori study using a closely-related data-based approach has been examined in Schoepplein et al. (2018) where Gene-Expression Programming (GEP) was employed. In this approach τ_{ij} was assumed to depend on the strain rate and the rotation rate tensors S_{ij} and Ω_{ij} respectively (as in Gamahara and Hattori 2017), but also on the filter width Δ and filtered density $\bar{\rho}$. GEP was then used to derive a best-fit function for the stress-tensor which showed good agreement against the DNS data.

The direct modelling approach for reacting flows was first examined in Nikolaou et al. (2021). A DNS database of a turbulent premixed hydrogen V-flame was used in order to train a network to predict all six components of the stress tensor using as inputs the filtered density $\bar{\rho}$, and the nine velocity gradients $\partial \bar{u}_i/\partial x_j$ (suitably normalised). In comparison to previous studies in the literature this DNS configuration was particularly challenging to model due to the strong inhomogeneity in the direction perpendicular to the mean stream-wise flow, the presence of a bluff body, and the presence of heat release modelled using detailed chemistry—the configuration is shown in Fig. 1. The lowest turbulence cases V60 and V60H ($Re_T = 220$) were used for training the networks while the highest turbulence level case V90 ($Re_T = 562.8$) for testing the networks. A 1(10)-2(40)-3(10)-4(18)-5(6) network structure was developed for each filter width considered, able to predict all six components of the stress tensor (Nikolaou et al. 2021). In contrast to previous studies employing fully connected layers in order to account for the strong inhomogeneity in the cross-stream directions it was found necessary to decouple layers 4 and 5 by introducing 3 to 1 connections rather than fully connected between these two layers.

A thorough a priori comparison against all models presented in Sect. 2 was conducted for all three filter widths considered i.e. at $\Delta/\delta_L = 1, 2$ and 3 where δ_L is the laminar thermal flame thickness. Figures 2 and 3 show the instantaneous predictions (normalised) of all models considered for the largest filter width for the dominant components τ_{11} and τ_{13} respectively. These results are quantified in terms of the Pearson correlation coefficient for each individual component of the stress tensor averaged over all filter widths in Fig. 4. The results show that the networks are able to outperform the predictions obtained using the classic models while the work in

Fig. 1 Averaged (in homogeneous y direction) instantaneous progress variable $c = (T - T_r)/(T_p - T_r)$ for all three cases (T_r =reactants temperature, T_p =products temperature). Note that the databases are 3D: cases V60 and V60H used for training and case V97 (highest turbulence level case) used for testing (Nikolaou et al. 2021)

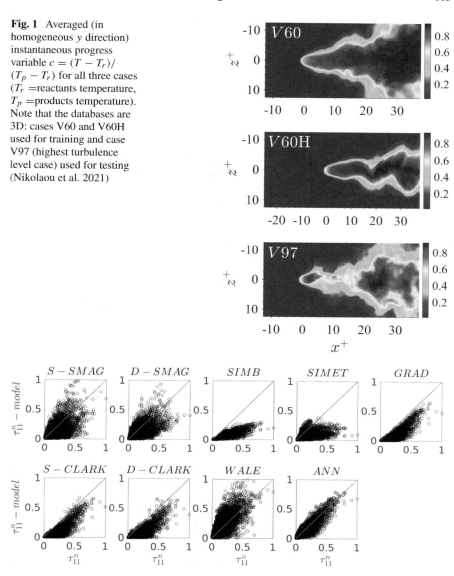

Fig. 2 Scatter plots of instantaneous values of DNS and modelled τ_{11} on the LES mesh, for $\Delta^+ = 3$ (Nikolaou et al. 2021)

Nikolaou et al. (2021) also confirmed the results found in Klein et al. (2015) on the poor performance of the Smagorinsky model (static and dynamic) for reacting flows.

Another important point to consider in the model evaluation step is the ability of a model to predict the correct relative magnitude between the different stress

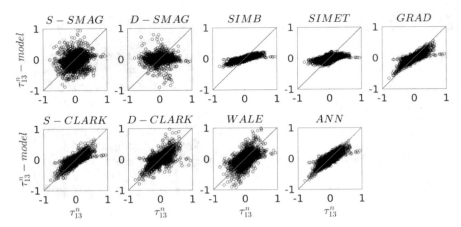

Fig. 3 Scatter plots of instantaneous values of DNS and modelled τ_{13} on the LES mesh, for $\Delta^+ = 3$ (Nikolaou et al. 2021)

Fig. 4 Pearson correlation coefficients averaged across all filter widths, for each stress tensor component for case V97 (Nikolaou et al. 2021)

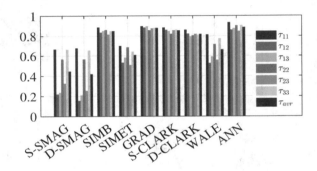

components which amounts to evaluating the alignment angle between the DNS and modelled resultant stress in a given direction. A perfect model would correspond to a zero alignment angle between the modelled and DNS stresses in a particular direction and the probability density function would approach a δ function at zero. This evaluation step is particularly important to do in flows with strong inhomogeneities since in such cases one must ensure that the model's predictions are not biased towards any of the dominant or non-dominant components of the stress tensor. Therefore, in a further evaluation step in Nikolaou et al. (2021) probability density functions of the alignment angle between the modelled and DNS stress tensor τ_{j1} were extracted and compared for each model. The results are shown in Fig. 5 where it is apparent that the ANN-based model shows an improved performance in comparison to the classical models.

Fig. 5 Probability density function of the angle θ between DNS τ_{j1} and modelled τ_{j1}^m for $\Delta^+ = 3$ (Nikolaou et al. 2021)

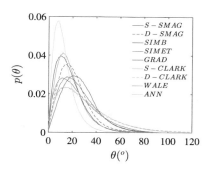

4.3 Type (c)

The first use of machine-learning in a deconvolution-based context dates to the work of Maulik and San (2017) where a single-layer network with 100 neurons was trained to recover estimates of the unfiltered velocity components u_i^* from their filtered counterparts $\bar{u}_i$. The inputs to the network consisted of the filtered velocity components in the neighbourhood of a given point. This enabled the direct modelling of the stress tensor using explicit filtering on the deconvoluted variables. The developed networks were tested a priori for different cases including 2D Kraichman, 3D Kolmogorov and compressible stratified turbulence with overall good results.

In the same spirit, a neural network was trained in Yuan et al. (2020) to reconstruct the unfiltered velocity components which was tested both against the DNS data and a posteriori in LES of forced isotropic turbulence. The inputs consisted of the filtered velocities in the region surrounding a given point as in Maulik and San (2017) and the outputs consisted of the three unfiltered velocity components which were then filtered explicitly to model the stress tensor as in classical deconvolution-based approaches. In a posteriori testing, the ANN-based models provided improved predictions over the dynamic Smagorinsky model.

5 A Note: Sub-grid Versus Sub-filter

It is important to note that the terms "sub-grid" and "sub-filter" are different. "Sub-grid" refers to scales not resolved by the mesh h while "sub-filter" refers to scales not resolved by the filter width Δ. In the majority of classic approaches $h/\Delta = 1$ and the terms are equivalent however in approaches which include deconvolution/machine-learning $h/\Delta < 1$ in which case the terms are not equivalent: in such cases "sub-filter" refers to scales between h and Δ which are resolved by the mesh and can potentially be recovered e.g. using deconvolution and/or suitably trained neural/convolutional networks.

6   Challenges of Data-Based Models

6.1   Universality

As the name suggests, data-based methods depend on data. One can view machine-learning methods such as ANNs and CNNs as a multi-dimensional data-fitting procedure. As a result, the predictive ability of a network depends on the dataset. For datasets not too dissimilar to the dataset used to train a network in the first place, the predictions are expected to be reasonably good since in such cases inference is equivalent to a form of high-dimensional interpolation. For datasets which are too dissimilar (which lie far from the multi-dimensional fitted surface) the predictions are expected to be poorer in comparison since in such cases inference is equivalent to extrapolation. For instance, a neural network trained solely on homogeneous decaying turbulence data to predict the stress tensor would probably perform poorly in shear-dominated flows and vice versa. Increasing the training data-size is always an option however this would lead to even more complex networks with increased computational cost. Another option would be to train case-specific networks and switch between them depending on the local flow configuration. In general, the universality of a network depends on the size, quality, and diversity of the databases used for training.

6.2   Choice and Pre-processing of Data

Any inputs to a data-driven model need to be appropriately scaled, and standardization is a commonly used procedure for this purpose. Usually in the turbulence modelling community, such standardization is performed on the input variables which are already appropriately normalized by using some physical quantities such as mean flow velocity and turbulence length scale. However, it is often the case that such reference quantities are not available or they do not necessarily represent flow phenomena in practical problems. For example, non-reacting flow DNS is often performed for non-dimensional quantities. One way to train a model is to use such non-dimensional quantities as they are with or without standardization. While such a strategy would not require normalization based on physical quantities for training, applying a model based on this strategy to practical LES problems, one would face an issue of finding appropriate parameters to non-dimensionalize the quantities.

6.3   Training, Validation, Testing

Developing a model based on machine-learning typically involves three steps namely training, validation, and testing. The validation step is typically performed during the training phase on a subset of the training data while the chosen testing dataset varies from study to study. In some studies for instance the testing dataset is also a subset of the training dataset albeit at different spatio-temporal coordinates within the computational domain. This approach is convenient as there is no need to perform additional and often expensive simulations to generate new data e.g. at a higher Re or Ma number. However this approach may introduce a bias in the predictive ability of the network since the testing dataset may be too similar to the training/validation datasets. Therefore careful thought is required on the most appropriate training and testing strategy.

6.4 Network Structure

The choice of network structure is typically performed on a trial and error basis and to date there is no formal/theoretical procedure to a priori obtain the best network structure (number of layers, number of nodes, type of activation function, type of loss function) for a given set of inputs and outputs which minimises the training error. In addition, increasing the number of layers and/or nodes does not always improve the predictive ability of the network. Furthermore, and perhaps there is no formal way of a priori choosing the best set of input variables for a given output set and for a given network structure-typically a range of inputs are tested based on intuition.

When it comes to practical LES, some networks are more difficult to implement and parallelise in LES solvers than others. For instance, point-wise inputs are very convenient for LES applications while inputs requiring the values of the surrounding mesh points are tricky to implement and parallelise in practice using MPI. This is often the case with CNNs and other types of networks utilizing plane and volumetric inputs on Cartesian mesh points. However most LES codes often employ non-uniform and unstructured meshes. Of course, the fields can be interpolated to generate CNN-like inputs at every iteration at every point, but this would result in increased computational cost and other associated issues (Kashefi et al. 2021). One potential strategy to circumvent this issue while keeping the important spatial information for the inputs is so-called "point-cloud deep learning" (Kashefi et al. 2021). Although this framework is not yet well established for modelling the stress tensor, the compatibility to arbitrary mesh geometry is something future machine-learning models should consider.

6.5 LES Mesh Size

The development of LES models using DNS data involves explicit filtering operations with a filter size Δ. An important question is then how does one choose h i.e. the LES mesh size? Typically in classic approaches $h/\Delta = 1$ but this choice does not ensure that the resolved fields such as velocity and scalar fields are well-resolved.

Consequently, the gradients of these variables as obtained on the LES mesh which are typically used as inputs to neural networks are also not well-resolved which introduces a bias in the predictive ability of the network-this is also the case when evaluating the performance of classic models which involve gradient terms.

In an effort to resolve this Nikolaou and Vervisch (2018) proposed a criterion for the LES mesh size, based on a scalar variation evolving from 0 to 1, which was originally proposed for a "reaction progress variable" (e.g. non-dimensional temperature) but which can also be regarded as a normalized fluctuating velocity component $\phi(x)$ ($0 \leq \phi \leq 1$).

$$\phi(x) = \frac{1}{2} \left(1 + erf \left(\frac{x\sqrt{\pi}}{\delta} \right) \right), \tag{29}$$

where δ is a length scale for the gradient defined as $\delta = 1/\max(d\phi/dx)$. Filtering Eq. (29) based on the filtering operation (Eq. (2)) with a Gaussian kernel, the filtered field $\bar{\phi}(x)$ can be obtained as,

$$\bar{\phi}(x) = \frac{1}{2} \left(1 + erf \left(\frac{1}{\sqrt{1 + \frac{\pi}{6}\frac{\Delta^2}{\delta^2}}} \frac{x\sqrt{\pi}}{\delta} \right) \right). \tag{30}$$

The length scale for the gradient of the filtered field can be obtained in the same manner as $\delta = 1/\max(d\bar{\phi}/dx)$, which leads to

$$\bar{\delta} = \delta \left(1 + \frac{\pi}{6}\frac{\Delta^2}{\delta^2} \right)^{1/2}, \tag{31}$$

ensuring $\bar{\delta}/\delta > 1$ i.e. that the length scale increases due to the filtering operation. It would be more useful to rewrite Eq. (31) in terms of $\bar{\delta}/\Delta$, since our interest here is how fine the mesh should be to capture the gradient information of the filtered field with Δ,

$$\frac{\bar{\delta}}{\Delta} = \left(\frac{\pi}{6} + \frac{\delta^2}{\Delta^2} \right)^{1/2}. \tag{32}$$

Usually, to resolve a filtered gradient n mesh points are required within $\bar{\delta}$ which results to,

$$\frac{h}{\Delta} = \frac{1}{n} \left(\frac{\pi}{6} + \frac{\delta^2}{\Delta^2} \right)^{1/2}. \tag{33}$$

In most turbulent flows, it is expected that $\delta/\Delta \sim 0$. Equation (33) yields $h/\Delta \simeq 0.36$ for $n = 2$ (two mesh points within the filtered slope), and $h/\Delta \simeq 0.18$ for $n = 4$, leading to the insight that the LES mesh required to capture the filtered gradient should have two to five mesh points within Δ. This consideration is required when generating filtered quantities from resolved fields such as DNS, especially for

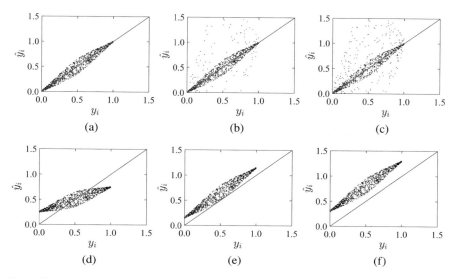

Fig. 6 Scatter plots of target values y_i and predicted values $\hat{y}_i$. **a–f**: scenarios (a) to (f), respectively

machine-learning with gradient-related inputs, but is also useful for conventional gradient model assessments.

6.6 Performance Metrics

The quantification of prediction accuracy is very important since in the modelling of the stress tensor a model assessment needs to be performed spatio-temporally and for all six components of the stress tensor—a comprehensive visual examination is just not enough. Amongst the possible quantification methods, the mean squared error (MSE) would be the most convenient to use since it is already incorporated in the loss function of most machine-learning algorithms. Another choice is the root mean squared error (RMSE). However, MSE and RMSE are considered to be sensitive to local outliers which are prevalent in non-linear phenomena. For this reason, the mean absolute error (MAE) may be more suitable for model assessment purposes.

In various model developments in the turbulent flow community, the cross-correlation coefficient is also used extensively. While this quantity is familiar to the community, relying on this coefficient alone can bias the model performance assessment significantly. This point is illustrated by using the following simulated target values y_i and predicted values $\hat{y}_i$ in scenarios (a) to (f), where i is the index of N-samples.

(a) Predicted values are scattered around the target values.
(b) Predicted values are scattered around the target values, but 15% of samples have much larger deviation (outliers).
(c) Predicted values are scattered around the target values, but 30% of samples have much larger deviation (outliers).
(d) Predicted values are scattered around a line $\hat{y}_i = 0.5y_i + 0.25$. The deviation from the line is the same as (a).
(e) Predicted values are scattered around a line $\hat{y}_i = y_i + 0.15$. The deviation from the line is the same as (a).
(f) Predicted values are scattered around a line $\hat{y}_i = y_i + 0.30$. The deviation from the line is the same as (a).

Scenario (a) represents perhaps a good model. In turbulent flow problems where the variables take a wide range of values however, such a good model may output a prediction with a large deviation for a limited number of samples, and such situations may correspond to scenarios (b) and (c). The situations where the trend of predicted values is close to the target values but there is some deviation between the two may correspond to scenarios (d), (e) and (f). Examples of such scenarios are shown in Fig. 6.

For the scenarios (a)–(f), the following metrics often used for model assessments are considered,

- Mean absolute error

$$\epsilon_{MAE} = \frac{\sum_{i=1}^{N} |y_i - \hat{y}_i|}{N}. \tag{34}$$

- Relative mean absolute error

$$\epsilon_{rMAE} = \frac{\epsilon_{MAE}}{\bar{y}}. \tag{35}$$

- Mean squared error

$$\epsilon_{MSE} = \frac{\sum_{i=1}^{N} (y_i - \hat{y}_i)^2}{N}. \tag{36}$$

- Root mean squared error

$$\epsilon_{RMSE} = \sqrt{\epsilon_{MSE}}. \tag{37}$$

- Relative root mean squared error

$$\epsilon_{rRMSE} = \frac{\epsilon_{RMSE}}{\bar{y}}. \tag{38}$$

- Pearson's cross-correlation coefficient

$$\rho_p = \frac{\sum_{i=1}^{N} (y_i - \bar{y}) \left(\hat{y}_i - \bar{\hat{y}}\right)}{\sqrt{\sum_{i=1}^{N} (y_i - \bar{y})^2 \left(\hat{y}_i - \bar{\hat{y}}\right)^2}} \tag{39}$$

Table 1 Scatter plots of generic target values y_i and predicted values $\hat{y}_i$

Scenario	ϵ_{MAE}	ϵ_{rMAE}	ϵ_{MSE}	ϵ_{RMSE}	ϵ_{rRMSE}	ρ_p	ρ_p^2	R^2	E_1
(a)	0.05	0.09	0.00	0.06	0.12	0.98	0.96	0.96	0.81
(b)	0.11	0.23	0.06	0.24	0.49	0.76	0.58	0.29	0.55
(c)	0.17	0.36	0.11	0.33	0.68	0.64	0.41	−0.39	0.29
(d)	0.13	0.26	0.02	0.14	0.30	0.98	0.96	0.74	0.48
(e)	0.15	0.31	0.03	0.16	0.33	0.98	0.96	0.67	0.38
(f)	0.30	0.62	0.09	0.31	0.63	0.98	0.96	−0.18	−0.23

- Coefficient of determination

$$R^2 = 1 - \frac{\sum_{i=1}^{N} \left(y_i - \hat{y}_i\right)^2}{\sum_{i=1}^{N} (y_i - \bar{y})^2} \tag{40}$$

- Coefficient of Legates and McCabe (2013)

$$E_1 = 1 - \frac{\sum_{i=1}^{N} \left|y_i - \hat{y}_i\right|}{\sum_{i=1}^{N} |y_i - \bar{y}|} \tag{41}$$

In the list above $\bar{}$ denotes the mean value. The metrics ρ_p, R^2 and E_1, yield 1 for a perfect model. All of the above metrics are computed and summarised in Table 1 for scenarios (a)–(f). Note that ρ_p^2 is also shown since it is often used as an alternative definition for the coefficient of determination. As clearly seen, the cross-correlation coefficient ρ_p shows relatively high values for all the scenarios except for (c) where $\rho_p = 0.64$, which may still be acceptable for certain purposes. However, there is substantial discrepancy between the intuitive interpretation of Fig. 6 and ρ_p in Table 1 for scenarios (d)–(f). For these cases the relative errors ϵ_{rMAE} and ϵ_{rRMSE}, vary from 25% to 63%, while $\rho_p = 0.98$ for these scenarios. Also, ϵ_{rRMSE} and R^2 tends to be more sensitive to large deviation of small number of samples respectively than ϵ_{rMAE} and E_1 (see the scenario (b)), and this is considered due to $(y_i - \hat{y}_i)^2$. These considerations suggest that model assessments based on ρ_p alone cannot thoroughly assess a model's performance accurately, and ρ_p should be used along with visual examination and/or another metric.

7 Summary

Machine-learning methods are increasingly being used by the fluid mechanics community for modelling purposes and in particular for the unresolved stress tensor. The applications are diverse while a large number of both a priori but also a posteriori assessments have shown data-based methods either to outperform the predictions of classic models or to at least match them. The developed networks are typically one to five layers deep with around one hundred neurons in each hidden layer with the structure of the networks varying from study to study. Overall, the best-performing inputs appear to be gradients of the filtered velocity components and functions of the velocity gradients such as the strain rate tensor and the rotation-rate tensor irrespective of the nature of the flow i.e. reacting or non-reacting. In terms of computational cost this depends on the structure of the networks with most of the developed networks in the literature, despite being slower than the classical algebraic models, exhibiting around the same order of magnitude cost. Despite however the success of the developed networks some important issues still remain which are discussed in the text. The most important in the authors view is universality. The predictive ability and versatility of a network is tightly coupled to the dataset used for training in the first place. At the time being, in the majority of studies in the literature these databases are restricted to small-scale DNS of often canonical flow problems such as decaying homogeneous turbulence, turbulent channel flow, statistically planar freely-propagating flames etc. while in practical LES the flows are significantly more complex but also at higher Re and Ma numbers. In order to overcome this issue and to eventually obtain a truly case-independent and parameter-free machine-learning-based model for the stress tensor, further research is required at conditions which are more relevant for practical flows including both a priori and a posteriori studies.

References

Ali Kashefi, Davis Rempe, Guibas Leonidas J (2021) A point-cloud deep learning framework for prediction of fluid flow fields on irregular geometries. Phys Fluids 33:027104

Anderson BW, Domaradzki JA (2012) A subgrid-scale model for Large Eddy simulation based on the physics of inter-scale energy transfer in turbulence. Phys Fluids 24:1–35

Aspden A, Day M, Bell J (2016) Three-dimensional direct numerical simulation of turbulent lean premixed methane combustion with detailed kinetics. Combust Flame 166:266–283

Bardina J, Ferziger JH, Reynolds WC (1983) Improved turbulence models based on Large Eddy Simulation of homogeneous, incompressible, turbulent flows. Technical report no. TF-19, Dep. Mech. Eng. Stanford University

Bray KNC, Libby PA, Masuya G, Moss JB (1981) Turbulence production in premixed turbulent flames. Combust Sci Technol 25:127–140

Chatzopoulos AK, Rigopoulos S (2013) A chemistry tabulation approach via rate-controlled constrained equilibrium (RCCE) and artificial neural networks (ANNs), with application to turbulent non-premixed CH4/H2/N2 flames. Proc Combust Inst 34:1465–1473

Chomiak J, Nisbet J (1995) Modelling variable density effects in turbulent flames-some basic considerations. Combust Flame 102:371–386

Clark RA (1979) Evaluation of sub-grid scalar models using an accurately simulated turbulent flow. J Fluid Mech 91:1–16

Deardoff JW (1970) A numerical study of three-dimensional turbulent channel flow at large Reynolds numbers. J Fluid Mech 41:453–480

Domaradzki JA, Saiki EM (1997) A sub-grid scale model based on the the estimation of unresolved scales of turbulence. Phys Fluids 9(2148)

Domaradzki JA, Liu W, Brachet ME (1993) An analysis of sub-grid scale interactions in numerically simulated isotropic turbulence. Phys Fluids 5(1747)

Domingo P, Vervisch L (2015) Large Eddy Simulation of premixed turbulent combustion using approximate deconvolution and explicit flame filtering. Proc Combust Inst 35:1349–1357

Domingo P, Vervisch L (2017) DNS and approximate deconvolution as a tool to analyse one-dimensional filtered flame sub-grid scale modelling. Combust Flame 177:109–122

Domingo P, Nikolaou Z, Seltz A, Vervisch L (2020) From discrete and iterative deconvolution operators to machine learning for premixed turbulent combustion modeling. In: Pitsch H, Attili A (eds) Data analysis for direct numerical simulations of turbulent combustion. Springer, Cham, pp 215–232

Fjortof R (1953) On the changes in the spectral distribution of kinetic energy for two-dimensional and non-divergent flow. Svenska Geophysica Foreningen, Tellus 5:225

Gamahara M, Hattori Y (2017) Searching for turbulence models by artificial neural network. Phys Rev Fluids 2:054604

Germano M, Piomelli U, Moin P, Cabot WH (1991) A dynamic sub-grid scale eddy viscosity model. Phys Fluids 3:1760–1765

Geurts BG (1997) Inverse modelling for Large Eddy simulation. Phys Fluids 9:3585–3587

Gicquel LYM, Staffelbach G, Poinsot T (2012) Large Eddy simulations of gaseous flames in gas turbine combustion chambers. Prog Energy Combust Sci 38:782–817

Hornik K (1991) Approximation capabilities of multi-layer feed-forward networks. Neural Netw 4:251–257

Ihme M, Schmitt C, Pitsch H (2009) Optimal artificial neural networks and tabulation methods for chemistry representation in LES of a bluff-body swirl-stabilized flame. Proc Combust Inst 32:1527–1535

Kanov K, Burns R, Lalescu C, Eyink G (2015) The John hopkins turbulence databases: an open simulation laboratory for turbulence research. Comput Sci Eng 17:10–17

Kerr RM, Domaradzki JA, Barbier G (1996) Small-scale properties of non-linear interactions and sub-grid scale energy transfer in isotropic turbulence. Phys Fluids 8(197)

Khan J, Wei JS, Ringer M, Saal LH, Ladanyi M, Westermann F, Berthold F, Schwab M, Antonescu CR, Peterson C, Meltzer PS (2001) Classification and diagnostic prediction of cancers using gene expression profiling and artificial neural networks. Nature 7:673–679

Klein M, Kasten C, Gao Y, Chakraborty N (2015) A priori direct numerical simulation assessment of sub-grid scale stress tensor closures for turbulent premixed combustion. Comput Fluids 122:1–1

Krizhevsky A, Sutskever I, Hinton G (2012) ImageNet classification with deep convolutional neural networks. Proc Adv Neural Inf Process Syst 25:1090–1098

Lapeyre CJ, Misdariis A, Cazard N, Veynante D, Poinsot T (2019) Training convolutional neural networks to estimate turbulent sub-grid scale reaction rates. Combust Flame 203:255–264

Legates DR, McCabe GJ (2013) A refined index of model performance: a rejoinder. Int J Climatol 33:1053–1056

Leonard A (1974) Energy cascade in Large Eddy simulation of turbulent fluid flows. Adv Geophys 18A:237–248

Lilly DK (1966) On the application of the eddy viscosity concept in the inerial sub-range of turbulence. Nation Center for Atmospheric Research (NCAR) report 5:1–19

Lilly DK (1992) A proposed modification of the Germano subgrid-scale closure method. Phys Fluids 4:633–635

Lodato G, Vervisch L, Domingo P (2009) A compressible wall-adapting similarity mixed model for Large Eddy simulation of the impinging round jet. Phys Fluids 21:1–21

Mason PJ, Callen NS (1986) On the magnitude of the subgrid-scale eddy coefficient in large-eddy simulations of turbulent channel flow. J Fluid Mech 162:439–462

Mathew J (2002) Large Eddy simulation of a premixed flame with approximate deconvolution modelling. Proc Combust Inst 29:1995–2000

Maulik R, San O (2017) A neural network approach for the blind deconvolution of turbulent flows. J Fluid Mech 831:151–181

Mehl C, Fiorina B (2017) Evaluation of deconvolution modelling applied to numerical combustion. Combust Th Model 22:38–70

Meneveau C, Katz J (2000) Scale invariance and turbulence models for Large-Eddy simulation. Ann Rev Fluid Mech 32:1–32

Minamoto Y, Fukushima N, Tanahashi M, Miyauchi T, Dunstan T, Swaminathan N (2011) Effect of flow geometry on turbulence-scalar interaction in premixed flames. Phys Fluids 23:125107

Mnih V, Kavukcuoglu K, Silver D, Rusu AA, Veness J, Bellemare MG, Graves A, Riedmiller M, Fidjeland AK, Ostrovski G, Petersen S, Beattie C, Sadik A, Antonoglou I, King H, Kumaran D, Wierstra D, Legg S, Hassabis D (2015) Human-level control through deep reinforcement learning. Nature 518:529–533

Moin P, Squires K, Cabot W, Lee S (1991) A dynamic sub-grid scale model for compressible turbulence and scalar transport. J Fluid Mech 3:2746–2757

Nikolaou ZM, Minamoto Y, Vervisch L (2019) Unresolved stress tensor modelling in turbulent premixed V-flames using iterative deconvolution: an a priori assesment. Phys Rev Fluids 4(063202)

Nikolaou ZM, Swaminathan N (2014) Evaluation of a reduced mechanism for turbulent premixed combustion. Combust Flame 161:3085–3099

Nikolaou ZM, Swaminathan N (2015) Direct numerical simulation of complex fuel combustion with detailed chemistry: physical insight and mean reaction rate modelling. Combust Sci Technol 187:1759–1789

Nikolaou ZM, Swaminathan N (2018) Assessment of FSD and SDR closures for turbulent flames of alternative fuels. Flow Turb Combust 101:759–774

Nikolaou ZM, Vervisch L (2018) A priori assessment of an iterative deconvolution method for LES sub-grid scale variance modelling. Flow Turb Combust 101:33–53

Nikolaou ZM, Vervisch L, Cant RS (2018) Scalar flux modelling in turbulent flames using iterative deconvolution. Phys Rev Fluids 3:043201

Nikolaou ZM, Chrysostomou C, Vervisch L, Cant S (2019) Progress variable variance and filtered rate modelling using convolutional neural networks and flamelet methods. Flow Turb Combust 103:485–501

Nikolaou ZM, Chrysostomou C, Minamoto Y, Vervisch L (2021) Evaluation of a neural network-based closure for the unresolved stresses in turbulent premixed V-flames. Flow Turb Combust 106:331–356

Park J, Choi H (2021) Toward neural-network-based Large Eddy Simulation: application to turbulent channel flow. J Fluid Mech 914(A16)

Perez FH, Mukhadiyev N, Xu X, Sow A, Li B, Sankaran R, Im H (2018) Direct numerical simulation of reacting flows with detailed chemistry using many-core CPU acceleration. Comput Fluids 173:73–79

Pfandler P, Beyrau F, Dinkelacker F, Leipertz A (2010) A priori testing of an eddy viscosity model for the density-weighted sub-grid scale stress tensor in turbulent premixed flames. Exp Fluids 49:839–851

Piomelli U, Cabot WH, Moin P, Lee S (1991) Sub-grid scale backscatter in turbulent and transitional flows. Phys Fluids 3(1747)

Pitsch H (2006) Large Eddy simulation of turbulent combustion. Ann Rev Fluid Mech 38:453–482

Sagaut P (2001) In Large Eddy simulation for incompressible flows: an introduction. Springer, Berlin

Salvetti MV (1994) A priori tests of a new dynamic sub-grid scale model for finite difference Large Eddy simulations. Phys Fluids 7:2831–2847

Sarghini F, de Felice G, Santini S (2003) Neural networks based subgrid scale modeling in Large Eddy simulations. Comput Fluids 32:97–108

Schmitt FG (2007) About Boussinesq's turbulent viscosity hypothesis: historical remarks and a direct evaluation of its validity. C R Mech 335:617–627

Schoepplein M, Weatheritt J, Sandberg R, Talei M, Klein M (2018) Application of an evolutionary algorithm to LES modelling of turbulent transport in premixed flames. J Comp Phys 374:1166–1179

Seltz A, Domingo P, Vervisch L, Nikolaou Z (2019) Direct mapping from LES resolved scales to filtered-flame generated manifolds using convolutional neural networks. Combust Flame 210:71–82

Sen BA, Menon S (2009) Turbulent premixed flame modeling using artificial neural networks based chemical kinetics. Proc Combust Inst 32:1605–1611

Sen BA, Hawkes ER, Menon S (2010) Large Eddy simulation of extinction and re-ignition with artificial neural networks based chemical kinetics. Combust Flame 157:566–578

Silver D, Huang A, Maddison CJ, Guez A, Sifre L, van den Driessche G, Schrittwieser J, Antonoglou I, Panneershelvam V, Lanctot M, Dieleman S, Grewe D, Nham J, Kalchbrenner N, Sutskever I, Lillicrap T, Leach M, Kavukcuoglu K, Graepel T, Hassabis D (2016) Mastering the game of go with deep neural networks and tree search. Nature 529:484–489

Smagorinsky J (1963) General circulation experiments with the primitive equations. Mon Weath Rev 91:99–164

Stolz S, Adams N (1999) An approximate deconvolution procedure for Large Eddy simulation. Phys Fluids 11:1699–1701

Stolz S, Adams N (2001) An approximate deconvolution model for Large Eddy simulation with application to incompressible wall-bounded flows. Phys Fluids 13:997–1015

Sutskever I, Vinyals O, Le QV (2014) Sequence to sequence learning with neural networks. Proc Adv Neural Inf Process Syst 27:3104–3112

Tao B, Katz J, Meneveau C (2000) Geometry and scale relationships in high Reynolds number turbulence determined from three-dimensional holographic velocimetry. Phys Fluids 12:941–944

Tao B, Katz J, Meneveau C (2002) Statistical geometry of subgrid-scale stresses determined from holographic velocimetry measurements. J Fluid Mech 457:35–78

Treichler S, Bauer M, Bhagatwala A, Borghesi G, Sankaran R, Kolla PMH, Slaughter E, Lee W, Aiken A, Chen J (2017) S3D-Legion: an exascale software for direct numerical simulation of turbulent combustion with complex multicomponent chemistry. Exascale Sci Appl 12:257–258

Vreman B, Geurts B, Kuerten H (1996) Large Eddy simulation of the temporal mixing layer using the Clark model. Theor Comput Fluid Dyn 8:309–324

Vreman B, Geurts B, Kuerten H (1997) Large Eddy simulation of the temporal mixing layer. J Fluid Mech 339:357–390

Wang Q, Ihme M (2017) Regularized deconvolution method for turbulent combustion modelling. Combust Flame 176:125–142

Wang H, Hawkes E, Chen J, Zhou B (2017) Direct numerical simulations of a high Karlovitz number laboratory premixed jet flame-an analysis of flame stretch and flame thickening. J Fluid Mech 815:511–536

Wang Z, Luo K, Li D, Tan J, Fan J (2018) Investigations of data-driven closure for subgrid-scale stress in large-eddy simulation. Phys Fluids 30:125101

Xie C, Wang J, Li H, Wan M (2019) Artificial neural network mixed model for Large Eddy simulation of compressible isotropic turbulence. Phys Fluids 31(085112)

Yoshizawa A (1986) Statistical theory for compressible turbulent shear flows, with the application to sub-grid modelling. Phys Fluids 29:2152–2164

Yuan Z, Xie C, Wang J (2020) Deconvolutional artificial neural network models for Large Eddy simulation of turbulence. Phys Fluids 32(115106)

Zhou Z, He G, Wang S, Jin G (2019) Sub-grid scale model for large-eddy simulation of isotropic turbulent flows using an artificial neural network. Comp Fluids 195(104319)

Machine Learning for Combustion Chemistry

T. Echekki, A. Farooq, M. Ihme, and S. M. Sarathy

Abstract Machine learning provides a set of new tools for the analysis, reduction and acceleration of combustion chemistry. The implementation of such tools is not new. However, with the emerging techniques of deep learning, renewed interest in implementing machine learning is fast growing. In this chapter, we illustrate applications of machine learning in understanding chemistry, learning reaction rates and reaction mechanisms and in accelerating chemistry integration.

1 Introduction and Motivation

Machine-learning (ML), a term associated with a range of data analysis and discovery methods, can provide enabling tools for effective data-based science in the analysis, reduction and acceleration of combustion chemistry. The tools associated with ML can carry out a variety of automated tasks that either serve as effective substitutes for modern data analysis and discovery techniques applied to combustion chemistry or additional tools for its effective integration in CFD codes.

The implementation of ML in combustion chemistry is not new. Several tools have been used for chemistry reduction or chemistry acceleration. Perhaps one of the earliest analysis tools used for combustion chemistry is principal component analysis (PCA) (Vajda et al. 2006). By identifying redundant species in a mechanism

T. Echekki (✉)
North Carolina State University, Campus Box 7910, Raleigh, NC, USA
e-mail: techekk@ncsu.edu

A. Farooq · S. M. Sarathy
King Abdullah University of Science and Technology, Thuwal 23955, Saudi Arabia
e-mail: aamir.farooq@kaust.edu.sa

S. M. Sarathy
e-mail: mani.sarathy@kaust.edu.sa

M. Ihme
Stanford University, Stanford, CA 94305, USA
e-mail: mihme@stanford.edu

© The Author(s) 2023
N. Swaminathan and A. Parente (eds.), *Machine Learning and Its Application to Reacting Flows*, Lecture Notes in Energy 44, https://doi.org/10.1007/978-3-031-16248-0_5

and eventually eliminating their reactions, PCA plays a similar role to more recent methods based on directed relations graphs (DRG) (Lu and Law 2005).

Artificial neural networks (ANN) also have been used in combustion chemistry. Since the early work of Christos et al. (1995), ANNs have been used as substitutes for the direct evaluation of the chemical source terms in combustion. Beside their use as generalized function evaluators, ANNs have been used in other contexts, as discussed below. More recent applications of ANNs in combustion chemistry addressed the integration of chemically-stiff systems of equations.

The premise of ML tools in combustion chemistry lies in the availability of an ever-expanding body of data from experiments and computations and the complexity of handling this chemistry in the presence of 10–1000s of chemical species and 100–10,000s chemical reactions. Some of the challenges associated with combustion chemistry and potential applications of ML are highlighted below.

First, chemistry integration represents the ultimate bottleneck in reacting flow simulations. This is partly attributed to the size of chemical systems, involving many species and reactions, and the stiffness of their chemistry. This stiffness is associated with the presence of disparate timescales for the different reactions in a chemical mechanism. Approaches to overcome the presence of such bottlenecks can rely on chemistry reduction, chemistry tabulation and strategies to remove the fast time scales in chemistry integration. This reduction can be implemented offline from detailed chemistry or *in situ* using adaptive chemistry techniques. Careful chemistry reduction can also achieve a significant reduction of the stiffness of the chemical systems through the elimination of fast reactions and associated species.

Second, another difficult challenge with combustion chemistry is the development of new chemical mechanisms for an expanding range of fuels. Detailed mechanism development is a complex and time-consuming process that usually represents a first step prior to chemistry reduction. Identifying the elementary reactions relevant to a particular fuel oxidation, then determining their rates and relative importance in the mechanisms are integral steps in this process. Such an effort cannot be sustained given the need to develop the important elementary reaction data, especially data critical for the low-temperature oxidation for these fuels. More importantly, practical fuels tend to be complex blends and mixtures of different molecules. Establishing the chemical description of 10 or 100s of molecules is very challenging and must include models for their transport and thermodynamic properties. Until recently, strategies to develop a reduced description of chemistry without access to detailed or skeletal descriptions of chemistry have been limited to *ad hoc* global chemistry approaches that optimize rate constant and stoichiometric coefficients for the global reactions by matching global observables, such as flame speeds, ignition delay times or extinction strain rates.

However, a growing body of data and detailed mechanisms is now available that can be exploited to develop "rules" for representing the chemistry of complex fuels (Buras et al. 2020; Ilies et al. 2021; Zhang and Sarathy 2021b, c; Zhang et al. 2021). Temporal measurements from shock tubes and rapid compression machines (RCMs), although may be limited to a subset of the chemical species present, which may be subject to experimental uncertainty, can provide important relief to detail mechanism development as discussed below.

The challenges listed above can lend themselves to applications of data science and the implementation of ML tools for combustion chemistry discovery, reduction and acceleration. The various ML methods in combustion chemistry and other applications can generally be classified as either supervised (e.g. classification, regression models) or unsupervised (e.g. clustering and PCA). Supervised models are a class of models in which both input and output are known and prescribed from the training data. This data is called labeled. For example, in a regression ANN for chemical source terms, we attempt to map the thermo-chemical state (i.e. pressure, temperature and composition) to a chemical source terms. During unsupervised learning, the output is not labeled. This approach may include for example identifying principal components (using PCA) from a thermo-chemical state or clustering of states based on the proximity of the thermo-chemical state vector.

Another class of models that have not been extensively used in combustion chemistry are the so-called semi-supervised models. In semi-supervised models, both labeled and unlabeled data are used for the training of these models. These models include for example generative models where available data is trained to generate new similar data. A popular such model is the generative adversarial network (GAN). As expected, ML approaches require data. The quality and quantity of the data is critical as discussed below. These approaches are trained on this data, while a portion can be used for either validation or testing.

In this chapter, we illustrate different implementations of ML tools in combustion chemistry. The goal is not to provide a comprehensive review of these tools or to address all studies involving ML for combustion chemistry. Instead, we attempt to provide an overview of various applications of ML in combustion chemistry. It is important to note that research in ML for combustion chemistry is a very active area of research and more progress is expected in the coming years. The chapter is divided into 3 general topics related to: (1) learning reaction rates, (2) learning reaction mechanisms and (3) chemistry integration and acceleration.

2 Learning Reaction Rates

The law of mass action and the Arrhenius model for the rate constant form the traditional representation of the rate of reaction of chemical species in combustion. This rate can be expressed in terms of a linear combination of rate of progress for each elementary reaction a species is involved in. The integration of chemistry is limited by the cost of this evaluation as well as the inherent stiffness of reaction mechanisms, exhibited by a wide range of timescales involved and the time-step size required to integrate chemistry in combustion simulations.

Artificial neural networks (ANNs) have been proposed as an alternative tool to the direct evaluation of reaction rates based on the law-of-mass-action and the Arrhenius law. Perhaps one of the earliest implementations of ANNs in combustion is through

Fig. 1 Illustration of the ANN-based matrix formulation for reaction rates

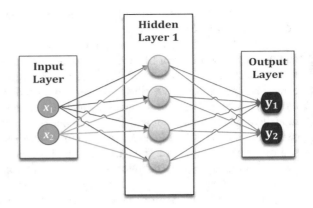

their use in their implementation as regression tools for species and temperature chemical source terms (Blasco et al. 1998, 1999, 2000; Chatzopoulos and Rigopoulos 2013; Chen et al. 2000; Christo et al. 1996; Christos et al. 1995; Flemming et al. 2005; Franke et al. 2017; Ihme 2010; Ihme et al. 2008, 2009; Sen and Menon 2010a, b; Sinaei and Tabejamaat 2017). The primary goal of representing chemical source terms with ANN is to accelerate the evaluation of chemistry. The different demonstrations of ANN for chemistry tabulation have shown that the ANN-based chemistry tabulation method is computationally efficient and accurate.

ANNs are perhaps the most versatile ML tools that have been used for combustion chemistry and other applications. Among these ANNs, one of the most popular ANN architectures are the so-called multi-layer perceptions (MLP). A representative MLP-ANN architecture is shown in Fig. 1. It is designed to construct a functional relation between a prescribed input vector $\mathbf{x}$ (x_1, x_2) and an output vector $\mathbf{y}$ (y_1, y_2). Within the context of a regression model, the ANN forms a function for $\mathbf{y}$ in terms of $\mathbf{x}$, i.e., $\mathbf{y} = \mathbf{f}(\mathbf{x})$. The input layer in the figure contains the input vector elements, which are represented by "neurons". A similar arrangement is present for the output layer where each element is represented by a neuron. The neurons carrying values are in the hidden layer, which separate the input and the output layers. In the illustration, there is only one hidden layer with 4 neurons shown. The illustrated MLP here is fully-connected, meaning that starting with the first hidden layer all the way to the output layer, the neurons carrying values are in the hidden layers, which separate the input and output layers. The strength of the connections are represented by "weights" and the value at the neurons at these layers is expressed in terms of the values of the neurons of the previous layers weighted the strength of the connections. Although not shown in the figure, additional "bias" neurons can be added to the input and all hidden layers. The role of the bias neurons is to provide more flexibility to train the model that relates the input to the output vectors.

To illustrate the relation between the input and the output layers, we use the network illustrated in Fig. 1. The output y_1, which corresponds to the value of the first neuron in the output layer, is expressed in terms of the hidden layer:

$$y_1 = f\left(\sum_{i=1}^{4} w_{1i}^{(1)} a_{1i}^{(1)} + b^{(1)}\right),\qquad(1)$$

where the superscript (1) corresponds the first hidden layer with weights $w_i^{(1)}$ and values $a_i^{(1)}$ at the ith neuron in the hidden layer. $b^{(1)}$ is the bias value at the hidden layer and f is the activation function. The bias neuron value serves as an additional parameters to fine-tune the network architecture and potentially reduce its complexity (i.e., less hidden layers or less neurons per hidden layer). The values of the ith neuron, $a_i^{(1)}$, in the hidden layer can be related to the input variables as follows:

$$a_i^{(1)} = f\left(w_{1i}^{(0)} x_1 + w_{2i}^{(0)} x_2 + b^{(0)}\right).\qquad(2)$$

Here $w_{1i}^{(0)}$ and $w_{2i}^{(0)}$ correspond to the weights of the connections between the input layer and the ith neuron in the first hidden layer associated with inputs x_1 and x_2, respectively. The network is trained to determine the weights of all connections from input to output layers and the bias values.

In matrix form, the output values for the hidden layer neurons and the output layer neurons can be expressed as follows:

$$\mathbf{a}^{(1)} = f\left(\mathbf{W}^{(0)} \mathbf{x} + \mathbf{b}^{(0)}\right)\qquad(3)$$

and

$$\mathbf{y} = f\left(\mathbf{W}^{(1)} \mathbf{a}^{(1)} + \mathbf{b}^{(1)}\right)\qquad(4)$$

where $\mathbf{W}^{(0)}$ and $\mathbf{W}^{(1)}$ are the weight matrices corresponding to the weights of the connections between the input and the first hidden layer and the first hidden layer and the output layer, respectively. $\mathbf{b}^{(0)}$ and $\mathbf{b}^{(1)}$ are the bias vectors for the input and the first hidden layers, respectively, with identical elements in each vector.

The expression above can be generalized to related on hidden layer or an output layer at a level $n + 1$ to the vector of values from the previous layer level n:

$$\mathbf{y}^{(n+1)} = f\left(\mathbf{W}^{(n)} \mathbf{y}^{(n)} + \mathbf{b}^{(n)}\right)\qquad(5)$$

MLPs vary in complexity as well as in purpose. Accommodating complexity can be achieved by increasing the number of hidden layers, the number of neurons per hidden layer and the activation functions, which can be varied from one layer to another. Prescribing the loss function can also improve the prediction of the target output. Although, there are usual choices for the activation functions, there is an inherent flexibility in the choice of network parameters, including the activation function to represent systems of equations representing physics, as illustrated below.

2.1 Chemistry Regression via ANNs

In this section, we briefly summarize key considerations for establishing efficient regression for chemical reaction rates using ANNs. Figure 2 illustrates a relatively deep network topology that constructs a regression of the reaction rates for 10 species and the heat release rate for the temperature equation from the work of Wan et al. (2020). This network has 5 fully connected dense layers between the input and output layers. In dense layers, neurons in a given layer are connected through weights to all neurons in the previous layer. As indicated, the number of neurons in the hidden layers is higher towards the input layer and decays towards the output layer. The rectified linear unit (ReLU) activation function is used. The network has approximately 180,000 weights to be optimized during the training stage, which required approximately 2.2 h on an Nvidia GeForce GTX 1089 Ti GPU. Other variants of the topology shown in Fig. 2 have been adopted in the literature (see for example, (Blasco et al. 1998, 1999, 2000; Chatzopoulos and Rigopoulos 2013; Chen et al. 2000; Christo et al. 1996; Christos et al. 1995; Flemming et al. 2005; Franke et al. 2017; Ihme 2010; Ihme et al. 2008, 2009; Sen and Menon 2010a, b; Sinaei and Tabejamaat 2017)).

Determining all these chemical source terms invariably requires more complex neural networks than those specialized to predict only one quantity. Within such complex networks, the weights from the input layer to the layer prior to the last layer are shared among all the input quantities; and the weights relating the last hidden layer to the output layer are the primary differentiators for the individual reaction rates. There are potentially 3 attractive features for the use of ANNs to model chemical source terms. The first feature is the potential acceleration in the evaluation of the chemical source through graphical processing units (GPUs) through integration of neural networks with existing accelerated packages designed to optimize ANN evaluations through mixed hardware frameworks.

A second attractive feature is that ANNs can be made simpler by adopting only a subset of the input. This is motivated by the inherent correlation of thermo-chemical

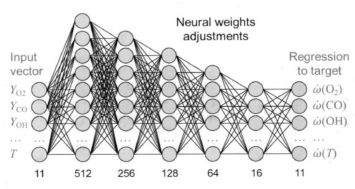

Fig. 2 Illustration of the ANN-based matrix formulation for reaction rates with multiple inputs and multiple outputs (from Wan et al. (2020))

scalars in a chemical mechanism, which lends itself to dimensionality reduction methods. Alternatively, low-dimensional manifold parameters, such as principal components (PCs) from PCA, or a choice of representative species, including major reactants, products and intermediates can be used.

A third feature of using ANNs for learning chemical reaction rates, related to the previous one, is that, if a subset of the inputs is used, then, the solution vector may also require only a subset of the thermo-chemical scalars to be transported, which corresponds primarily to the thermo-chemical scalars in the input vector. This can reduce the computational cost. It follows that, if species and associated reactions that represent a bottle-neck in chemistry integration are eliminated, then, the stiffness of the chemical system is significantly reduced, further accelerating chemistry integration.

Implementing the regression for chemical source terms within a single ANN has a number of advantages. First, constraints can be built in the training for the chemical source terms, for example to enforce the conservation of elements, mass or energy. Moreover, a single network with a number of shared weights may be exploited for computational efficiency, since the contributions to the individual source terms occur primarily at the connections between the last hidden layer and the output layer.

However, accommodating all species chemical source terms in a single layer may also require a more complex ANN architecture. Alternative strategies to reduce this complexity have been used. One approach relies on adapting different ANNs for different clusters of data, such as different networks for the reacting and the non-reacting zones in the mixture. This approach has been implemented by Blasco et al. (2000), Chatzopoulos and Rigopoulos (2013) and Franke et al. (2017) using self-organizing maps (SOM) (Kohonen 2013). In these studies, chemistry tabulation was implemented in conjunction with closure models for turbulent combustion and SOM was used as an adaptive tool to cluster similar conditions of the composition space to establish a single ANN regression tables for them.

SOMs are a popular method and an unsupervised ML technique for clustering and model reduction as stated earlier. They are single-layer neural networks that connect inputs, which corresponds to data to be clustered, to a (generally 2D) map of nodes or clusters. The clustering of the input data is based on their weights relative to the different nodes, which are determined iteratively by measuring their "proximity" to the node measures. The outcome of this iterative procedure is a mapping of the original data into a lower-dimensional space represented by the 2D map of nodes. The versatility of SOM in addressing how data is grouped is established through the choice of measures of similarity that are used to identify the mapping. For tabulation, these measures can be related to the proximity in thermo-chemical space (e.g. similar temperatures and compositions); while, for identifying different phases, these measures may rely on the evolution of marker thermo-chemical scalars in time and their correlations with other scalars.

Alternatively, clustering was implemented to group thermo-chemical scalars of similar behavior, such as the construction of an ANN for intermediates and another for reactants and products (Owoyele et al. 2020). This approach attempts to construct a minimum set of neural networks that are also less complex than the ones that accommodate all thermo-chemical scalars.

Additional consideration for constructing ANNs for reaction rate regression is related to the high variability of the input, the thermo-chemical scalars, and the output data, their chemical source terms, resulting in strongly nonlinear regressions, which may require, unnecessarily, complex and deeper ANNs. A potential way of "taming" the data variability is to pre-process the input and the output data. Sharma et al. (2020) used log-normalization to pre-process free radicals, which tend to skew towards zero.

Finally, determining an optimum topology for a chemistry regression network is not a trivial task. A shallow (one hidden layer) to a moderately deep network may not be sufficient to capture the functional complexity of the chemical source terms and may result in "under-fitting". Meanwhile, a much deeper network with numerous neurons in their hidden layers may achieve better predictions with an increased cost of evaluating the networks and the associated storage needed for the trained weights. It can also result in "over-fitting" when data is sparse or does not represent the true variability of the accessed composition space.

Ihme et al. (2008), Ihme (2010); Ihme et al. (2009) proposed an approach to determine an optimum artificial neural network (OANN) using the generalized pattern search (GPS) method (Torczon 1997). The GPS method is a derivative-free optimization that generates a sequence of iterates with a prescribed objective functions. The optimum network in this method is designed to determine the choice of network parameters (number of hidden layers, number of neurons in hidden layers) that minimize the memory requirements, the computational cost and the approximation error of the network.

Nowadays, other automated tools can be used to help optimize a given network. These include the so-called automated machine learning (or AutoML) tools, such as the Keras Tuner, Auto-PyTorch and the AutoKeras tools (Hutter et al. 2019). However, special attention must be paid to the choice of the measure of convergence of the training schemes.

3 Learning Reaction Mechanisms

Machine learning tools are set to provide greater insight into (1) the discovery of chemical pathways and key reactions in a mechanism, and (2) the reduction and representation of chemical mechanisms. In this section, we review a number of applications in which ML tools have been used for learning reaction mechanisms.

3.1 Learning Observables in Complex Reaction Mechanisms

Although, for many, the ultimate goal of understanding chemical mechanisms is to develop ways to reduce them, developing a qualitative and quantitative understanding of important pathways for reaction and the various stages of oxidation and identifying the main species and reactions important to this oxidation are important crucial steps towards mechanism reduction. ML offers powerful tools to achieve these goals.

Clustering methods have been used in a different context by Blurock and co-workers Blurock (2004), Blurock (2006), Tuner et al. (2005), Blurock et al. (2010) to identify the different mechanistic phases of fuel oxidation, which can be helpful in devising reduced chemistry schemes for these different phases. In Blurock (2004, 2006) clustering based on reaction sensitivity is used to identify the different phases of oxidation of aldehyde combustion and the ignition stages of ethanol, respectively. These studies exploit the presence of "similarity" between chemical states to identify the phases were the associated species are dominant. Identifying such phases can be important in several respects. For example, during the high-temperature oxidation of complex hydrocarbon fuels, identifying the two distinct phases of fuel pyrolysis and subsequent oxidation have enabled pathways to the development of hybrid chemistry approaches (Wang et al. 2018) (see Sect. 3.4). A less obvious distinction between the different phases of the low-temperature oxidation of the same complex fuels, can also reveal similar strategies to construct hybrid chemistry descriptions by identifying representative or marker species for each phase.

Insight to the physics from simulations or experiments can also provide a pathway towards generalizing observations, such as among different fuel functional groups. A recent study by Buras et al. (2020) used convolutional neural networks (CNNs) to construct correlations between the time scales of the low-temperature fuel spontaneous oxidation and chemical species profiles, primarily for OH, HO_2, CH_2O and CO_2 from plug-flow reactors (PFRs) and the first stage autoignition delay time (IDT). In their study, the authors relied on PFR simulation of 23 baseline fuels (18 pure fuels and 5 fuel blends) spanning a range of functional groups, including alkanes, alkenes/aromatics, oxygenates and fuel blends. They used existing mechanisms and perturbations of the parameters of these mechanisms to construct a wide database of species profiles. Emphasis on OH and HO_2 is motivated by their role during the onset of spontaneous fuel oxidation. These intermediates exhibit different behaviors for two general fuels that show different propensities to form OH and HO_2 during their oxidation cycle resulting in different correlations between the time scales for spontaneous fuel oxidation and the first stage IDT, one showing comparable values between the two quantities and another exhibiting a much slower first stage IDT. These different propensities are exhibited in the temporal profiles of these 2 intermediates as shown by Buras et al. (2020).

CNNs are a different class of neural networks compared to the fully-connected multi-layer perceptrons shown in Figs. 1 and 2. They are specialized for multi-dimensional inputs, such as 2D images and include intermediate processing layers, convolutional and pooling layers, that are designed to dissect patterns in multi-

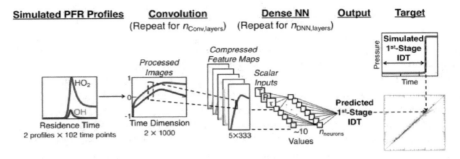

Fig. 3 A schematic of the CNN architecture used by Buras et al. (2020) to construct correlations between profiles of OH, HO_2, CH_2O and CO_2 from PFR simulations of the low-temperature oxidation of a range of fuels and the first stage ignition delay times (IDTs). Reproduced with permission from Buras et al. (2020)

Fig. 4 Correlations of predictions of first stage IDT from CNN to the simulated values using OH and HO_2 profiles as inputs to the CNN. The inset also shows the histogram of the percent error indicating a relatively narrow distribution. Reproduced with permission from Buras et al. (2020)

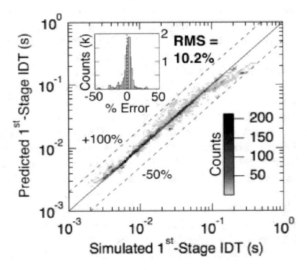

dimensional and structured input data. Within the context of the work by Buras et al. (2020), the CNN architecture captures the different patterns with the profiles of the intermediates, OH and HO_2.

Figure 3 shows a schematic of the CNN architecture used by Buras et al. (2020) to construct correlations between profiles of OH and HO_2 from PFR simulations of the low-temperature oxidation for a range of fuels and the first stage ignition delay times (IDTs). The input data corresponds to 1D profiles of both OH and HO_2; while the output (or target) is represented by the first stage IDT. By using a CNN, Buras et al. (2020) show that they can generate adequate predictions of the first stage IDT as shown in Fig. 4.

3.2 *Chemical Reaction Neural Networks*

One of the more recent developments in ML learning for chemical kinetics is the representation of reaction rates with prescribed inputs as the thermo-chemical state in terms of neural networks (Barwey and Raman 2021; Ji and Deng 2021). Such a representation enables the use of various tools to both accelerate the evaluation of reaction rates and develop skeletal descriptions of detailed mechanisms.

The rate of progress of a global reaction, $\nu_A A + \nu_B B \rightarrow \nu_C C + \nu_D D$, can be expressed as:

$$r = k \, C_A^{\nu_A} \, C_B^{\nu_B}, \tag{6}$$

where the rate constant k is expressed in terms of the Arrhenius law:

$$k = A \, T^b \, \exp\left(-\frac{E_a}{\mathcal{R}T}\right) \tag{7}$$

In this expression, A, b and E_a correspond to the frequency factor, the pre-exponential temperature power and the activation energy. This expression can be re-written as follows:

$$r = \exp\left(\ln k + \nu_A \, \ln C_A + \nu_B \, \ln C_B\right) \tag{8}$$

$$= \exp\left(\ln A + b \, \ln T - \frac{E_a}{\mathcal{R}T} + \nu_A \, \ln C_A + \nu_B \, \ln C_B\right) \tag{9}$$

This expression can be formulated as an artificial neural network as illustrated in Fig. 5a for a single reaction and Fig. 5b for multi-step reactions. In Fig. 5a, the network emulates the structure of an ANN with no hidden layers. In this network, the input layer corresponds to the natural logs of the concentrations for A, B, C and D. The output layer corresponds to their rate of change, $-\nu_A \, r, -\nu_B \, r, \nu_C \, r$ and $\nu_D \, r$, respectively. The activation function is the exponential functions and the bias is $\ln k$. The stoichiometric coefficients, ν_A, ν_B, ν_C and ν_D correspond to the weights of the network. The bias $\ln k$, which represents the temperature-dependent rate constant, incorporates to the contributions of the rate parameters, A, b and E_a. The illustrated CRNN can be generalized to accommodate more reactions and more species, as shown in Fig. 5b, thus enabling a neural network description of a set of global reactions to be optimized via ANNs. However, perhaps the main advantages of CRNN beyond the ability to frame reaction mechanisms within a neural network are the potential implications for such network for chemistry reduction and acceleration. Ji and Deng (2021) demonstrated a framework where the CRNN can be learned in the context of neural ODEs as discussed in Sect. 4 below.

An additional advantage of the CRNN is the potential for chemistry reduction via threshold pruning where input and output weights are clipped below a certain threshold. This pruning enhances the sparsity of the CRNN, which in turn can help speed up the evaluation of reaction rates. Ji and Deng (2021) showed that this prun-

Fig. 5 Illustration of the CRNN network by Ji and Deng (From Ji et al. (2021)). In the figure, the symbols "[ ]" denote concentrations while the "dots" over the concentrations in the output layer denote reaction rates. Reproduced with permission from Ji et al. (2021)

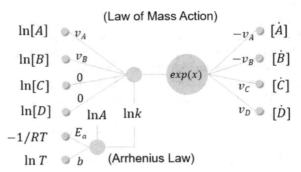

(a) A Neuron for A Single-step Reaction

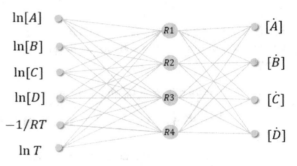

(b) A CRNN Network for Multi-step Reactions

ing can still recover accurately the reaction rates in the CRNN by re-balancing the remaining weights.

A similar formulation was proposed by Barwey and Raman (2021). These authors also recast Arrhenius kinetics as a neural network using matrix-based formulations. By this process, the evaluation of the neural network can exploit specially optimized libraries for machine learning that are also optimized for use with graphical processing units (GPUs).

3.3 PCA-Based Chemistry Reduction and Other PCA Applications

As indicated earlier, PCA has been one of the earliest ML tools implemented for combustion chemistry. From the earlier work of Turány and co-workers (see for example (Vajda et al. 2006)) PCA was used to identify the most influential reactions in a mechanism through an eigen decomposition related to the sensitivity matrix. Their analysis is based on identifying the contributions to a "response function":

$$Q\left(\alpha\right)=\sum_{j=1}^{l}\sum_{i=1}^{m}\left[\frac{f_i(x_j,\alpha)-f_i(x_j,\alpha^{\circ})}{f_i(x_j,\alpha^{\circ})}\right]^2 \qquad (10)$$

which evaluates the cumulative contribution of the normalized deviations of perturbed kinetic model response parameters relative to the original non-perturbed kinetic model. Here, f_i can correspond to temperature, a measure of species concentrations, or both and other global parameters, such as flame speeds or extinction strain rates. α_j is a reaction rate kinetic parameter, which is normally adopted as the rate constants for the reaction in a mechanism. Also, l and m in the sum correspond to the total number of analysis point (in space or time) and the number of target functions (e.g. species concentrations, temperatures). x_j corresponds to positions or times that involve all the samples in the calculation of Q.

PCA is implemented on the matrix $\mathbf{S}^T\mathbf{S}$, where $\mathbf{S}$ is the matrix of normalized sensitivity coefficients whose component i, j can be expressed as $\partial \ln f_i/\partial \ln \alpha_j$. An eigen-decomposition of the matrix yields a set of eigenvalues λ_i (ordered from high to lower magnitudes) and associated eigenvectors (which form an orthonormal set) and principal components (PCs), ϕ, which can be expressed in terms of the kinetic parameters as:

$$\phi = \mathbf{Q}^T \psi, \qquad (11)$$

where ψ is the vector logarithmic parameters $\psi_j = \ln \alpha_j$. The eigen-decomposition can be used to approximate the response function Q as follows (Vajda et al. 2006):

$$Q(\alpha) \simeq= \sum_{i=1}^{r}\lambda_i \left(\Delta\psi_i\right)^2 \qquad (12)$$

By ordering the eigenvalues, the PCs corresponding to the largest eigenvalues determine the influential part of the mechanism.

PCA can also be implemented within the context of a neural network using autoencoders. Figure 6 shows the architecture of an autoencoder with an input and an output layer and 3 hidden layers. The hidden layers are implemented with a decreasing number of neurons to a bottleneck layer, then an increasing number of neurons to the output. The dimensional of the output is identical to the input and the values of its neurons is designed to reproduce the corresponding values at the input layer. Therefore, the goal of an autoencoder is to reproduce the original data (at the input) by representing the data through a reduced dimension corresponding to the number of neurons in that hidden layer.

An autoencoder with one hidden layer, the bottleneck layer, a linear activation function and a penalty function that is the mean squared error (MSE) is designed to reproduce the PCA space from a prescribed input dimension to a dimension that corresponds to the number of neurons in the hidden layer. Additional steps are needed to reproduce the PCs from PCA analysis given that PCA also requires an orthonormal set of eigevenvectors for the PCs.

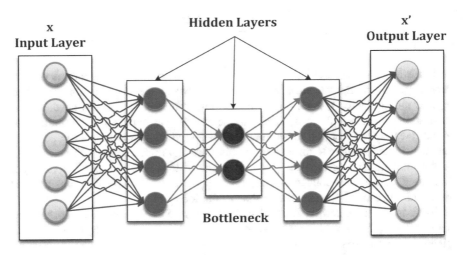

Fig. 6 Illustration of the network architecture of an autoencoder

Recently, Zhang et al. (2021) proposed the use of autoencoders as a tool for chemistry reduction. These autoencoders exploit the dimensionality reduction at the bottleneck to construct a reduced description of chemistry. Given the inherent risk of extrapolation when the autoencoder attempts to access out-of-distribution (OOD) regions via extrapolation, Zhang et al. (2021) proposed the coupling of the autoencoder with either a deep ensemble (DE) method (Lakshminarayanan et al. 2017) or the so-called PI3NN method (Zhang et al. 2021). Within an autoencoder structure, the DE method accounts for a predicted mean (the predicted values) as well as the output variance to assess uncertainty (Lakshminarayanan et al. 2017). While in the PI3NN method, two additional neural networks are introduced to estimate the upper and lower bounds of the data reconstruction, again as a measure to assess the uncertainty in the autoencoder performance.

Figure 7 illustrates the two OOD-aware autoencoder configurations investigated by Zhang et al. (2021). The authors showed that by using these configurations, the number of input species is reduced from 12 to 2 at the bottleneck. This reduction can translate into a reduction in the number of transported scalars.

Finally, another implementation of PCA in combustion chemistry has been proposed by D'Alessio et al. (2020a, b). In their recent studies, they proposed an adaptive reduced chemistry scheme in which the composition space is partitioned into different clusters where appropriate and efficient reduced chemistry models can be implemented. The partitioning is implemented, instead of using a standard clustering approach such as K-Means or SOM, using local PCA (or LPCA) (Kambhatla and Leen 1997). The main difference between the use of LPCA vs K-Means, for example, is in the criteria established to partition the composition space. Instead of minimizing the Euclidean error between data of a given cluster and its centroid, the criteria is to miminize the reconstruction error of the PCA within a given cluster. D'Alessio et al.

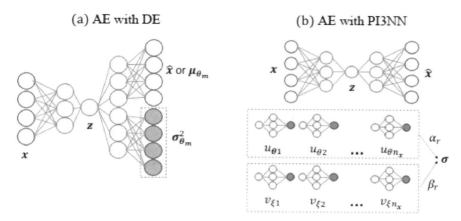

Fig. 7 Illustration of two OOD-aware autoencoder architectures with DE (left) and PI3NN (right). The input layer, **x** corresponds to the full chemistry description; while the bottleneck **z** represents the reduced chemical description. The autoencoder is designed to reproduce the input in the output; and the DE and PI3NN modifications attempt to assess the uncertainty of the predictions, especially when extrapolation is needed. Reproduced with permission from Zhang et al. (2021)

(2020b) showed that superior performance is established by adopting LPCA as part of the clustering algorithm instead of a hybrid clustering approach based on the coupling of self-organizing maps (SOMs) and K-Means in an unsteady laminar co-flow diffusion flame of methane in air. Within the context of a CFD simulation, LPCA is used as a classifier to determine the cluster to which a given cell state belongs. In each cluster, an *a priori* chemistry reduction is implemented using the training data, which in the studies of D'Alessio et al. (2020a, b) correspond to a series of unsteady 1D flames or data from 2D simulations of the same configuration, respectively.

3.4 Hybrid Chemistry Models and Implementation of ML Tools

The oxidation chemistry of a typical transportation fuel poses severe computational challenges for multi-dimensional reacting flow simulations. These challenges may be attributed primarily to the sheer size of associated chemical mechanisms when available. However, and oftentimes, the chemical kinetic data may not be available. While chemistry reduction strategies have been reasonably successful in overcoming the challenge of handling chemical complexity (Battin-Leclerc 2008; Turányi and Tomlin 2014), such strategies can only be used when reliable detailed mechanisms for the fuels of interest are available.

Experimental data-based chemistry reduction is one viable strategy for modeling the chemistry of complex fuels. Recently, the hybrid chemistry (HyChem) approach

Fig. 8 Species time-history measurements for n-dodecane oxidation. Initial mixture conditions: 1410 K, 2.3 atm, 457 ppm n-dodecane/O2/argon, $\phi = 1$. (Reproduced with permission from Davidson et al. (2011))

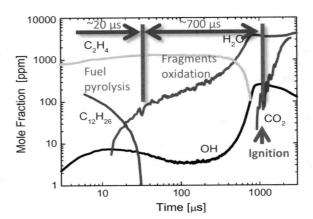

was proposed by Wang and co-workers Wang et al. (2018), Xu et al. (2018), Tao et al. (2018), Wang et al. (2018), Saggese et al. (2020), Xu et al. (2020), Xu and Wang (2021) as a chemistry reduction approach for the high-temperature oxidation of transportation fuels starting from time-series measurements of fuel fragments (and other relevant species) to capture the pyrolysis stage of these fuels. Such measurements can be achieved primarily using shock tubes and a variety of optical diagnostic techniques and sampling methods.

The approach is based on the premise that, at high temperatures, fuel oxidation undergoes: (1) a fast fuel pyrolysis step resulting in the formation of smaller fuel fragments, followed by (2) a longer oxidation step for these fragments. Figure 8 shows experimental observations by Davidson et al. (2011), which illustrate the 2 stages of n-dodecane oxidation through time-history measurements of the fuel, a fuel fragment, C_2H_4, and oxidation species, OH, H_2O and CO_2. The figure shows that the fuel is depleted in the first 30μs and it is replaced by pyrolysis fragments, which eventually oxidize towards simpler hydrocarbons.

In HyChem, a hybrid chemistry model represented by a set of lumped fuel pyrolysis steps is augmented by foundational C_0–C_4 chemistry for the fragments-oxidation. With experimental measurements of the key fragments, the stoichiometric coefficients and rate constants for the global reactions are determined through an optimization approach. The lumped reactions for the fuel pyrolysis is modeled using the following two reaction steps for a fuel C_mH_n:

- **Unimolecular decomposition reaction**

$$
\begin{aligned}
C_mH_n \rightarrow &e_d \, (C_2H_4 + \lambda_3 \, C_3H_6 + \lambda_4 \, C_4H_8) \\
&+ b_d \, [\chi C_6H_6 + (1 - \chi)] + \alpha \, H + (2 - \alpha) \, CH_3
\end{aligned}
\tag{13}
$$

- **H-atom abstraction and β-scission reactions of fuel radicals**

$$C_mH_n + R \rightarrow RH + \gamma\, CH_4 + e_a\, (C_2H_4 + \lambda_3\, C_3H_6 + \lambda_4\, C_4H_8)$$
$$+ b_a\, [\chi C_6H_6 + (1 - \chi)] + \beta\, H + (1 - \beta)\, CH_3 \tag{14}$$

where R represents the following species: H, CH_3, O, OH, O_2 and HO_2. In these reactions, α, β, λ_3, λ_4 and χ are the stoichiometric parameters that need to be determined for each fuel chemistry. More specifically, α and β correspond to the number of H atoms per C_mH_n in the two reactions, respectively. The remaining parameters, e_d, e_a, b_d and b_a can be expressed in terms of the stoichiometric parameters using elemental conservation principles across each reaction (Wang et al. 2018).

The HyChem approach relies on the ability to measure some key fuel fragments, CH_4, C_2H_4 (in shock tubes), C_3H_6, C_4H_8 isomers, C_6H_6 and C_7H_8 (in flow reactors). Therefore, these fuel fragments represent much less complex species than the original fuel and their oxidation can be modeled using a simpler foundational chemistry model as the subsequent oxidation stage. More importantly, the fragments' measurements can be used to determine the stoichiometric parameters and the rate constants of the lumped reactions needed to model the pyrolysis stage.

Hybrid chemistry approaches, such as the HyChem ML can play useful roles to formulate robust chemistry descriptions for complex fuels. In two recent studies, Ranade and Echekki (2019a, b) proposed an ANN-based implementation of HyChem. In a first step, a shallow regression ANN is implemented on the temporal species measurements to evaluate directly their rate of change, which directly measures their rate of reaction. In the second step, deep regression ANNs are trained to relate fragments' concentrations to their rate of reaction. This network, as in the HyChem approach, is used to evaluate the fragments' chemical source terms during the pyrolysis stage. Ranade and Echekki (2019b) showed that the procedure can be extended beyond the pyrolysis stage to enable the use of a simpler foundational chemistry.

More recently, Echekki and Alqahtani (2021) proposed a data-based hybrid chemistry approach to accelerate chemistry integration during the high-temperature oxidation of complex fuels. The approach is based on the ANN regression of representative species, which may or may not include the pyrolysis fragments, during the pyrolysis stage. These representative C_0–C_4 species are determined using reactor simulation data and PCA on all species reaction rates. This PCA is used to determine the most important species to represent the evolution of the oxidation process. Beyond the pyrolysis stage, these species can be modeled with a foundational chemistry model like the remaining species.

Since the representative species are not tied to a particular list of fragments, the approach can be extended to the modeling of low-temperature oxidation where some of the initial intermediates are fuel-dependent. The work of Alqahtani (2020) demonstrated the feasibility of this extension to low-temperature fuel oxidation.

The approaches implemented in Ranade and Echekki (2019a, b), Echekki and Alqahtani (2021) or Alqahtani (2020) rely on ANN for the regression of the fragments or representative species in terms of the species concentrations. These studies suggest that the associated architectures of the ANN can be further simplified by using a subset of these species as inputs. This choice is motivated by the inherent correlations of the fragments/representative species and rely on the same motivation for using PCA in combustion modeling. However, ANNs may have limited interpretability unless they are implemented in the context of CRNN, as presented in Sect. 3.2.

The CRNNs (Ji and Deng 2021) offer an alternative optimization of the global reactions of the pyrolysis stage using the law of mass action and the Arrhenius form for the rate constants. Zanders et al. (2021) implemented a stochastic gradient descent (SGD) approach to optimize the lumped global reactions of pyrolysis starting with data of ignition delay times. Their approach was implemented within their Arrhenius.jl open-source software (Ji and Deng 2021) and by implementing the lumped reaction steps of pyrolysis within a CRNN. Their evaluation of the rate parameters of the lumped pyrolysis reactions yielded both an enhanced computational efficiency compared to approaches based on genetic algorithms and an improved predictions of IDT for ranges of temperature and equivalence ratios.

3.5 Extending Functional Groups for Kinetics Modeling

Functional group information has recently been used for the bottom-up development of chemical kinetic models. This approach was developed following the initial insight that AI models can predict combustion properties from several key functional group features of a fuel mixture. Recently, the team led by Zhang et al. advanced lumped fuel chemistry modeling approach using functional groups for mechanism development (FGMech) (Zhang and Sarathy 2021b; Zhang et al. 2021). They created a functional group-based approach, which can account for mixture variability and predict stoichiometric parameters of chemical reactions without the need for any tuning against experiments on the real fuel.

Figure 9 presents an overview of the functional group approach for kinetic model development. The effects of functional groups on the stoichiometric parameters and/or yields of key pyrolysis products were identified and quantified based on previous modeling of pure components (Zhang and Sarathy 2021a; Zhang et al. 2022; Zhang and Sarathy 2021c). A quantitative structure-yield relationship was developed by a multiple linear regression (MLR) model, which was used to predict the stoichiometric parameters and/or yields of key pyrolysis products based on ten input features (eight functional groups, molecular weight, and branching index). The approach was then extended to predict thermodynamic data, lumped reaction rate parameters and transport data based on the functional-group characterization of real fuels. FGMech is fundamentally different in that no parameters need to be tuned to match actual real-fuel pyrolysis/oxidation data, and all the model parameters were derived only from functional group data. It was shown that the FGMech approach can make good predictions on the reactivity of various aviation, gasoline, and diesel fuels (Zhang and Sarathy 2021b; Zhang et al. 2021).

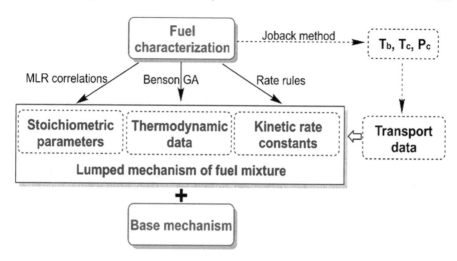

Fig. 9 Overview of the functional group approach for kinetic model development

3.6   Fuel Properties' Prediction Using ML

The properties of fuels are carefully controlled to enable engines to operate at their optimal conditions and to ensure that fuels can be safely handled and stored. Important properties include those that can be easily determined based on simple thermo-physical models and linear blending (e.g., density, viscosity, heating values) to more complex properties that cannot be easily determined from physical modeling (e.g., octane number, cetane number, and sooting tendency). For the latter, ML techniques may be used to predict these fuel properties.

The first requirement for fuel property prediction is a suitable input descriptor for model training. Various molecular 1-3D representations such as SMILES (Simplified Molecular Input Line Entry Specification), InChI (International Chemical Identifier) or connectivity matrices can be used to obtain molecular descriptors for AI-based quantitative structure-property relationships (QSPR). Table 1 illustrates the use of different ML approaches to evaluate fuel properties.

Abdul Jameel et al. have demonstrated significant progress in the use of ANNs to predict various fuel properties including octane numbers (Jameel et al. 2018), derived cetane number (Jameel et al. 2016, 2021), flash point (Aljaman et al. 2022), and sooting indices (Jameel 2021). In general, they used functional groups derived from ^{1}H NMR spectra of pure hydrocarbons and real fuel mixtures as input descriptors for model training, as illustrated in Fig. 10. The functional groups used include nine structural descriptors (paraffinic primary to tertiary carbons, olefinic, naphthenic, aromatic and ethanolic OH groups, molecular weight and branching index). Ibrahim and Farooq (2020, 2021) utilized the methodology proposed by Abdul Jameel et al.

Table 1 Example of fuel properties predicted by AI and associated descriptors

Species property	ML approach	Reported descriptors
Octane/Cetane Number, autoignition metrics	ANN based group contributions, SVM based on Boruta features elimination, CNN, Graph NN, ANN, k-NN, RF, HDMR/CNN	Molecular weight, critical volume, Balaban/Kier-Hall/Wiener index, water/octanol partitioning
HHV, LHV, HoV	GC, SVM, ANN, Ant colony—PLS-MLR, MLR, GA-SVM	Van der Waals surface area, number of carbons
Soot index, exhaust after-treatment activity	Bayesian inference of GC, PCA-ANN, SVM, RF, PLS	LUMO-HOMO energy, functional groups

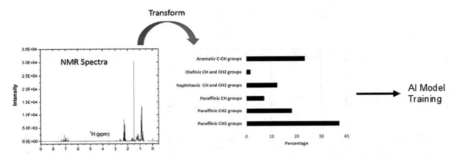

Fig. 10 Conversion of NMR spectra to functional groups followed by training for ML model for property prediction

for fuel property (RON, MON, DCN, H/C ratio) prediction based on infrared (IR) absorption spectra rather than NMR shifts.

3.7 Transfer Learning for Reaction Chemistry

Chemical kinetic modelling is an indispensable tool for our understanding of the formation and composition of complex mixtures. These models are routinely used to study pollution, air quality, and combustion systems. Recommendations from kinetic models often help shape and guide environmental policies and future research directions. There are two essential data feeds for such models: species thermochemistry and rate coefficients of elementary reactions. Uncertainties in these feeds directly affect the predictive accuracy of chemical kinetic models. Historically, these data were measured experimentally and/or estimated from simple rules, such as group-additivity and structure-activity-relations. Ab-initio quantum chemistry based theoretical models have been developed over the years to calculate thermochemistry and reaction rate coefficients, and the accuracies of these calculations have been increas-

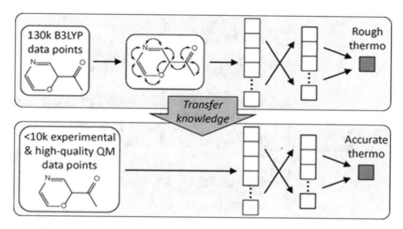

Fig. 11 Transfer learning model architecture to learn molecular embedding and neural network parameter initialization for application to small datasets. Reproduced with permission from Grambow et al. (2019)

ing steadily. These methods, however, require significant computational power and are challenging to apply to large molecular systems. In recent times, machine-learning based methods have attracted significant attention for the prediction of thermochemistry and reaction rate coefficients. In particular, inspired by the success of transfer learning approach in image processing, researchers have applied it in the domain of reaction chemistry. Transfer learning applies the knowledge (or model) learned in one task to another task. One of the benefits of transfer learning is that it can overcome the lack of large datasets, which are generally needed for machine learning algorithms.

Grambow et al. (2019) trained three base models, one each for enthalpy of formation, entropy and heat capacity, on a large dataset (≈130,000) generated from low-level (high uncertainty) theoretical calculations. These based models were then used as the starting models for the prediction of more accurate values of those thermochemistry properties by using a much smaller (<10,000) dataset of experimental values and high-accuracy theoretical calculations (see Fig. 11). Bhattacharjee and Vlachos (2020) implemented a 'data fusion' methodology to map thermo-chemical quantities, calculated at various levels of theory, to a higher level of theory. Zhong et al. (2022) overcame the challenge of small datasets by transferring knowledge among them for predictions with higher accuracy (see Fig. 12). The authors also compared their results with two other similar approaches, namely multitask learning and image-based transfer learning. Likewise, Han and Choi (2021) presented a framework of leveraging the learning from a large simulated database (with high uncertainty) to a small experimental database (with small uncertainty) for reliably predicting NMR (nuclear magnetic resonance) chemical shifts over a wide range of chemical space.

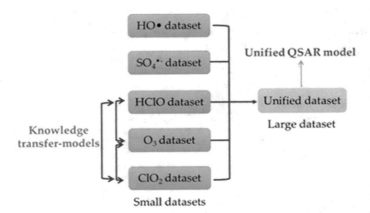

Fig. 12 Transfer learning approach for combining small datasets. Reproduced with permission from Zhong et al. (2022)

More recently, Ibrahim and Farooq (2022) showcased a temperature-dependent multi-target model with a custom-made Arrhenius loss applied to the AtmVOCkin reaction rate dataset. The Arrhenius loss dictates physically sound temperature-dependence which reduces overfitting, makes use of all available data in literature, and it outputs the three Arrhenius parameters which are compatible with modern automated chemical mechanism generator inputs. The graph-based D-MPNN was used for transfer learning from the publicly available QM9 dataset which stretches the applicability domain and supplements fixed molecular descriptors. Multi-target predictions were also implemented to enable cross-reaction learning which can enhance predictive capability for reactions with small datasets. Tuning was done using Bayesian optimization which gives robust/automatic predictions and a fair comparison among various models. The model was used to predict the three modified-Arrhenius parameters for the temperature-dependent reactions of OH, O_3, NO_3 and Cl with a wide range of hydrocarbons (see Fig. 13).

4 Chemistry Integration and Acceleration

Chemistry integration represents a true bottleneck in combustion simulations involving both transport and chemistry. Measures to accelerate chemistry have adopted different strategies that are often combined with an initial step of chemistry reduction to global or skeletal mechanisms. Such strategies include chemistry tabulation, such as the use of *in situ* adaptive tabulation (ISAT) (Pope 1997), regression (such as ANN-based regression discussed in Sect. 2) and the piecewise reusable implementation of solution mapping (PRISM) (Tonse et al. 2003), adaptive chemistry, including dynamic approaches (see for example Liang et al. (2009), Continuo et al. (2011), Sun

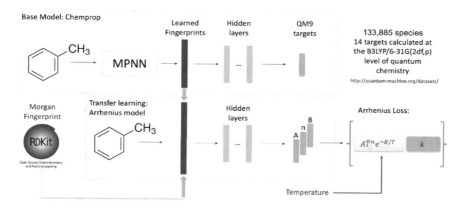

Fig. 13 Reaction rate prediction scheme (with toluene shown as a representative molecule). (Courtesey of Ibrahim and Farooq (2022))

and Ju (2017) and D'Alessio et al. (2020a)), manifold-based methods, such as intrinsic low-dimensional manifolds (ILDM) (Maas and Pope 1992) and computational singular perturbation (CSP) methods (Lam and Goussis 1994). Chemistry acceleration primarily relies on operator splitting of the chemical source terms resulting in the solution for ordinary differential equations (ODEs).

In the last few years, there has been a growing excitement about the potential of neural ODE (NODE) solutions (Chen et al. 2018; Rackauckas et al. 2020). NODEs construct solutions for ODEs using neural networks and ODE solvers where model parameters (i.e. weights) are evaluated by a backward solution of the adjoint state. Implementing NODEs for combustion reaction presents numerous challenges associated with the inherent stiffness of the ODEs and the requirement for the simultaneous solutions of multiple ODEs for species and energy (Kim et al. 2021).

However, there have been several attempts in recent years to implement chemistry integration with neural networks. Owoyele and Pal (2022) proposed the so-called ChemNODE approach. The implementation of ChemNODE is summarized in Fig. 14. In ChemNODE, a stiff ODE solver is used to advance the solution of a thermo-chemical state at different time increments. These solutions constitute the observations that are used to train for the reaction rates implemented on the right column of the figure. These ANN-based reaction rates are integrated as well using the same ODE solver. The loss function to be minimized is the mean squared error comparing the solutions at the various observation points based on integration with the Arrhenius law and integration with the ANN-based reaction rates. Recognizing the difficulty of learning chemical sources within the proposed ChemNODE approach, Owoyele and Pal (2022) used a progressive approach for training these terms where each species is trained sequentially while the remaining species' source terms are modeled with the solution from the ODE solver based on the Arrhenius law. Moreover, the optimization process involves the evaluation of derivatives of the neural

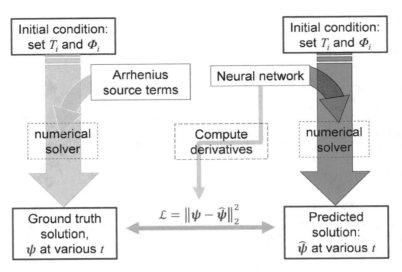

Fig. 14 Illustration of the ChemNODE algorithm. Reproduced with permission from Owoyele and Pal (2022)

network solution with respect to the network parameters, Owoyele and Pal (2022) adopted a forward-mode continuous sensitivity analysis using packages available for the Julia language.

An alternative procedure for accelerating chemistry integration is proposed by Galassi et al. (2022). Their acceleration strategy is built on the use of CSP to remove the fast time scales from the chemistry integration. CSP usually requires the evaluation of a Jacobian matrix for the local chemical source terms and its eigen-decomposition. This decomposition is needed to identify the fast and slow timescales of the chemical system. By projecting the fast time scales out of the chemistry integration, the inherent stiffness of this chemical system is significantly reduced. However, there is an inherent cost to the evaluation of the Jacobian and the process of its eigen-decomposition, which scales strongly with the size of the chemical mechanism. Galassi et al. (2022) proposed the use of ANN regression as a cheaper surrogate to the local projection basis. Otherwise the CSP procedure shown in Fig. 15 is adopted. Figure 15 shows the general algorithm used to integrate chemistry within the proposed CSP-ANN framework. Given a current chemical state, the CSP basis is retrieved using ANN. The training for this basis is implemented offline, which was carried out in the Galassi et al. (2022) study using 0D ignition data for hydrogen-air mixtures. The procedure, then, involves an implementation of "radical correction" to account for the fast time scales, an explicit integration using the projection into the slow invariant manifold, then another radical correction. For the 9-species mechanism, 7 neural networks are trained in the Galassi et al. (2022) study. They each feature 2 hidden layers with 128 neurons each in each layer.

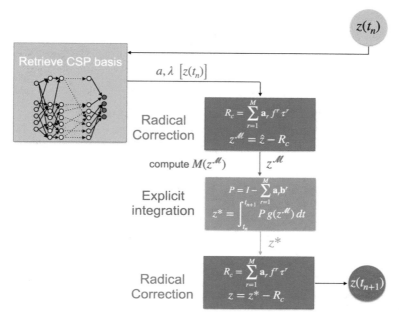

Fig. 15 Illustration of the CSP-ANN algorithm. Reproduced with permission from Galassi et al. (2022)

Zhang et al. (2021) proposed a different scheme for chemistry integration, which is based on training a deep neural network (DNN) to project a solution of the thermochemical state vector at a given time (i.e. the input) to the corresponding solution after a small time increment (i.e. the output). Figure 16 illustrates the structure of the DNN, which was implemented for a dimethyl ether (DME) mechanism with 54 species. The input solution at a given time includes 56 neurons for the species, pressure and temperature. The DNN features two independent, fully-connected branches for the low- and high-temperature oxidation for DME. Each branch has 3 hidden layers with 1600, 400 and 400 neurons. The output corresponds to the projection of the solution at a later time with 56 neurons in the output layer. The approach adopted by Zhang et al. (2021) is very reminiscent of the ISAT approach (Pope 1997), except for relying on DNNs to project solutions instead of a tree-based storage and tabulation.

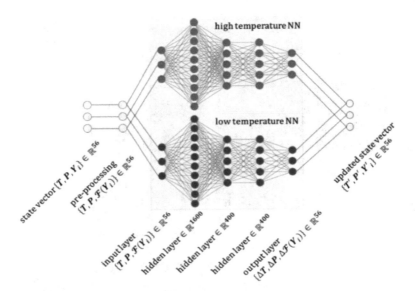

Fig. 16 Illustration of the deep neural network for DLODE. Reproduced with permission from Zhang et al. (2021)

5 Conclusions

In this chapter, we have illustrated a number of applications of ML tools in combustion chemistry. These applications span the scopes of understanding, the reduction and the acceleration of chemistry in combustion applications. Based on the material presented, we anticipate important advances in the following areas:

- The development of experiment-based HyChem-style mechanisms for a broad range of fuels that also rely on rules extracted for fuels of the same functional group.
- The implementation of novel ML tools for chemistry reduction either by developing skeletal mechanisms, global mechanisms or hybrid chemistry models combining empirical global steps coupled with foundational chemistry.
- The development of chemistry acceleration schemes that exploit dimensionality reduction of the composition space and features of stiffness removal.

Acknowledgements The authors would like to acknowledge the support of King Abdullah University of Science and Technology under grant: 4351-CRG9.

References

Al Ibrahim E, Farooq A (2020) Octane prediction from infrared spectroscopic data. Energy Fuels 34(1):817–826

Al Ibrahim E, Farooq A (2021) Prediction of the derived cetane number and carbon/hydrogen ratio from infrared spectroscopic data. Energy Fuels 35:8141–8152

Al Ibrahim E, Farooq A (2022) A transfer learning approach to multi-target temperature-dependent reaction rate prediction. Submitted

Aljaman B, Ahmed U, Zahid U, Reddye VM, Sarathy SM, Jameel AGA (2022) A comprehensive neural network model for predicting flash point of oxygenated fuels using a functional group approach. Fuel 317:123428

Alqahtani SSH (2020) Machine learning methods for chemistry reduction in combustion. PhD thesis, North Carolina State University

Barwey S, Raman V (2021) A neural network-inspired matrix formulation of chemical kinetics for acceleration on gpus. Energies 14(9)

Battin-Leclerc F (2008) Detailed chemical kinetic models for the low-temperature combustion of hydrocarbons with application to gasoline and diesel fuel surrogates. Prog Energy Combust Sci 34:40–498

Bhattacharjee H, Vlachos DG (2020) Thermochemical data fusion using graph representation learning. J Chem Info Model 60:4673–4683

Blasco JA, Fueyo N, Dopazo C, Ballester J (1998) Modelling the temporal evolution of a reduced combustion chemical system with an artificial neural network. Combust Flame 113:38–52

Blasco JA, Fueyo N, Larroya JC, Dopazo C, Chen J-Y (1999) A single-step time-integrator of a methane-air chemical system using artificial neural networks. Comput Chem Eng 23:1127–1133

Blasco JA, Fueyo N, Dopazo C, Chen J-Y (2000) A self-organizing-map approach to chemistry representation in combustion applications. Combust Theo Model 4:61–76

Blurock ES (2004) Characterizing complex reaction mechanisms using machine learning clustering techniques. Int J Chem Kin 36:107–118

Blurock ES (2006) Automatic characterization of ignition processes with machine learning clustering techniques. Int J Chem Kin 38:621–633

Blurock ES, Tuner M, Mauss F (2010) Phase optimized skeletal mechanisms for engine simulations. Combust Theo Model 14:295–313

Buras ZJ, Safta C, Zádor J, Sheps L (2020) Simulated production of OH, HO_2, CH_2O, and CO_2 during dilute fuel oxidation can predict 1st-stage ignition delays. Combust Flame 216:472–484

Chatzopoulos AK, Rigopoulos S (2013) A chemistry tabulation approach via rate-controlled constrained equilibrium (RCCE) and artificial neural networks (ANNs), with application to turbulent non-premixed $CH_4/H_2/N_2$ flames. Proc Combust Inst 34:1465–1473

Chen RT, Rubanova Y, Bettencourt J, Duvenaud DK (2018) Neural ordinary differential equations. Adv Neural Info Sys 6571–6583

Chen J-Y, Blasco JA, Fueyo N, Dopazo C (2000) An economical strategy for storage of chemical kinetics: fitting in situ adaptive tabulation with artificial neural networks. Proc Combust Inst 28:115–121

Christo FC, Masri AR, Nebot EM (1996) Artificial neural network implementation of chemistry with PDF simulation of H_2/CO_2 flames. Combust Flame 106:406–427

Christos FC, Masri AR, Nebot EM, Turanyi T (1995) Utilising artificial neural network and repro-modelling in turbulent combustion. In: 1995 IEEE international conference on neural networks proceedings, pp 911–916

Continuo F, Jeanmart H, Lucchini T, D'Errico G (2011) Coupling of in situ adaptive tabulation and dynamic adaptive chemistry: an effective method for solving combustion in engine simulations. Proc Combust Inst 33:3057–3064

D'Alessio G, Cuoci A, Aversano G, Bracconi M, Stagni A, Parente A (2020a) Impact of the partitioning method on multidimensional adaptive-chemistry simulations. Energies 13

D'Alessio G, Parente A, Stagni A, Cuoci A (2020b) Adaptive chemistry via pre-partitioning of composition space and mechanism reduction. Combust Flame 211:68–82

Davidson DF, Hong Z, Pilla G, Farooq A, Cook R, Hanson RK (2011) Multi-species time-history measurements during n-dodecane oxidation behind reflected shock waves. Proc Combust Inst 33:151–157

Echekki T, Alqahtani S (2021) A data-based hybrid model for complex fuel chemistry acceleration at high temperatures. Combust Flame 223:142–152

Flemming F, Sadiki A, Janicka J (2005) LES using artificial neural networks for chemistry representation. Prog Comput Fluid Dyn 5:375–385

Franke LLC, Chatzopoulos AK, Rigopoulos S (2017) Tabulation of combustion chemistry via artificial neural networks (ANNs): methodology and application to LES-PDF simulation of Sydney flame L. Combust Flame 185:245–260

Galassi RM, Ciottoli PP, Valorani M, Im HG (2022) An adaptive time-integration scheme for stiff chemistry based on computational singular perturbation and artificial neural networks. J Comput Phys

Grambow CA, Li Y-P, Green WH (2019) Accurate thermochemistry with small data sets: a bond additivity correction and transfer learning approach. J Phys Chem A 123(27):5826–5835

Han H, Choi S (2021) Transfer learning from simulation to experimental data: NMR chemical shift predictions. J Phys Chem Lett 12:3662–3668

Hutter F, Kotthoff L, Vanschoren J (2019) Automated machine learning methods, systems, challenges. In: Hutter F, Kotthoff L, Vanschoren J (eds) Automated machine learning methods, systems, challenges. Springer Series on Challenges in Machine Learning

Ihme M (2010) Topological optimization of artificial neural networks using a pattern search method. NOVA Science Inc., USA, pp 323–343

Ihme M, Marsden AL, Pitsch H (2008) Generation of optimal artificial neural networks using a pattern search algorithm: application to approximation of chemical systems. Neural Comput 20:573–601

Ihme M, Schmidt C, Pitsch H (2009) Optimal artificial neural networks and tabulation methods for chemistry representation in LES of a bluff-body swirl-stabilized flame. Proc Combust Inst 32:1527

Ilies BD, Khandavilli M, Li Y, Kukkadapu G, Wagnon SW, Jameel AGA, Sarathy SM (2021) Probing the chemical kinetics of minimalist functional group gasoline surrogates. Energy Fuels 35(4):3315–3332

Jameel AGA (2021) Predicting sooting propensity of oxygenated fuels using artificial neural networks. Proc 9

Jameel AGA, van Oudenhoven VCO, Naser N, Emwas AH, Gao X, Sarathy SM (2021) Predicting ignition quality of oxygenated fuels using nuclear magnetic resonance spectroscopy and artificial neural networks. SAE Int J Fuels Lubr

Jameel AGA, Naser N, Emwas A-H, Dooley S, Sarathy SM (2016) Predicting fuel ignition quality using h-1 NMR spectroscopy and multiple linear regression. Energy Fuels 30(11):9819–9835

Jameel AGA, Van Oudenhoven VCO, Emwas A-H, Sarathy SM (2018) Predicting octane number using nuclear magnetic resonance spectroscopy and artificial neural networks. Energy Fuels 32(5):6309–6329

Ji W, Deng S (2021) Autonomous discovery of unknown reaction pathways from data by chemical reaction neural networks. J Phys Chem A 125:1082–1092

Ji W, Qiu W, Shi Z, Pan S, Deng S (2021) Stiff-pinn: physics-informed neural network for stiff chemical kinetics. J Phys Chem A 125:8098–8106

Ji W, Deng S (2021) Arrhenius.jl: a differentiable combustion simulation package

Ji W, Zanders J, Park J-W, Deng S (2021) Machine learning approaches to learn HyChem models. In: Proceedings of the ASME 2021 international combustion conference, number paper ICEF2021/69657

Kambhatla N, Leen TK (1997) Dimension reduction by local principal component analysis. Neural Comput 9:1493–1516

Kim S, Ji W, Deng S, Ma Y, Rackauckas C (2021) Stiff neural ordinary differential equations. Chaos 31(093122)

Kohonen T (2013) Essentials of the self-organizing map. Neural Netw 37:52–65

Lakshminarayanan B, Pritzelnd A, Blundell C (2017) Simple and scalable predictive uncertainty estimation using deep ensembles. Proc Adv Neural Inf Process Syst 6402–6413

Lam SH, Goussis DA (1994) The CSP method for simplifying kinetics. Int J Chem Kin 26:461–486

Liang L, Stevens JG, Raman S, Farrell JT (2009) The use of dynamic adaptive chemistry in combustion simulation of gasoline surrogate fuels. Combust Flame 156:1493–1502

Lu TF, Law CK (2005) A directed relation graph method for mechanism reduction. Proc Combust Inst 30:1333–1341

Maas U, Pope SB (1992) Simplifying chemical kinetics: intrinsic low-dimensional manifolds in composition space. Combust Flame 88:239–264

Owoyele O, Kundu P, Ameen MM, Echekki T, Som S (2020) Application of deep artificial neural networks to multi-dimensional flamelet libraries and spray flames. Int J Engine Res 21(1, SI):151–168

Owoyele O, Pal P (2022) ChemNODE: a neural ordinary differential equations framework for efficient chemical kinetic solvers. Energy AI 7

Pope SB (1997) Computationally efficient implementation of combustion chemistry using in situ adaptive tabulation. Combust Sci Tech 1:41–63

Rackauckas C, Ma Y, Martensen J, Warner C, Zubov K, Supekar R, Skinner D, Ramadhan A, Edelman A (2020) Universal differential equations for scientific machine learning. arXiv:2001.04385

Ranade R, Echekki T (2019a) A framework for data-based turbulent combustion closure: a priori validation. Combust Flame 206:490–505

Ranade R, Echekki T (2019b) A framework for data-based turbulent combustion closure: a posteriori validation. Combust Flame 210:279–291

Saggese C, Wan K, Xu R, Tao Y, Bowman CT, Park JW, Lu T, Wang H (2020) A physics-based approach to modeling real-fuel combustion chemistry—V. NO_x formation from a typical jet A. Combust Flame 212:270–278

Sen BA, Menon S (2010a) Linear eddy mixing based tabulation and artificial neural networks for large eddy simulations of turbulent flames. Combust Flame 157:62–74

Sen BA, Menon S (2010b) Large eddy simulation of extinction and reignition with artificial neural networks based chemical kinetics. Combust Flame 157:566–578

Sharma AJ, Johnson RF, Kessler DA, Moses A (2020) Deep learning for scalable chemical kinetics. In: AIAA scitech 2020 forum, number AIAA paper 2020-0181

Sinaei P, Tabejamaat S (2017) Large eddy simulation of methane diffusion jet flame with representation of chemical kinetics using artificial neural network. Proc Inst Mech Eng Part E: J Process Mech Eng 231:147–163

Sun W, Ju Y (2017) TA multi-timescale and correlated dynamic adaptive chemistry and transport (CO-DACT) method for computationally efficient modeling of jet fuel combustion with detailed chemistry and transport. Combust Flame 184:297–311

Tao Y, Xu R, Wang K, Shao J, Johnson SE, Movaghar A, Han X, Park JW, Lu T, Brezinsky K, Egolfopoulos FN, Davidson DF, Hanson RK, Bowman CT, Wang H (2018) A physics-based approach to modeling real-fuel combustion chemistry—III: reaction kinetic model of JP10. Combust Flame 198:466–476

Tonse SR, Moriarty NW, Frenklach M, Brown NJ (2003) Computational economy improvements in PRISM. Int J Chem Kin 35:438–452

Torczon V (1997) On the convergence of pattern search algorithms. SIAM J Opt 7(1):1–25

Tuner M, Blurock ES, Mauss F (2005) Phase optimized skeletal mechanisms in a stochastic reactor model for engine simulation. SAE, USA

Turányi T, Tomlin AS (2014) Reduction of reaction mechanisms. Springer, pp 183–312

Vajda S, Valko P, Turányi T (2006) Principal component analysis of kinetic models. Int J Chem Kin 17:55–81

Wan K, Barnaud C, Vervisch L, Domingo P (2020) Chemistry reduction using machine learning trained from non-premixed micro-mixing modeling: application to DNS of a syngas turbulent oxy-flame with side-wall effects. Combust Flame 220:119–129

Wang K, Xu R, Parise T, Shao J, Movaghar A, Lee DJ, Park JW, Gao Y, Lu T, Egolfopoulos FN, Davidson DF, Hanson RK, Bowman CT, Wang H (2018) A physics based approach to modeling real-fuel combustion chemistry—IV: HyChem modeling of combustion kinetics of a bio-derived jet fuel and its blends with a conventional jet A. Combust Flame 198:477–489

Wang H, Xu R, Wang K, Bowman CT, Hanson RK, Davidson DF, Brezinsky K, Egolfopoulos FN (2018) A physics-based approach to modeling real-fuel combustion chemistry—I: evidence from experiments, and thermodynamic, chemical kinetic and statistical considerations. Combust Flame 193:502–519

Xu R, Saggese C, Lawson R, Movaghar A, Parise T, Shao J, Choudhary R, Park JW, Lu T, Hanson RK, Davidson DF, Egolfopoulos FN, Aradi A, Prakash A, Raja V, Mohan R, Cracknell R, Wang H (2020) A physics-based approach to modeling real-fuel combustion chemistry—VI: predictive kinetic models of gasoline fuels. Combust Flame 220:475–487

Xu R, Wang H (2021) A physics-based approach to modeling real-fuel combustion chemistry—VII: relationship between speciation measurement and reaction model accuracy. Combust Flame 224(SI):126–135

Xu K, Wang R, Banerjee S, Shao J, Parise T, Zhu Y, Wang S, Movaghar A, Lee DJ, Zhao R, Han X, Gao Y, Lu T, Brezinsky K, Egolfopoulos FN, Davidson DF, Hanson RK, Bowman CT, Wang H (2018) A physics-based approach to modeling real-fuel combustion chemistry—II: reaction kinetic models of jet and rocket fuels. Combust Flame 193:520–537

Zhang X, Sarathy SM (2021a) High-temperature pyrolysis and combustion of C_5–C_{19} fatty acid methyl esters (FAMEs): a lumped kinetic modeling study. Energy Fuels 35(23):19553–19567

Zhang X, Sarathy SM (2021b) A functional-group-based approach to modeling real-fuel combustion chemistry—II: kinetic model construction and validation. Combust Flame 227:510–525

Zhang X, Sarathy SM (2021c) A lumped kinetic model for high-temperature pyrolysis and combustion of 50 surrogate fuel components and their mixtures. Fuel 286

Zhang X, Yalamanchi KK, Sarathy SM (2021) A functional-group-based approach to modeling real-fuel combustion chemistry—I: prediction of stoichiometric parameters for lumped pyrolysis reactions. Combust Flame 227:497–509

Zhang P, Liu S, Lu D, Sankaran R, Zhang G (2021) An out-of-distribution-aware autoencoder model for reduced chemical kinetics. Disc Contin Dyn Syst—Series S

Zhang P, Liu S, Lu D, Zhang G, Sankaran R (2021) A prediction interval method for uncertainty quantification of regression models. In: Conference: ninth international conference on learning representations (ICLR), Virtual, Austria—5/7/2021

Zhang X, Li W, Xu Q, Zhang Y, Jing Y, Wang Z, Sarathy SM (2022) A decoupled modeling approach and experimental measurements for pyrolysis of C_6–C_{10} saturated fatty acid methyl esters (FAMEs). Combust Flame, page in press

Zhang T, Zhang Y, E W, Ju Y (2021) DLODE: a deep learning-based ode solver for chemical kinetics (AIAA paper 2021-1139)

Zhong S, Zhang Y, Zhang H (2022) Machine learning-assisted QSAR models on contaminant reactivity toward four oxidants: combining small data sets and knowledge transfer. Env Sci Tech 56:681–692

Deep Convolutional Neural Networks for Subgrid-Scale Flame Wrinkling Modeling

V. Xing and C. J. Lapeyre

Abstract Subgrid-scale flame wrinkling is a key unclosed quantity for premixed turbulent combustion models in large eddy simulations. Due to the geometrical and multi-scale nature of flame wrinkling, convolutional neural networks are good candidates for data-driven modeling of flame wrinkling. This chapter presents how a deep convolutional neural network called a U-Net is trained to predict the total flame surface density from the resolved progress variable. Supervised training is performed on a database of filtered and downsampled direct numerical simulation fields. In an *a priori* evaluation on a slot burner configuration, the network outperforms classical dynamic models. In closing, challenges regarding the ability of deep convolutional networks to generalize to unseen configurations and their practical deployment with fluid solvers are discussed.

1 Introduction

As the effects of human activities become increasingly visible on the planet's climate, the combustion of fossil fuels is in need of renewal. Many ambitious carbon reduction scenarios, e.g. the IEA's "Net Zero by 2050" (International Energy Agency 2021), suggest a growing reliance on non-carbon fuels such as hydrogen and ammonia in the next decade. The large expected increase in intermittent renewable power notably solar and wind is well complemented by these means of storing, transporting, and distributing energy. While some applications will require fuel cells, it seems that combustion still has a large role to play in consuming these energy sources whether *via* adapted gas turbines for power generation, in heaters for homes and offices, in engines for propulsion, and even in some industrial processes such as iron or glass production. Additionally, the manipulation, storage and transport of these fuels can

V. Xing · C. J. Lapeyre (✉)
CERFACS, 42 avenue Gaspard Coriolis, 31000 Toulouse, France
e-mail: lapeyre@cerfacs.fr

V. Xing
e-mail: xing@cerfacs.fr

© The Author(s) 2023
N. Swaminathan and A. Parente (eds.), *Machine Learning and Its Application to Reacting Flows*, Lecture Notes in Energy 44, https://doi.org/10.1007/978-3-031-16248-0_6

lead to various safety issues that must be assessed and accounted for in the design phases. This is particularly true for hydrogen, which is hard to contain, hard to keep in a liquid phase, and has a low flammability limit, meaning leaks can easily arise and lead to unwanted fires and explosions. Overall, many new design problems might arise for turbulent combustion systems in this upcoming energy transition.

The relentless increase in computational power enables the use of large eddy simulations (LES) to capture fine, unsteady combustion phenomena in ever more complex premixed combustion configurations (Vermorel et al. 2017; Carlos et al. 2021a, b). The main challenge lies in the separation of scales between the finest combustion structures—typically of the order of the laminar flame thickness—and the extent of the computational domain. This is exacerbated in the aforementioned example of hydrogen which burns at higher speeds and in thinner reaction zones than hydrocarbon fuels. As a result, one of the major challenges in LES of premixed turbulent combustion is the modeling of subgrid-scale (SGS) reaction source terms. Turbulent reaction source terms are highly dependent on unresolved interactions between fine turbulent scales and the flame front. To first order, this results in the increase of the total flame surface *via* wrinkling of the flame front at resolved and unresolved scales, leading to an increased consumption rate of the unburnt gases. Inspired by this observation, many premixed turbulent combustion models have been built under the *flamelet* assumption, where the reaction rate is proportional to the flame surface area (Poinsot and Veynante 2011). As a result, correctly capturing the turbulent combustion rate is contingent on accurate modeling of SGS flame wrinkling.

This chapter will begin in Sect. 2 with an overview of existing SGS wrinkling models, with a specific focus on algebraic fractal approaches. The success of dynamic approaches (Charlette et al. 2002b; Ronnie et al. 2004) suggests that the inclusion of contextual data leads to significant improvements in model accuracy. In this light, a promising opportunity for wrinkling modeling is to use convolutional neural networks, which have been at the forefront of recent major advances in computer vision and are presented in Sect. 3. The full supervised training and *a priori* evaluation of a deep convolutional neural network wrinkling model is presented in Sect. 4. Finally, issues that need to be addressed on the path towards the deployment of neural network-based wrinkling models in practical LES computations are discussed in Sect. 5.

2 Wrinkling Models

Turbulent fully premixed flames are commonly modeled using the flamelet assumption, under which chemical reactions take place in thin layers that are wrinkled but not fragmented by turbulence (Peters 1988). Chemical timescales are assumed to be fast compared to turbulent processes so that the effects of turbulence can be treated independently from the chemistry. Under these assumptions, the evolution of thermochemical variables can be tracked by a single scalar quantity, the progress variable

c, which increases monotonically from 0 in the unburnt state to 1 in the burnt state. Flamelet models often assume that the structure of local flame elements measured in the progress variable space is identical to that of a one-dimensional laminar flame propagating in the normal direction to the flame element, making tabulated chemistry an effective method to model the thermochemical state of the flamelet (Benoît 2015). Traditional turbulent combustion diagrams (Borghi 1985; Peters 1988, 1999) posit that flamelets exist as long as the Kolmogorov lengthscale is larger than the laminar flame thickness, δ_L, and turbulent eddies cannot penetrate inside the flame front. This limitation is challenged by a growing body of work (Skiba et al. 2018; Driscoll et al. 2020) that reports experimental and numerical evidence of the existence of flamelet structures even for highly turbulent premixed flames (turbulent Reynolds number $Re_t \approx 10^5$, Karlovitz number $Ka \approx 500$) and supports the validity of flamelet models for a much wider range of turbulent flames than previously assumed.

Under the flamelet assumption, the wrinkling of the reaction layer induced by turbulence leads to an increase of the turbulent flame speed s_T proportional to the total flame area A_T (Driscoll 2008):

$$\frac{s_T}{s_L} = I_0 \frac{A_T}{A_L},\qquad(1)$$

where s_L, I_0, A_L are the unstretched laminar flame speed, stretch factor, and unwrinkled flame area, respectively. I_0 accounts for the effect of differential diffusion, and although accurate modeling of this factor is still elusive, experimental and DNS measurements consistently report I_0 values close to unity even for highly turbulent flames (Driscoll et al. 2020). The main obstacle to determining the turbulent flame speed is therefore the evaluation of the wrinkled flame front surface area. Since LES of practical turbulent premixed flames typically cannot afford to resolve the smallest wrinkling scales, the unresolved flame area must be recovered by SGS models.

Following Boger et al. (1998), the transport equation for c is given by:

$$\frac{\partial \overline{\rho c}}{\partial t} + \nabla \cdot (\overline{\rho \mathbf{u} \widetilde{c}}) + \nabla \cdot (\overline{\rho \mathbf{u} c} - \overline{\rho \mathbf{u} \widetilde{c}}) = \overline{\rho w |\nabla c|} = \langle \rho w \rangle_s \overline{|\nabla c|},\qquad(2)$$

where $\rho, \mathbf{u}, w$ are the density, velocity vector, and flamelet displacement speed, and $\overline{Q}$, $\widetilde{Q} = \overline{\rho Q}/\overline{Q}$, $\langle Q \rangle_s$ denote filtered, density-weighted filtered, and surface-averaged versions of a quantity Q, respectively. For laminar flame elements that propagate at the laminar flame speed s_L ($I_0 \approx 1$), the first term of the right hand side can be simplified as $\langle \rho w \rangle_s = \rho_u s_L$ using the unburnt gas density ρ_u. The second term of the right hand side is the generalized flame surface density (FSD) noted $\overline{\Sigma} = \overline{|\nabla c|}$ and represents the total surface area per unit volume of the flame front, including unresolved wrinkles. $\overline{\Sigma}$ is often connected to the resolved FSD $|\nabla \bar{c}|$ through the wrinkling factor:

$$\Xi = \overline{\Sigma}/|\nabla \bar{c}|.\qquad(3)$$

Ξ is equal to one when flame wrinkling is fully resolved, like in the case of a laminar flame.

Equation 2 forms the basis of flame surface density models, which typically determine $\overline{\Sigma}$ or Ξ using a transport equation (Weller et al. 1998; Hawkes and Cant 2000; Richard et al. 2007) or algebraic models (Boger et al. 1998; Wang et al. 2012; Mouriaux et al. 2017). For instance, Boger et al. (1998) propose an algebraic expression for $\overline{\Sigma}$ in the limit of a thin flame front relative to the filter size Δ:

$$\overline{\Sigma} = 4\sqrt{\frac{6}{\pi}}\,\Xi\frac{\tilde{c}(1-\tilde{c})}{\Delta}\,, \tag{4}$$

where Ξ remains to be modeled.

The wrinkling factor is also an essential component of LES reaction rate closures that use filtering or artificial thickening to deal with insufficient flame resolution. In the F-TACLES formalism (Fiorina et al. 2010), unclosed terms are pre-computed on filtered 1D laminar flames and tabulated as a function of $\tilde{c}$ and Δ. The turbulent reaction rate is expressed as $\overline{\dot{\omega}} = \Xi\,\overline{\dot{\omega}}_{1D}$. Alternatively, the thickened flame model (TFLES) (Butler and O'Rourke 1977; Colin et al. 2000) artificially thickens the flame front by a factor F by multiplying the thermal diffusivity and dividing the reaction rate by F. This operation does not affect the flame speed and enables the computation of the reaction rate from a set of well-resolved thermochemical variables $\bar{\phi}$. An efficiency factor E compensates the reduced sensitivity of the thickened flame front to turbulent wrinkling:

$$\overline{\dot{\omega}} = \frac{E}{F}\dot{\omega}(\bar{\phi}) = \frac{\Xi(\delta_L^0)}{F\,\Xi(F\delta_L^0)}\,\dot{\omega}(\bar{\phi})\,, \tag{5}$$

where $\Xi(\delta_L^0)$ and $\Xi(F\delta_L^0)$ are the wrinkling factors associated with the unthickened and thickened flame, respectively.

The rest of this chapter will focus on algebraic models for Ξ which have seen extensive developments over the years and have been comparatively reviewed in the literature (Chakraborty and Klein 2008; Ma et al. 2013). They are divided into two families:

- Models based on correlations of the turbulent flame speed (Weller et al. 1998; Colin et al. 2000; Muppala et al. 2005). These models leverage Eq. 1 to express Ξ as a function of turbulence parameters such as u'/s_L, l_t/δ_L. For instance, Colin et al. (2000) propose the expression:

$$\Xi = 1 + \alpha\Gamma_{\Delta_e}\frac{u'_{\Delta_e}}{s_L}\,, \tag{6}$$

where Γ_{Δ_e} accounts for the net straining effect of all vortices smaller than Δ_e, and α is a model parameter prescribed by the user.

- Models based on a fractal description of the flame front (Gouldin 1987; Gouldin et al. 1989; Charlette et al. 2002a, b; Ronnie et al. 2004; Fureby 2005; Wang et al. 2011; Hawkes et al. 2012; Keppeler et al. 2014). These will be detailed in the following.

Building from the seminal work of Gouldin (1987); Gouldin et al. (1989), fractal models assume that in a range of physical scales bounded by an inner cutoff η and an outer cutoff L, the flame front is a fractal surface of dimension D such that $2 \leq D \leq 3$. As a result, the wrinkling factor is given by:

$$\Xi = \left(\frac{L}{\eta}\right)^{D-2} .$$

(7)

Theoretical scaling arguments based on Damköhler's small and large-scale limits (Peters 2000) indicate that D ranges from $7/3$ in flamelets to $8/3$ in high Karlovitz flames (Hawkes et al. 2012). Experimental measurements lean towards the lower end of this range, with recent results on highly turbulent flames reporting $2.1 \leq D \leq 2.3$ (Skiba et al. 2021a). L corresponds to the size of the largest unresolved wrinkles, which is roughly the turbulence integral lengthscale l_t in RANS (Gouldin 1987) and the combustion filter size Δ in LES (Knikker et al. 2002; Charlette et al. 2002b). η is the size of the smallest wrinkles which scales with the inverse of Ka (Gülder and Smallwood 1995; Skiba et al. 2021a) and is the subject of careful modeling endeavors in fractal models.

In Charlette et al. (2002a), the inner cutoff scale η is chosen as the inverse mean curvature of the flame $|\langle \nabla \cdot \mathbf{n} \rangle_s|$ with $\mathbf{n}$ the normal vector to the flame front. It is modeled by assuming an equilibrium of the production and destruction of SGS flame surface density, and lower bounded by the laminar flame thickness. The resulting model is expressed as Wang et al. (2011):

$$\Xi = \left(1 + \min\left[\frac{\Delta}{\delta_L} - 1, \Gamma_\Delta \frac{u'_\Delta}{s_L}\right]\right)^\beta .$$

(8)

where Γ_Δ is a vortex efficiency function that serves the same purpose as in the Colin model of Eq. 6. While the Colin model introduced a multiplicative model parameter α, the Charlette model uses a power-law exponent β which is linked to the fractal dimension by $\beta = D - 2$. A constant value $\beta = 0.5$ ($D = 2.5$) is proposed in the original paper and leads to a *static* version of the Charlette model. When u'_Δ is sufficiently large, Eq. 8 takes on a *saturated* form:

$$\Xi = \left(\frac{\Delta}{\delta_L}\right)^\beta ,$$

(9)

where the wrinkling does not depend on the turbulence intensity.

The power-law parameter β can also be determined by a dynamic procedure (Charlette et al. 2002b) where it becomes a spatially and temporally evolving quantity. This avoids the delicate and arbitrary choice of one single value for β, which is often only justified *post hoc* by comparison to DNS or experimental data. It is also supported by empirical evidence highlighting significant spatial and temporal variations of the fractal dimension in turbulent flames (Keppeler et al. 2014; Skiba et al. 2021a).

The dynamic procedure introduces a filtering operation $\hat{Q}$ at a test-filter size $\hat{\Delta} = \gamma \Delta > \Delta$ and an averaging operation $\langle Q \rangle$ over a size $\Delta_m > \hat{\Delta}$. By equating two expressions of the averaged test-filtered total FSD:

$$\langle \widehat{\Xi_\Delta |\nabla \bar{c}|} \rangle = \langle \Xi_{\hat{\Delta}} |\nabla \hat{\bar{c}}| \rangle , \tag{10}$$

and assuming that β is uniform over the averaging volume, a closed-form formula for β can be found. The high levels of turbulence seen in practical turbulent configurations mean that Eq. 8 often takes its saturated form (Veynante and Moureau 2015) and in this case, the dynamic expression for β is:

$$\beta = \frac{\ln \left(\langle \widehat{|\nabla \bar{c}|} \rangle / \langle |\nabla \hat{\bar{c}}| \rangle \right)}{\ln \gamma} . \tag{11}$$

The dynamic Charlette model has been applied to LES of jet flames (Wang et al. 2011; Schmitt et al. 2015; Volpiani et al. 2016), ignition kernels (Wang et al. 2012; Mouriaux et al. 2017), stratified non-swirling burners (Mercier et al. 2015; Proch et al. 2017), the PRECCINSTA swirled burner (Veynante and Moureau 2015; Volpiani et al. 2017), explosions in semi-confined domains (Volpiani et al. 2017), and light-around in an annular combustor (Puggelli et al. 2021). It has also seen numerous incremental improvements over the years (Wang et al. 2011; Mouriaux et al. 2017; Proch et al. 2017) and stands today as a strong model for the SGS wrinkling factor.

3 Convolutional Neural Networks

This section gives a primer for uninitiated combustion physicists on deep learning. It explores what neural networks are, what the adjective "convolutional" refers to in that context, and how Convolutional Neural Networks, a workhorse of the deep learning revolution of the past decade, can be put to use for SGS problems.

3.1 *Artificial Neural Networks*

As early as the 1940s, attempts to model the behavior of biological neural networks have led to a simple function representing the action of a neuron (McCulloch and Pitts 1943). In its simplest form, a neuron sums all of its weighted electrical inputs via its dendrites, and the result is fed to a threshold function: if the sum of the input signals is high enough, an electrical impulse is sent through the axon to other neurons. Formally:

$$y = \sigma(\mathbf{w}^T \mathbf{x} + b),\tag{12}$$

where $\mathbf{x}$ is the vector of inputs received by the dendrites, $\mathbf{w}$ the vector of weights that it applies to each, b is a bias value, σ some threshold-like function called the *activation function*, and y the resulting signal sent *via* the axon to other connected neurons. Several of these neurons can be connected together, side by side as well as front to back, to form a *neural network*. Networks are part hand-designed, part automatically optimized, but in their most simple form they are *feedforward*, i.e. there are no information loops in the network.

The understanding of neural biology has advanced well beyond these simple models today, but the terminology "neural" has persisted. Modern neural networks have moved away from a strict analogy with biological neurons, towards a more abstract formalism. A *network* is composed of a succession of *layers* that perform operations on their input *feature map*, and pass on the resulting output feature map to the next layer.

Another important choice concerns the activation functions: if σ is linear, then so is each neuron, and stacking several linear neurons successively would be equivalent to composing several linear functions. The result would still be a linear function that a single neuron can represent. σ is therefore usually non-linear, and is an empirical trade-off between the non-linearity and the computational complexity it introduces, as well as some considerations on ease of training. The most common example is the *ReLU* or *REctified Linear Unit* function: $\sigma(x) = \max(0, x)$. For binary classification tasks, the last activation function is usually a sigmoid function:

$$\sigma(x) = \frac{1}{1 + e^{-x}}\tag{13}$$

taking values from 0 to 1 that can be interpreted as a class probability.

Once a network architecture is chosen, it is time to *train* it. Essentially, training means finding the optimal weights $\mathbf{w}$ and biases b, called *trainable parameters*, for all the neurons in the network so as to minimize a given *loss function*. To this end, the gradient of the loss function on given training samples with respect to all of the trainable parameters can be computed. This error can then be minimized by updating the trainable parameters via an optimization procedure, usually a form of iterative gradient descent.

In practice however, this gradient often proves highly non-convex and high-dimensional, and the error minimization process is too challenging for many standard gradient descent techniques. Instead, the minimization process is usually performed using *backpropagation* and *stochastic gradient descent* (SGD). Backpropagation (Rumelhart et al. 1986) is simply the process of computing progressively the gradient of the error with respect to the trainable parameters in each layer of the neural network, working backwards (hence the name) from the output to the input. This is a special case of reverse automatic differentiation, which is now the standard framework in deep learning libraries to efficiently perform backpropagation on complex neural networks. SGD is another trick used by most deep learning strategies (Goodfellow et al. 2016). Ideally, the gradient of error with respect to trainable parameters should be estimated over the entire training set. However, training databases are very large in deep learning, and this is computationally intractable. But in many situations, approximating this gradient with a small subset (called a *mini-batch*) of the training database gives a sufficiently good estimate of the overall gradient to advance an iterative gradient descent algorithm. This mini-batch-based gradient descent is called SGD.

Machine learning models are trained to capture all the meaningful features of the training dataset that are relevant to their learning task. If a model is underparametrized, it can fail to fit the training dataset adequately, leading to a behavior named *underfitting*. For this reason, modern neural networks contain a very large number of parameters, more than hundreds of billions in recent architectures (Brown et al. 2020). This can however lead them to learn too much, eventually learning the full dataset entirely by heart, a process called *overfitting*. Although this results in a very low loss function during training, an overfitted network performs poorly on data outside of the training dataset, meaning that it fails to *generalize*. To guard against this, overfitting must be monitored during training. This is done by reserving part of the dataset as a separate *validation* set, which can never be used to optimize the networks weights directly. The quality of predictions on this validation set is evaluated regularly during training, demonstrating when the generalization performance starts to degrade, and suggesting that the network has started to learn the specific noise of the data, and is no longer improving on the general task. The compromise between underfitting and overfitting is called the *bias-variance trade-off* (Goodfellow et al. 2016) and is central to any machine learning task.

3.2 Convolutional Layers

Neural networks built only with *fully connected* (FC) layers, where each neuron is connected to every neuron of the previous layer are called *multi-layer perceptrons* (MLPs). MLPs are simple stacks of successive FC layers. While this gives some choice in the design of the network (number of dense layers, number of neurons in each layer, activation functions...), other more specialized layers have been proposed for specific tasks. For image data, where the pixels have a matrix structure, *convolutional* layers (ConvLayers) are usually used. For the purpose of physical modeling, it is believed that a direct analogy between pixels in images and discretized physical fields can be made. The output of a ConvLayer is obtained by the convolution of its *kernel*, containing its trainable parameters, with its input feature map, as illustrated in Fig. 1. Multiple independent channels, each with its own kernel, are usually used to enhance the expressiveness of the layer. Each kernel (here of size 3×3, in gray) is convolved with the input matrix, producing a new matrix at the output.

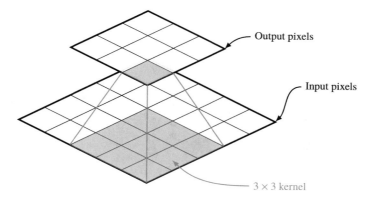

Fig. 1 Convolutional layer on a 2D matrix (e.g. an image). Input pixels (bottom) are convolved with a 3×3 kernel to produce the output pixels one by one

These convolutional kernels are the basis of many image treatment methods, where the kernel weights are prescribed to perform tasks such as contour detection, gaussian blur, denoising, etc. In a ConvLayer, the weights of the kernel (here 9 values) are the learnable parameters that are to be adjusted by the learning process instead of being explicitly prescribed. ConvLayers are well-adapted to dealing with spatial grids because of their translation equivariance and local consistency inductive bias (Battaglia et al. 2018). Since the same kernel is used for all input locations, the number of parameters of a ConvLayer is typically lower than in an FC layer. Moreover, unlike an FC layer, the number of parameters in a ConvLayer does not depend on the size of the input feature map, making it a good choice to process inputs of large dimensions like 3D computational domains.

Adding the ConvLayer to the layer arsenal leads to new network architectures, called convolutional neural networks (CNNs). Interestingly, shallow ConvLayers of a CNN have been observed to learn Gabor filters, which naturally occur in the visual cortex of mammals and are often chosen to extract image features in handmade image classifiers (Goodfellow et al. 2016). CNNs have been applied with great success for image-based tasks since the 1990s (LeCun et al. 1998), and have fueled the deep learning craze since the early 2010s successes (Krizhevsky et al. 2012) on the ImageNet classification challenge (Deng et al. 2009). Empirical evidence has shown that stacking small convolutional kernels leads to better performance than a single equivalent large kernel (Simonyan and Zisserman 2015; Szegedy et al. 2015). Depth is thus an important hyperparameter in CNNs, and deep CNNs have been universally used in recent breakthroughs in computer vision (He et al. 2015; Brock et al. 2019; Tan and Quoc 2019; Chen et al. 2020). Two of the most common learning tasks in computer vision, specifically when dealing with images, are *classification* and *segmentation*.

Image classification (Fig. 2a) is a task where a discrete label must be determined for an image. In the simple case of classifying of cat and dog images, the probability

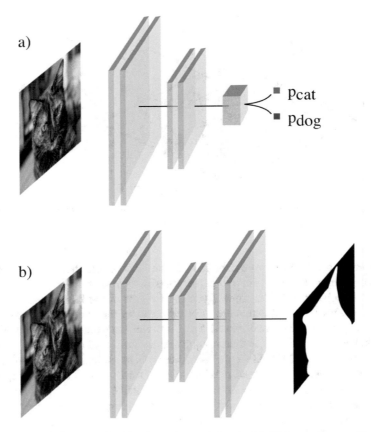

Fig. 2 Typical CNN tasks: **a** classification, where an image is classified according to a discrete list of labels; and **b** segmentation, where each pixel is classified according to a discrete label

that the image contains a cat p_{cat} is predicted by the network, and $p_{dog} = 1 - p_{cat}$ is inferred. If $p_{cat} > 0.5$, the label for the image is determined to be cat. Otherwise, it is dog. This prediction can then be compared to a truth value in the training database, and the network weights can be updated as described in Sect. 3.1. More generally, there can be more than 2 classes to choose from, and more than one class can be present at the same time. CNNs designed for classification tend to have a funnel-like shape, with a high-dimensional input (several thousand pixels, possibly in color) and a low-dimensional output (only 2 in our example, 1000 in the ImageNet dataset (Deng et al. 2009)).

Image segmentation (Fig. 2b) consists in identifying and classifying meaningful instances in an image by outlining them with labeled *masks*. Continuing with the previous example, the precise pixels belonging to the cat are sought. This changes the architecture of the network, which no longer needs to reduce the dimension of its output. Instead, the output has the same shape as the input, and each pixel is classified

as cat (1) or not (0). As a result, the layers chosen in the network must ensure that
the problem dimensionality is preserved at the output.

3.3   From Segmentation to Predicting Physical Fields with CNNs

A specific neural network architecture initiated a series of excellent results on image
segmentation tasks: the so-called *U-Net* (Ronneberger et al. 2015). This network,
introduced to detect tumors in medical images, can now be found in a variety of
projects, in its original form or in one of numerous variations (Çiçek et al. 2016;
Falk et al. 2019; Oktay et al. 2018), including in fluid dynamics (Wandel et al.
2021). Its structure is that of a "double funnel", one encoding the image into small
but numerous feature maps, and the other upscaling back to the input dimension
(Fig. 3). Compared to simple linear architectures (Fig. 2), the U-Net introduces *skip
connections* between some of the blocks, meaning data flows both to the lower
blocks (with deeper encoding of the features) and directly to the same-size output.
The intuition behind this is that in order to perform a segmentation decision on a
given pixel, a multi-scale analysis is needed. The influence of neighbouring pixels
informs on local textures. Further pixels (equivalent to a "zoomed-out" view of the
image) give information about the general shapes in the vicinity. Further pixels still
(seen by the deepest levels of the U-Net) offer an analysis of the position of the

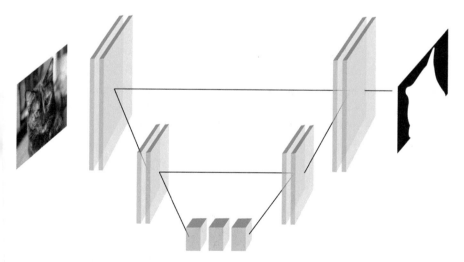

Fig. 3 Architecture of a U-Net neural network. Convolutional layers operate in an "double funnel" fashion, first reducing the feature map size, than increasing it again to match the input. Skip
connections are used between matching-size layers

shapes relative to each other. In the second (right in Fig. 3) half of the network, these levels of analysis coalesce gradually to form the final decision.

This process has analogies with the dynamic procedure of Eq. 11. Indeed, the dynamic estimation of β relies on observing the field of c at the resolved scale and the test-filter scale. Similarly, the first layer of a U-Net learns to detect structures on a 3-pixel wide stencil, and deeper layers aggregate features coming from several of these patches, effectively working at a larger scale. The U-Net can therefore be seen as a generalization of the concept introduced by dynamic models, where the effect of multiple scales on the target prediction is learned from the data, instead of only the resolved and test-filtered scales. This motivates the application of this type of network to the problem of predicting sub-grid scale wrinkling.

Some adaptations are needed to use a traditional U-Net on LES fields:

- The U-Net performs a regression task by predicting specific SGS values instead of a segmentation task. The final activation function should thus be a ReLU or an identity function.
- The U-Net must handle 3D data instead of 2D images. This poses very little challenge, as most modern implementations of neural network libraries natively offer 3D convolutional layers with the same functionality as classical 2D ones.
- Because the CNN is designed to work on structured data (pixels in image applications), it must operate on a *homogeneous, isotropic* mesh. This might mean that the field from a CFD mesh must be interpolated onto such a mesh. This limitation is due to the use of a "vanilla" U-Net, with no adaptations to more complex meshes. However, modern implementations with graph neural networks (Pfaff et al. 2021) could perform operations directly on an unstructured mesh if needed.

4 Training CNNs to Model Flame Wrinkling

This section presents the complete process of training and evaluating the CNN as a wrinkling model by following the steps described in Lapeyre et al. (2019). Full details are contained in the original paper, and code and data are available online.[1]

4.1 Data Preparation

The training and evaluation datasets are generated from the DNS of a slot burner configuration simulated with the AVBP unstructured compressible code (Schönfeld and Rudgyard 1999; Selle et al. 2004). A fully premixed stoichiometric mixture of methane-air unburnt gases is injected in a central rectangular inlet section at $U = 10$ m/s and surrounded by a slow coflow of burnt gases. The domain is a rectangular box meshed with a homogeneous grid containing $512 \times 256 \times 256$ hexahedral

---

[1] https://gitlab.com/cerfacs/code-for-papers/2018/arXiv_1810.03691.

elements of size $\Delta x = 0.1$ mm which resolve the reaction zone of the flame front on 4–5 points. A turbulent velocity field generated from a Passot-Pouquet spectrum (Passot and Pouquet 1987) is superimposed to the unburnt gas inlet velocities. Three separate DNS simulations are run:

- DNS1: inlet turbulence fluctuation intensity u' chosen such that $u'/s_L = 1.25$,
- DNS2: increased inlet turbulence, $u'/s_L = 2.5$,
- DNS3: starting from a steady-state snapshot of DNS2, the inlet velocity U is doubled for 1 ms, then set back to its initial level for 2 ms. This triggers the formation of a detached pocket of unburnt gases as evidenced in Fig. 4.

The training dataset is built from 50 snapshots of DNS1 and 50 snapshots of DNS2 extracted at 0.2 ms intervals in the steady-state regime. Similarly, the evaluation dataset is made up of 15 snapshots of DNS3. The slightly different large-scale flow dynamics and flame front geometry make it a good choice to assess the generalization of the CNN on a distribution close to that of the training set.

For each snapshot, the DNS field of c is filtered with a Gaussian kernel and downsampled to a coarse $64 \times 32 \times 32$ grid with a coarse cell size $8\Delta x$ to generate $\bar{c}$ and $\overline{\Sigma} = \overline{|\nabla c|}$. The network is trained to predict $\overline{\Sigma}^+ = \overline{\Sigma}/\overline{\Sigma}_{\text{lam}}^{\text{max}}$ corresponding to an input field of $\bar{c}$. $\overline{\Sigma}^+$ is the total FSD normalized by its maximum value measured on a laminar flame discretized on the same grid. While the values of $\overline{\Sigma}$ are specific to a given flame and coarse grid, $\overline{\Sigma}^+$ is a generic quantity that reflects the amount of unresolved wrinkling and should be more amenable to generalization. Normalizing the target value around 1 is also beneficial for the convergence of the early phase of SGD, since the output of the CNN resulting from inputs $\bar{c}$ and initial weights of the order of 1 will also be of the order of 1.

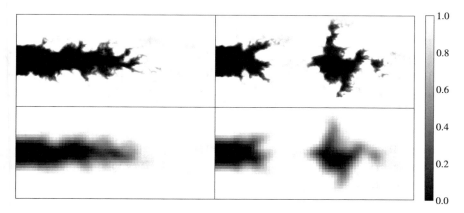

Fig. 4 Slices of progress variable field at $t = 0$ ms (left) and $t = 1$ ms (right) into DNS3. Top: DNS fields, bottom: filtered fields downsampled on the coarse mesh. The transient inlet velocity step leads to the separation of a pocket of unburnt gases

4.2 Building and Analyzing the U-Net

The U-Net architecture of Lapeyre et al. (2019) is detailed in Fig. 5. It follows a fully convolutional, symmetrical, three-stage encoder–decoder structure. Each stage is composed of two successive combinations of

- a 3D convolution with a $3 \times 3 \times 3$ kernel,
- a batch normalization layer (Ioffe and Szegedy 2015),
- a rectified linear unit (ReLU) nonlinear activation unit,

followed by $2 \times 2 \times 2$ pooling operations. In the encoder, maxpooling operations decrease the spatial dimensions of the feature maps by a factor of 2. The shape of the input field is then recovered by upsampling pooling operations in the decoder.

The network contains a total of 1.4 million trainable parameters. In cases where a smaller network would be preferrable, the parameter count could be reduced by using simpler neural network architectures (Shin et al. 2021) or by investigating architecture search and pruning methods (Frankle and Carbin 2019). On an Nvidia Tesla V100 GPU, training the network to convergence in 150 epochs takes 20 min, and inference on a single snapshot of DNS3 only requires 12 ms.

A key property of vision-based neural networks is their receptive field (RF), which corresponds to the input region that can influence the prediction on a single output point (Goodfellow et al. 2016). In practice, due to the distribution of the hidden layer connections inside the network, points located at the center of the receptive field contribute more to the prediction than those at the periphery. This leads to the notion of *effective* receptive field (ERF) (Luo et al. 2016) which measures the extent of the receptive field that is actually meaningful to the prediction, and can be quantified by counting the number of connections originating from each input

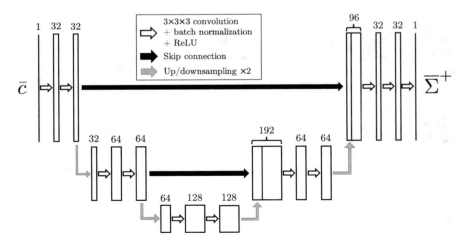

Fig. 5 Diagram of the U-Net architecture. Feature maps are represented by rectangles with their number of channels above. Arrows represent the hidden layers connecting the feature maps

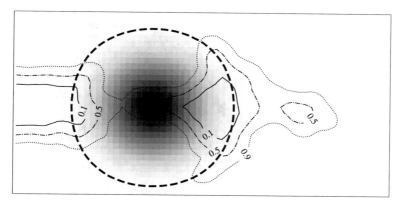

Fig. 6 ERF superimposed on iso-lines of $\bar{c}$ on a slice of a DNS3 snapshot ($t = 0.8$ ms). Grayscale intensity in the ERF is proportional to the impact of the input voxel location on the output prediction at the center of the ERF. Dashed circular line: edge of the ERF

location. Figure 6 compares the extent of the ERF of the U-Net with the DNS3 flame. The size (Luo et al. 2016) of the ERF is approximately 7.6 times the filtered laminar flame thickness and is large enough to encompass all of the large-scale structures of the flame front. In comparison, the context size of the Charlette dynamic model can be estimated as the averaging filter size which is typically 2−6 times the filtered laminar flame thickness (Veynante and Moureau 2015; Volpiani et al. 2016). Increasing the context size of the dynamic model may lead to numerical issues caused by flame/boundary and flame front interactions (Mouriaux et al. 2017) and greatly impacts the computational cost of the procedure (Volpiani et al. 2016), whereas for CNNs it can simply be achieved by using a deeper network.

4.3 A Priori Validation

After training the CNN on snapshots of DNS1 and DNS2, it is evaluated *a priori* on snapshots of DNS3 which are fully separate from the training dataset. The values of the trained weights of the CNN are frozen, and the model behaves like a large parametric function mapping $\bar{c}$ to $\overline{\Sigma}^{+}$. In Fig. 7a, the Charlette and CNN models are compared by plotting the downstream evolution of the total flame surface area that they predict on the DNS3 snapshot with the largest DNS total flame surface. For reference, target flame surface values from the DNS and values obtained without any SGS modeling are also shown. In this snapshot, the flame contains three distinct regions: a weakly turbulent flame base attached to the inlet ($x \approx 0$–15 mm), followed by a detached pocket of unburnt gases ($x \approx 15$–45 mm) and a postflame region of combustion products with no flame front.

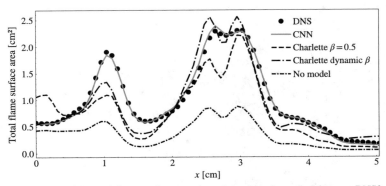

(a) Evolution of the total flame surface area along the streamwise x direction on a DNS3 snapshot ($t = 0.8$ ms). The flame surface values are computed by integrating the total FSD on cross-section slices of the width of a coarse cell.

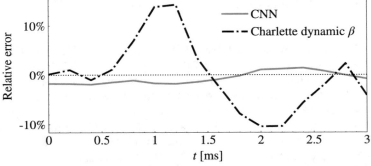

(b) Time evolution of the error on the domain-integrated total flame surface area relative to the target values on DNS3.

Fig. 7 *A priori* evaluation of a selection of wrinkling models

The static Charlette model with constant $\beta = 0.5$ finds the correct trend but consistently fails to accurately match the DNS flame surface values. The dynamic Charlette model with local β ($\hat{\Delta} = 1.5\Delta$, $\Delta_m = 2\hat{\Delta}$) using the corrections from Wang et al. (2011) and Mouriaux et al. (2017) performs very well in the detached pocket and close to the inlet, but still struggles near the tip of the attached flame which features prominent flame front interactions. Finally, the CNN agrees nearly perfectly with the target values in all regions of the domain. Figure 7b shows that this behavior is consistent throughout the whole duration of DNS3, whereas the error made by the Charlette dynamic model fluctuates in time.

5 Discussion

Deep CNNs trained to model SGS wrinkling show excellent modeling accuracy and consistency when compared to existing algebraic models on evaluation configurations that are similar to their training database. To move towards applications to practical complex configurations, some key questions still need to be addressed:

1. What information should be provided to the model? The U-Net presented above only used the field of $\bar{c}$ as input, but algebraic wrinkling models usually incorporate additional parameters like u'/s_L, l_t/δ_l, Ka, ...
2. To what extent can the model generalize to unseen configurations? Currently, the training dataset is built from DNS data which is rarely available in practice. If the model cannot reliably generalize well enough beyond its training distribution, this would severely limit its range of application.
3. Can the model be coupled to a fluid solver for on-the-fly predictions?

These questions apply broadly to any neural network model trained to predict an LES SGS quantity, not only to wrinkling models. Question 1 comes down to isolating the essential physical and numerical quantities that drive SGS wrinkling. A first meaningful quantity is the spatial distribution of $\bar{c}$ which identifies the location and thickness of the flame front in a premixed flame. Deep CNNs like the U-Net are presumably able to extract all the contextual information they need from the entire field of $\bar{c}$, and indeed experiments have indicated that providing gradients of $\bar{c}$ as additional inputs does not improve their accuracy. Other works that opt to use simpler architectures with fewer trainable parameters do include gradient information in the input of the network. Shin et al. (2021) train a shallow MLP combined with a mixture density network that captures the stochastic distribution of $\overline{\Sigma}$. Since the MLP only processes local data, $|\nabla\bar{c}|$ and $|\nabla^2\bar{c}|$ fields are used as additional inputs to provide some spatial context. Ren et al. (2021) use a network composed of a shallow 2D convolutional base followed by five fully connected layers. Local predictions are computed from 3×3 box stencils of the filtered fields of $\bar{c}$, $|\nabla\bar{c}|$ and the subgrid turbulence intensity u'_Δ discretized on the fine DNS grid.

Another relevant parameter is u'_Δ/s_L, which controls the amount of total flame surface wrinkling and is a crucial quantity in many wrinkling models covered in Sect. 2. Nonetheless, the challenges inherent to modeling u'_Δ from LES quantities (Colin et al. 2000; Veynante and Moureau 2015; Langella et al. 2017, 2018) have made the saturated Charlette dynamic model (Eq. 9) an attractive solution that does not directly depend on u'_Δ.

Finally, the proportion of unresolved flame wrinkling in the total flame surface is determined by the filter size Δ. Since CNNs work on grid data with no explicit distance embedding, Δ/δ_L sets the resolution of the filtered flame structures that are processed by the network. Figure 8 illustrates the ambiguity that may arise if Δ is not known by the network. There is an infinite number of combinations (c, Δ) that can lead to a given $\bar{c}$ field, each corresponding to a different amount of SGS wrinkling, and the sole knowledge of $\bar{c}$ is not sufficient to discriminate between them. Additionally,

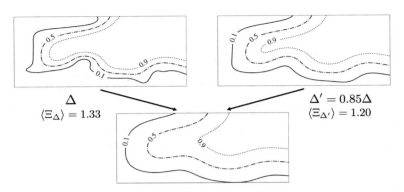

Fig. 8 Illustration of the filtering ambiguity. A filtered flame front (bottom) outlined by iso-lines of c can correspond to several unfiltered flames (top), each with a different filter size and mean wrinkling factor

CNNs are known to be sensitive to resolution discrepancies between the training and evaluation datasets (Touvron et al. 2019). This issue was avoided in Lapeyre et al. (2019) by training and evaluating the U-Net at the same Δ/δ_L but should be considered when generalizing to arbitrary flame resolutions.

To move towards generalizable SGS neural network models, u'_Δ/s_L and Δ/δ_L should henceforth be accounted for in the model either implicitly, in the choice of the training and evaluation datasets, or explicitly, by incorporating them in the model inputs or feature maps. Xing et al. (2021) started to investigate this by evaluating a U-Net trained on a statistically planar turbulent flame to predict the SGS variance of the progress variable $\overline{c'^2}$. A jet flame evaluation configuration (Luca et al. 2019) was chosen to test the ability of the network to generalize to a case featuring major differences from the training dataset regarding the large-scale flow and flame structures, thermophysical, and chemical parameters. The U-Net was observed to generalize better than existing dynamic approaches when u'_Δ/s_L and Δ/δ_L were chosen to match between the training and generalization configuration. Its performance dropped when either of these parameters did not match the unique values of the training set. However, when trained on a dataset containing a range of filter sizes, the U-Net was able to discriminate between the various Δ/δ_L values without explicitly providing Δ/δ_L as an input parameter. Apart from u'_Δ/s_L and Δ/δ_L, the inclusion of other relevant physical quantities can be investigated through feature importance analysis (Yellapantula et al. 2020).

The limits to generalization of SGS neural network models are still not well understood. Generalization is usually assessed by evaluating the model on the training distribution sampled at different spatial (Henry de Frahan et al. 2019; Wan et al. 2020) or temporal (Bode et al. 2021; Cellier et al. 2021; Chen et al. 2021) locations, or through minor parametric variations (Nikolaou et al. 2019; Lapeyre et al. 2019; Yao et al. 2020; Yellapantula et al. 2020; Chen et al. 2021). For wrinkling models specifically, Ren et al. (2021) study highly turbulent statistically stationary planar flames at $Ka = 38$ (case L), 390 (case M), and 1710 (case H). Cases M and H are located in the broken reaction zone regime, where the flamelet assumption may not hold. Snapshots show a highly fragmented reaction front and the authors point out that the resolved and total FSD fields have large discrepancies for these cases. After training on case H, the model performs well on case M and at larger filter sizes, beating a selection of static wrinkling models. It is interesting to note that it performs relatively poorly on case L which belongs to the thin reaction zone regime and features an intact reaction zone. This result highlights the model's sensitivity to changes in the turbulent combustion regime. Attili et al. (2021) draw similar conclusions after training the U-Net from Lapeyre et al. (2019) on four DNS of jet flames with increasing Reynolds numbers (Luca et al. 2019). Their results show that generalization to unseen turbulent levels works better between high Reynolds number flames, which they suggest is due to the asymptotic behavior of high Reynolds turbulence. In addition, models trained on a specific region of the flame (flame base, fully turbulent region, or flame tip) perform noticeably worse when tested on a different region, thus highlighting the spatial variations of the wrinkling distribution in a given flame.

Supervised training of neural networks is a form of inductive learning, for which generalization depends on the inductive biases of the model (Griffiths et al. 2010). These are the factors outside of the observed data that intrinsically steer the model towards learning a specific representation. Generalization is largely driven by how well the model's inductive biases fit the properties of the data representation it is trained to learn. The inductive biases of neural networks are heavily influenced by their architecture. MLPs have weak inductive biases, whereas CNNs have strong locality and translation equivariance inductive biases (Battaglia et al. 2018) which explains their success in generalization of computer vision tasks (Zhang et al. 2020). Since locality and translation equi-variance are also desirable properties of an SGS model, CNNs seem better suited than MLPs to generalize on SGS modeling tasks.

On the other hand, coupling CNNs with a fluid solver for on-the-fly predictions and *a posteriori* validation comes with numerous implementation challenges. In the case of the U-Net, its field-to-field nature allows it to output predictions in the entire domain in a single inference of the network, which is a strong asset for computations on large meshes. However, the input field needs to be built by gathering LES data points from the whole domain, and the prediction of the model has to be scattered back. For massively parallel solvers which perform domain decomposition, this requires dedicated message-passing communications between the solver and the CNN instances. Additionally, since the CNN can only process structured data, if the LES is performed on an unstructured mesh, the input and prediction fields must be interpolated between the solver mesh and a structured mesh that can be read by the CNN. Coupling interfaces such as OpenPALM (Duchaine et al. 2015) have successfully been used to manage these operations and perform fully coupled simulations using the AVBP solver (Lapeyre et al. 2018). The computational overhead due to the coupling and the neural network prediction is less than 5%. As a reference, the filtering operations used in the Charlette dynamic model typically induce overheads of 20–30% (Volpiani et al. 2016; Puggelli et al. 2021). Finally, given the large number of parameters of the U-Net, inference is preferably performed on a GPU. This requires additional care in the coupling implementation, but should not limit the deployment of the model given the growing adoption of hybrid CPU-GPU supercomputer infrastructures.

6 Conclusion

The intersection of LES subgrid-scale modeling and machine learning is a promising and rapidly growing field in numerical combustion. The large modeling capacity of deep neural networks is a strong asset to model complex SGS flame-turbulence phenomena in a data-rich environment fueled by high-fidelity simulation results. Taking inspiration from the computer vision community, a deep CNN U-Net architecture is trained to predict the total—resolved and unresolved—flame surface density field from the LES resolved progress variable field. The U-Net is built to aggregate multi-scale spatial information on the flame front, ranging from the coarse mesh resolution to large flame structures, thanks to its wide receptive field. In this sense, it can be viewed as an extension of existing dynamic models that combine information at the filtered and test-filtered scales. DNS snapshots are filtered and downsampled to generate the training and evaluation datasets that are used to evaluate the U-Net in an *a priori* context. On the evaluation set of a slot burner configuration, the U-Net consistently matches the target flame surface density distribution, beating the static and dynamic versions of the Charlette wrinkling model. More generally, the modeling methodology outlined in this chapter can be applied to any SGS quantity, such as the SGS variance of the progress variable. These results open the way to many compelling directions for future work. Coupling a deep CNN with a massively parallel fluid solver is a key step towards *a posteriori* validation. Graph neural networks could be explored as alternatives that could handle on arbitrary meshes and complex geometries. Finally, an issue at the core of the practical deployment of any machine learning combustion model is to assess whether it can robustly generalize outside of its training distribution, a feature that will need to be demonstrated if these models are to replace traditional models in CFD solvers.

References

Arroyo CP, Dombard J, Duchaine F, Gicquel L, Martin B, Odier N, Staffelbach G (2021a) Towards the large-eddy simulation of a full engine: integration of a 360 azimuthal degrees fan, compressor and combustion chamber. Part ii: comparison against stand-alone simulations. J Glob Power Propuls Soc Spec Issue (May):1–16. https://doi.org/10.33737/jgpps/133116

Arroyo CP, Dombard J, Duchaine F, Gicquel L, Martin B, Odier N, Staffelbach G (2021b) Towards the large-eddy simulation of a full engine: Integration of a 360 azimuthal degrees fan, compressor and combustion chamber. Part ii: comparison against stand-alone simulations. J Glob Power Propuls Soc (May)1–16. https://doi.org/10.33737/jgpps/133116

Attili A, Sorace N, Nista L, Schumann C, Karimi A (2021) Investigation of the extrapolation performance of machine learning models for les of turbulent premixed combustion. In: Proceedings European combustion meeting, pp 349–354

Battaglia PW, Hamrick JB, Bapst V, Sanchez-Gonzalez A, Zambaldi V, Malinowski M, Tacchetti A, Raposo D, Santoro A, Faulkner R, Gulcehre C, Song F, Ballard A, Gilmer J, Dahl G, Vaswani A, Allen K, Nash C, Langston V, Dyer C, Heess N, Wierstra D, Kohli P, Botvinick M, Vinyals O, Li Y, Pascanu R (2018) Relational inductive biases, deep learning, and graph networks. CoRR. arXiv:abs/1806.01261

Bode M, Gauding M, Lian Z, Denker D, Davidovic M, Kleinheinz K, Jitsev J, Pitsch H (2021) Using physics-informed enhanced super-resolution generative adversarial networks for subfilter modeling in turbulent reactive flows. Proc Combust Inst 38(2):2617–2625. https://doi.org/10.1016/j.proci.2020.06.022

Boger M, Veynante D, Boughanem H, Trouvé A (1998) Direct numerical simulation analysis of flame surface density concept for large eddy simulation of turbulent premixed combustion. In: Symposium (International) on combustion, vol 27, no 1, pp 917–925. https://doi.org/10.1016/S0082-0784(98)80489-X

Borghi R (1985) On the structure and morphology of turbulent premixed flames. In: Casci C, Bruno C (eds) Recent advances in the aerospace sciences. Springer, Boston, MA, pp 117–138. https://doi.org/10.1007/978-1-4684-4298-4_7

Brock A, Donahue J, Simonyan K (2019) Large scale GAN training for high fidelity natural image synthesis. In: International conference on learning representations

Brown T, Mann B, Ryder N, Subbiah M, Kaplan JD, Dhariwal P, Neelakantan A, Shyam P, Sastry G, Askell A, Agarwal S, Herbert-Voss A, Krueger G, Henighan T, Child R, Ramesh A, Ziegler D, Wu J, Winter C, Hesse C, Chen M, Sigler E, Litwin M, Gray S, Chess B, Clark J, Berner C, McCandlish S, Radford A, Sutskever I, Amodei D (2020) Language models are few-shot learners. Adv Neural Inf Process Syst 33:1877–1901

Butler TD, O'Rourke PJ (1977) A numerical method for two dimensional unsteady reacting flows. In: Symposium (International) on combustion, vol 16, no 1, pp 1503–1515. https://doi.org/10.1016/S0082-0784(77)80432-3

Cellier A, Lapeyre CJ, Öztarlik G, Poinsot T, Schuller T, Selle L (2021) Detection of precursors of combustion instability using convolutional recurrent neural networks. Combust Flame 233:111558. https://doi.org/10.1016/j.combustflame.2021.111558

Chakraborty N, Klein M (2008) A priori direct numerical simulation assessment of algebraic flame surface density models for turbulent premixed flames in the context of large eddy simulation. Phys Fluids 20(8):085108. https://doi.org/10.1063/1.2969474

Charlette F, Meneveau C, Veynante D (2002a) A power-law flame wrinkling model for les of premixed turbulent combustion part ii: dynamic formulation. Combust Flame 131(1–2):181–197. https://doi.org/10.1016/S0010-2180(02)00401-7

Charlette F, Meneveau C, Veynante D (2002b) A power-law wrinkling model for les of premixed turbulent combustion part i: non-dynamic formulation and initial tests. Combust Flame 131(1–2):159–180. https://doi.org/10.1016/S0010-2180(02)00400-X

Chen ZX, Iavarone S, Ghiasi G, Kannan V, D'Alessio G, Parente A, Swaminathan N (2021) Application of machine learning for filtered density function closure in mild combustion. Combust Flame 225:160–179. https://doi.org/10.1016/j.combustflame.2020.10.043

Chen T, Kornblith S, Norouzi M, Hinton G (2020) A simple framework for contrastive learning of visual representations. In: Proceedings of the 37th international conference on machine learning, pp 1597–1607

Çiçek O, Abdulkadir A, Lienkamp SS, Brox T, Ronneberger O (2016) 3D u-net: learning dense volumetric segmentation from sparse annotation. In: International conference on medical image computing and computer-assisted intervention, pp 424–432

Colin O, Ducros F, Veynante D, Poinsot T (2000) A thickened flame model for large eddy simulations of turbulent premixed combustion. Phys Fluids 12(7):1843. https://doi.org/10.1063/1.870436

Deng J, Dong W, Socher R, Li L-J, Li K, Fei-Fei L (2009) Imagenet: a large-scale hierarchical image database. In: 2009 IEEE conference on computer vision and pattern recognition. IEEE, pp 248–255

Driscoll James F (2008) Turbulent premixed combustion: flamelet structure and its effect on turbulent burning velocities. Prog Energy Combust Sci 34(1):91–134. https://doi.org/10.1016/j.pecs.2007.04.002

Driscoll James F, Chen Jacqueline H, Skiba Aaron W, Carter Campbell D, Hawkes Evatt R, Wang Haiou (2020) Premixed flames subjected to extreme turbulence: some questions and recent answers. Prog Energy Combust Sci 76:100802. https://doi.org/10.1016/j.pecs.2019.100802

Duchaine F, Jauré S, Poitou D, Quémerais E, Staffelbach G, Morel T, Gicquel L (2015) Analysis of high performance conjugate heat transfer with the openpalm coupler. Comp Sci Discov 8(1):15003. https://doi.org/10.1088/1749-4699/8/1/015003

Falk T, Mai D, Bensch R, Çiçek O, Abdulkadir A, Marrakchi Y, Böhm A, Deubner J, Jäckel Z, Seiwald K et al (2019) U-net: deep learning for cell counting, detection, and morphometry. Nat Methods 16:67–70

Fiorina B, Vicquelin R, Auzillon P, Darabiha N, Gicquel O, Veynante D (2010) A filtered tabulated chemistry model for les of premixed combustion. Combust Flame 157(3):465–475. https://doi.org/10.1016/j.combustflame.2009.09.015

Fiorina B, Veynante D, Candel S (2015) Modeling combustion chemistry in large eddy simulation of turbulent flames. Flow, Turbul Combust 94(1):3–42. https://doi.org/10.1007/s10494-014-9579-8

Frankle J, Carbin M (2019) The lottery ticket hypothesis: finding sparse, trainable neural networks. In: International conference on learning representations

Fureby C (2005) A fractal flame-wrinkling large eddy simulation model for premixed turbulent combustion. Proc Combust Inst 30(1):593–601. https://doi.org/10.1016/j.proci.2004.08.068

Goodfellow I, Bengio Y, Courville A (2016) Deep learning. MIT Press

Gouldin FC (1987) An application of fractals to modeling premixed turbulent flames. Combust Flame 68(3):249–266. https://doi.org/10.1016/0010-2180(87)90003-4

Gouldin FC, Bray KNC, Chen JY (1989) Chemical closure model for fractal flamelets. Combust Flame 77(3–4):241–259. https://doi.org/10.1016/0010-2180(89)90132-6

Griffiths TL, Chater N, Kemp C, Perfors A, Tenenbaum JB (2010) Probabilistic models of cognition: exploring representations and inductive biases. Trends Cogn Sci 14(8):357–364. https://doi.org/10.1016/j.tics.2010.05.004

Gülder OL, Smallwood GJ (1995) Inner cutoff scale of flame surface wrinkling in turbulent premixed flames. Combust Flame 103(1–2):107–114. https://doi.org/10.1016/0010-2180(95)00073-F

Hawkes ER, Cant RS (2000) A flame surface density approach to large-eddy simulation of premixed turbulent combustion. Proc Combust Inst 28(1):51–58. https://doi.org/10.1016/S0082-0784(00)80194-0

Hawkes ER, Chatakonda O, Kolla H, Kerstein AR, Chen JH (2012) A petascale direct numerical simulation study of the modelling of flame wrinkling for large-eddy simulations in intense turbulence. Combust Flame 159(8):2690–2703. https://doi.org/10.1016/j.combustflame.2011.11.020

Henry de Frahan MT, Yellapantula S, King R, Day MS, Grout RW (2019) Deep learning for presumed probability density function models. Combust Flame 208:436–450. https://doi.org/10.1016/j.combustflame.2019.07.015

He K, Zhang X, Ren S, Sun J (2015) Delving deep into rectifiers: surpassing human-level performance on imagenet classification. In: Proceedings of the IEEE international conference on computer vision, pp 1026–1034

International Energy Agency (2021) Net zero by 2050. Technical report, International Energy Agency

Ioffe S, Szegedy C (2015) Batch normalization: Accelerating deep network training by reducing internal covariate shift. In: Proceedings of the 32nd international conference on machine learning, vol 1, pp 448–456

Keppeler R, Tangermann E, Allaudin U, Pfitzner M (2014) Les of low to high turbulent combustion in an elevated pressure environment. Flow, Turbul Combust 92(3):767–802. https://doi.org/10.1007/s10494-013-9525-1

Knikker R, Veynante D, Meneveau C (2002) A priori testing of a similarity model for large eddy simulations of turbulent premixed combustion. Proc Combust Inst 29(2):2105–2111. https://doi.org/10.1016/S1540-7489(02)80256-5

Knikker R, Veynante D, Meneveau C (2004) A dynamic flame surface density model for large eddy simulation of turbulent premixed combustion. Phys Fluids 16(11):91–95. https://doi.org/10.1063/1.1780549

Krizhevsky A, Sutskever I, Hinton GE (2012) Imagenet classification with deep convolutional neural networks. Adv Neural Inf Process Syst 25:1097–1105

Langella I, Swaminathan N, Gao Y, Chakraborty N (2017) Large eddy simulation of premixed combustion: sensitivity to subgrid scale velocity modeling. Combust Sci Technol 189(1):43–78. https://doi.org/10.1080/00102202.2016.1193496

Langella I, Doan NAK, Swaminathan N, Pope SB (2018) Study of subgrid-scale velocity models for reacting and nonreacting flows. Phys Rev Fluids 3(5):054602

Lapeyre CJ, Misdariis A, Cazard N (2018) A-posteriori evaluation of a deep convolutional neural network approach to subgrid-scale flame surface estimation. In: Proceedings of the summer program, center for turbulence research, pp 349–358

Lapeyre CJ, Misdariis A, Cazard N, Veynante D, Poinsot TJ (2019) Training convolutional neural networks to estimate turbulent sub-grid scale reaction rates. Combust Flame 203:255–264. https://doi.org/10.1016/j.combustflame.2019.02.019

LeCun Y, Bottou L, Bengio Y, Haffner P (1998) Gradient-based learning applied to document recognition. Proc IEEE 86(11):2278–2324

Luo W, Li Y, Urtasun R, Zemel R (2016) Understanding the effective receptive field in deep convolutional neural networks. Adv Neural Inf Process Syst 29

Luca S, Attili A, Schiavo EL, Creta F, Bisetti F (2019) On the statistics of flame stretch in turbulent premixed jet flames in the thin reaction zone regime at varying Reynolds number. Proc Combust Inst 37(2):2451–2459. https://doi.org/10.1016/j.proci.2018.06.194

Ma T, Stein OT, Chakraborty N, Kempf AM (2013) A posteriori testing of algebraic flame surface density models for les. Combust Theory Model 17(3):431–482. https://doi.org/10.1080/13647830.2013.779388

McCulloch WS, Pitts W (1943) A logical calculus of the ideas immanent in nervous activity. B Math Biophys 5(4):115–133

Mercier R, Schmitt T, Veynante D, Fiorina B (2015) The influence of combustion SGS submodels on the resolved flame propagation. application to the les of the Cambridge stratified flames. Proc Combust Inst 35(2):1259–1267. https://doi.org/10.1016/j.proci.2014.06.068

Mouriaux S, Colin O, Veynante D (2017) Adaptation of a dynamic wrinkling model to an engine configuration. Proc Combust Inst 36(3):3415–3422. https://doi.org/10.1016/j.proci.2016.08.001

Muppala SR, Aluri NK, Dinkelacker F, Leipertz A (2005) Development of an algebraic reaction rate closure for the numerical calculation of turbulent premixed methane, ethylene, and propane/air flames for pressures up to 1.0 MPa. Combust Flame 140(4):257–266. https://doi.org/10.1016/j.combustflame.2004.11.005

Nikolaou ZM, Chrysostomou C, Vervisch L, Cant S (2019) Progress variable variance and filtered rate modelling using convolutional neural networks and flamelet methods. Flow, Turbul Combust 103(2):485–501. https://doi.org/10.1007/s10494-019-00028-w

Oktay O, Schlemper J, Folgoc LL, Lee M, Heinrich M, Misawa K, Mori K, McDonagh S, Hammerla NY, Kainz B et al (2018) Attention u-net: learning where to look for the pancreas. Med Imaging Deep Learn

Passot T, Pouquet A (1987) Numerical simulation of compressible homogeneous flows in the turbulent regime. J Fluid Mech 181:441–466. https://doi.org/10.1017/S0022112087002167

Peters N (1988) Laminar flamelet concepts in turbulent combustion. In: Symposium (International) on combustion, vol 21, no 1 pp 1231–1250. https://doi.org/10.1016/S0082-0784(88)80355-2

Peters N (1999) The turbulent burning velocity for large-scale and small-scale turbulence. J Fluid Mech 384:107–132. https://doi.org/10.1017/S0022112098004212

Peters N (2000) Turbulent combustion. Cambridge University Press

Pfaff T, Fortunato M, Sanchez-Gonzalez A, Battaglia PW (2021) Learning mesh-based simulation with graph networks. In: International conference on learning representations

Poinsot T, Veynante D (2011) Theoretical and numerical combustion. 3rd edn. www.cerfacs.fr/elearning

Proch F, Domingo P, Vervisch L, Kempf AM (2017) Flame resolved simulation of a turbulent premixed bluff-body burner experiment. Part i: analysis of the reaction zone dynamics with tabulated chemistry. Combust Flame 180:321–339. https://doi.org/10.1016/j.combustflame.2017.02.011

Puggelli S, Veynante D, Vicquelin R (2021) Impact of dynamic modelling of the flame subgrid scale wrinkling in large-eddy simulation of light-round in an annular combustor. Combust Flame 230:111416. https://doi.org/10.1016/j.combustflame.2021.111416

Ren J, Wang H, Luo K, Fan J (2021) A priori assessment of convolutional neural network and algebraic models for flame surface density of high karlovitz premixed flames. Phys Fluids 33(3):036111. https://doi.org/10.1063/5.0042732

Richard S, Colin O, Vermorel O, Benkenida A, Angelberger C, Veynante D (2007) Towards large eddy simulation of combustion in spark ignition engines. Proc Combust Inst 31 II(2):3059–3066

Ronneberger O, Fischer P, Brox T (2015) U-net: convolutional networks for biomedical image segmentation. In: International conference on medical image computing and computer-assisted intervention. Springer, pp 234–241

Rumelhart DE, Hinton GE, Williams RJ (1986) Learning representations by back-propagating errors. Nature 323(6088):533–536

Schmitt T, Boileau M, Veynante D (2015) Flame wrinkling factor dynamic modeling for large eddy simulations of turbulent premixed combustion. Flow, Turbul Combust 94(1):199–217. https://doi.org/10.1007/s10494-014-9574-0

Schönfeld T, Rudgyard M (1999) Steady and unsteady flow simulations using the hybrid flow solver AVBP. AIAA J 37(11):1378–1385. https://doi.org/10.2514/2.636

Selle L, Lartigue G, Poinsot TJ, Koch R, Schildmacher KU, Krebs W, Prade B, Kaufmann P, Veynante D (2004) Compressible large eddy simulation of turbulent combustion in complex geometry on unstructured meshes. Combust Flame 137(4):489–505. https://doi.org/10.1016/j.combustflame.2004.03.008

Shin J, Ge Y, Lampmann A, Pfitzner M (2021) A data-driven subgrid scale model in large eddy simulation of turbulent premixed combustion. Combust Flame 231:111486. https://doi.org/10.1016/j.combustflame.2021.111486

Simonyan K, Zisserman A (2015) Very deep convolutional networks for large-scale image recognition. In: International conference on learning representations

Skiba AW, Wabel TM, Carter CD, Hammack SD, Temme JE, Driscoll JF (2018) Premixed flames subjected to extreme levels of turbulence part i: flame structure and a new measured regime diagram. Combust Flame 189:407–432. https://doi.org/10.1016/j.combustflame.2017.08.016

Skiba AW, Carter CD, Hammack SD, Driscoll JF (2021a) High-fidelity flame-front wrinkling measurements derived from fractal analysis of turbulent premixed flames with large reynolds numbers. Proc Combust Inst 38(2):2809–2816. https://doi.org/10.1016/j.proci.2020.06.041

Skiba AW, Carter CD, Hammack SD, Driscoll JF (2021a) Experimental assessment of the progress variable space structure of premixed flames subjected to extreme turbulence. Proc Combust Inst 38(2):2893–2900. https://doi.org/10.1016/j.proci.2020.06.129

Szegedy C, Liu W, Jia Y, Sermanet P, Reed S, Anguelov D, Erhan D, Vanhoucke V, Rabinovich A (2015) Going deeper with convolutions. In: Proceedings of the IEEE conference on computer vision and pattern recognition, pp 1–9

Tan M, Le QV (2019) Efficientnet: rethinking model scaling for convolutional neural networks. In: International conference on machine learning, pp 6105–6114

Touvron H, Vedaldi A, Douze M, Jégou H (2019) Fixing the train-test resolution discrepancy. Adv Neural Inf Process Syst 32:8252–8262

Vermorel O, Quillatre P, Poinsot T (2017) Les of explosions in venting chamber: a test case for premixed turbulent combustion models. Combust Flame 183:207–223. https://doi.org/10.1016/j.combustflame.2017.05.014

Veynante D, Moureau V (2015) Analysis of dynamic models for large eddy simulations of turbulent premixed combustion. Combust Flame 162(12):4622–4642. https://doi.org/10.1016/j.combustflame.2015.09.020

Volpiani PS, Schmitt T, Veynante D (2016) A posteriori tests of a dynamic thickened flame model for large eddy simulations of turbulent premixed combustion. Combust Flame 174:166–178. https://doi.org/10.1016/j.combustflame.2016.08.007

Volpiani PS, Schmitt T, Veynante D (2017a) Large eddy simulation of a turbulent swirling premixed flame coupling the TFLES model with a dynamic wrinkling formulation. Combust Flame 180:124–135. https://doi.org/10.1016/j.combustflame.2017.02.028

Volpiani PS, Schmitt T, Vermorel O, Quillatre P, Veynante D (2017b) Large eddy simulation of explosion deflagrating flames using a dynamic wrinkling formulation. Combust Flame 186:17–31. https://doi.org/10.1016/j.combustflame.2017.07.022

Wan K, Hartl S, Vervisch L, Domingo P, Barlow RS, Hasse C (2020) Combustion regime identification from machine learning trained on Raman/Rayleigh line measurements. Combust Flame 219:268–274. https://doi.org/10.1016/j.combustflame.2020.05.024

Wandel N, Weinmann M, Klein R (2021) Learning incompressible fluid dynamics from scratch—towards fast, differentiable fluid models that generalize. In: International conference on learning representations

Wang G, Boileau M, Veynante D (2011) Implementation of a dynamic thickened flame model for large eddy simulations of turbulent premixed combustion. Combust Flame 158(11):2199–2213. https://doi.org/10.1016/j.combustflame.2011.04.008

Wang G, Boileau M, Veynante D, Truffin K (2012) Large eddy simulation of a growing turbulent premixed flame kernel using a dynamic flame surface density model. Combust Flame 159(8):2742–2754. https://doi.org/10.1016/j.combustflame.2012.02.018

Weller HG, Tabor G, Gosman AD, Fureby C (1998) Application of a flame-wrinkling les combustion model to a turbulent mixing layer. In: Symposium (International) on combustion, vol 27, no 1, pp 899–907. https://doi.org/10.1016/S0082-0784(98)80487-6

Xing V, Lapeyre C, Jaravel T, Poinsot TJ (2021) Generalization capability of convolutional neural networks for progress variable variance and reaction rate subgrid-scale modeling. Energies 14(16):5096. https://doi.org/10.3390/en14165096

Yao S, Wang B, Kronenburg A, Stein OT (2020) Modeling of sub-grid conditional mixing statistics in turbulent sprays using machine learning methods. Phys Fluids 32(11). https://doi.org/10.1063/5.0027524

Yellapantula S, Perry BA, Grout RW (2020) Deep learning-based model for progress variable dissipation rate in turbulent premixed flames. Proc Combust Inst 38:2929–2938. https://doi.org/10.1016/j.proci.2020.06.205

Zhang C, Bengio S, Hardt M, Mozer MC, Singer Y (2020) Identity crisis: memorization and generalization under extreme over parameterization. In: International conference on learning representations

Machine Learning Strategy for Subgrid Modeling of Turbulent Combustion Using Linear Eddy Mixing Based Tabulation

R. Ranjan, A. Panchal, S. Karpe, and S. Menon

Abstract This chapter describes the use of machine learning (ML) algorithms with the linear-eddy mixing (LEM) based tabulation for modeling of subgrid turbulence-chemistry interaction. The focus will be on the use of artificial neural network (ANN), particularly, supervised deep learning (DL) techniques within the finite-rate kinetics framework. We discuss the accuracy and efficiency aspects of two different strategies, where LEM based tabulation is used in both of them. While in the first approach, referred to as LANN-LES, the subgrid reaction-rate term is obtained efficiently using ANN in the conventional LEMLES framework, in the other approach referred to as TANN-LES, the filtered reaction rate terms are obtained using ANN. First, we assess the implications of the employed network architecture, and the associated hyperparameters, such as the amount of training and test data, epoch, optimizer, learning rate, sample size, etc. Afterward, the effectiveness of the two strategies is examined by comparing with conventional LES and LEMLES approaches by considering canonical premixed and non-premixed configurations. Finally, we describe the key challenges and future outlook of the use of ML based subgrid modelling within the finite-rate kinetics framework.

1 Introduction

Combustion within energy conversion and propulsion devices such as internal combustion engines, gas turbines, rocket engines, etc., usually occurs under turbulent conditions. The turbulence-chemistry interaction in such devices is characterized by

R. Ranjan
Department of Mechanical Engineering, University of Tennessee at Chattanooga, 615 McCallie Ave, Chattanooga, TN 37403, USA
e-mail: reetesh-ranjan@utc.edu

A. Panchal · S. Karpe · S. Menon (✉)
School of Aerospace Engineering, Georgia Institute of Technology, 270 Ferst Drive, Atlanta, GA 30332, USA
e-mail: suresh.menon@ae.gatech.edu

© The Author(s) 2023
N. Swaminathan and A. Parente (eds.), *Machine Learning and Its Application to Reacting Flows*, Lecture Notes in Energy 44, https://doi.org/10.1007/978-3-031-16248-0_7

highly nonlinear, unsteady, multi-scale, and multi-physics processes, which makes its investigation a challenging task. Although advancements in experimental diagnostics and computational tools have enabled some detailed studies, there are still challenges that need to be addressed. For example, while experiments under extreme operating conditions are often limited to measurements of fewer quantities, computational studies using high-fidelity approaches such as direct numerical simulation (DNS) and large-eddy simulation (LES) usually tend to be computationally expensive, and limited to a few simpler problems. Specifically, DNS, where all relevant spatial and temporal scales are resolved, is used to carry out fundamental studies, but it requires simplifications in the geometry, flow conditions, or chemistry to address the computational cost concerns. On the other hand, although LES, where only large-scales are captured and the effects of small-scales are parameterized using the subgrid-scale (SGS) closure models, is considered a promising strategy (Fureby and Möller 1995; Gonzalez-Juez et al. 2017; Pitsch 2006), to obtain statistical convergence, its computational cost is also not trivial. While subgrid-scale (SGS) closure for reacting LES remains an ongoing research effort for many approaches, the computational cost is a key challenge when employing finite-rate chemistry (FRC) approach with detailed chemical mechanisms. Here, we discuss past strategies to develop machine learning (ML) tools for LES of reacting flows, with a particular focus on finite-rate kinetics.

In recent years, rapid advancements in computing resources and data storage capabilities have led to increased usage of supervised deep learning (DL) using artificial neural network (ANN) (Goodfellow et al. 2016; LeCun et al. 2015) to tackle challenging problems from several fields such as computer vision (Krizhevsky et al. 2012), speech, image and text recognition (Bishop 2006), natural language processing (Collobert and Weston 2008), health-care (Leung et al. 2014), genetic sequencing (Libbrecht and Noble 2015), materials discovery (Pilania et al. 2013), complex game playing (Silver et al. 2017), high-energy physics (Baldi et al. 2014), etc. This is primarily due to the ability of the DL to effectively deal with high-dimensional data and the modeling of complex and nonlinear relationships. DL techniques are essentially representational learning methods that employ multiple levels of representation. These techniques transform the representation at one level starting with the raw input to an abstract representation at a higher level, which allows learning complex nonlinear relationships. The layers of features are learned from huge datasets using general-purpose learning procedures. Such a representational learning approach enables the discovery of intricate structures in high-dimensional data and is therefore amenable to different domains of science and engineering. Furthermore, the recent advancements in the back-propagation algorithm, mini-batch stochastic gradient, novel architectures such as convolutional neural network (CNN), and recurrent neural network (RNN) have also accelerated a wider adoption of DL techniques in different domains of science and engineering (LeCun et al. 2015).

To apply this approach to LES of reacting flows, the data-driven modeling through DL must focus on performance improvements via generalizing a model that captures all variations within the data. A conventional deep neural network (DNN) for modeling of the reaction-rate term is shown in Fig. 1, which is a multilayer fully connected feed-forward network where the information flows in a forward direction from input

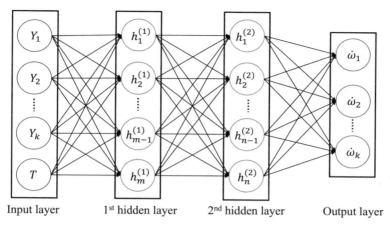

Fig. 1 Schematic of a multi-layer perceptron (MLP) for modeling of the reaction-rate term with two hidden layers having the vector $x = (Y_1, Y_2, \ldots, Y_k, T)$ as an input and the vector $y = (\dot{\omega}_1, \dot{\omega}_2, \ldots, \dot{\omega}_k)$ as an output

to output. Here, the input comprise of species mass fraction (Y_i with $i = 1, 2, \ldots, k$) and temperature (T), and the output comprise of the corresponding reaction-rate term ($\dot{\omega}_i$). Here, k denotes the total number of chemical species. Mathematically, a DNN defines the mapping $\mathcal{A} : x \rightarrow y$, where x and y denote input and output variables, respectively, and $\mathcal{A}$ represents a composition of many different functions, which can be represented through a network structure. A typical DNN comprises an input layer, an output layer, and more than one hidden layer. Each layer consists of several nodes, which are connected to all the nodes in the previous and the following layers. The complexity of a DNN increases with an increase in the number of hidden layers and the number of nodes per hidden layer. Such a basic network is also referred to as a multilayer perceptron (MLP). It has been shown that MLPs can yield universal function approximations (Hornik et al. 1989). Therefore, with enough layers and nodes, MLPs can be used to model arbitrarily complex and highly nonlinear functional forms, such as those needed for closure of the SGS terms while performing LES.

ANN algorithms have been used for SGS closure models in the context of Reynolds-averaged Navier-Stokes (RANS) and LES in past studies of both non-reacting (Beck et al. 2019; Duraisamy et al. 2015, 2019; Ling et al. 2016; Maulik and San 2017; Vollant et al. 2017) and reacting (Christo et al. 1995, 1996; Lapeyre et al. 2019; Seltz et al. 2019; Sen et al. 2010; Yellapantula et al. 2020) flows. In the context of LES of turbulent combustion, there are two key areas of relevance (a) the need to use detailed chemical kinetics for an accurate representation of the thermochemical state space, and (b) the modeling of the filtered reaction-rate term to account for the SGS turbulence-chemistry interaction. Over several years, past studies have focused on tackling both of these challenges, and further research is still underway.

To address the challenge related to thermochemical representation, detailed chemical kinetics can be used for accurate predictions over a wide range of operating conditions, In contrast, while the use of simplified chemical mechanisms is computationally expedient, they are known to affect the quality of predictions (Bilger et al. 2005). For several reacting flow conditions, the use of flamelet (Peters 2000; Pitsch 2006) and other low-dimensional manifold based approaches (Maas and Pope 1992; Bradley et al. 1988; Van Oijen and De Goey 2000) have been popular for their computational tractability. ANN has also been used to store flamelet libraries to reduce the computational storage requirements (Kempf et al. 2005; Ihme et al. 2009; Zhang et al. 2020). Additionally, it has also been used to model SGS source and transport terms (Seltz et al. 2019). Although low-dimensional manifold formulation can be used for some problems, often detailed finite-rate chemical mechanism is needed to accurately capture the flame dynamics and other features such as extinction, re-ignition, lean blowout, pollutant emissions, etc. However, FRC-based LES, referred here onwards as FRC-LES, becomes computationally intractable for simulation of practical applications when using a detailed chemical mechanism. The higher computational cost of FRC-LES is associated with the need to solve a highly stiff ODE system resulting from a wide range of time scales associated with different chemical species in a complex chemical mechanism, and the need to transport a large number of chemical species. In addition to approaches for computational cost reduction such as hybrid transported-tabulated chemistry (HTTC) (Ribert et al. 2014) and dynamic adaptive chemistry (DAC) (Yang et al. 2017) to name a few, ANN algorithms have also been used to address the computational cost concerns of FRC-LES (Christo et al. 1995; Christo et al. 1996; Sen et al. 2010; Sen and Menon 2010; Zhou et al. 2013; Franke et al. 2017; Sinaei and Tabejamaat 2017; Ranade et al. 2021).

A major challenge for LES of turbulent combustion is the need for accurate modeling of the filtered reaction-rate term. It has led to numerous physics-based SGS closure models for both low-dimensional manifold and FRC-based approaches. The reader is referred to the review articles (Pitsch 2006; Fureby 2009; Gonzalez-Juez et al. 2017), where challenges of different modeling paradigms and strengths and limitations of several modeling approaches are discussed. The modeling of the SGS turbulence-chemistry interaction is key for the accurate prediction of the flame dynamics. ANN-based strategies have been employed for computational cost reduction of filtered reaction-rate term modeling within both low-dimensional manifold (Nikolaou et al. 2019; Lapeyre et al. 2019; Seltz et al. 2019; Yellapantula et al. 2020) and FRC (Sen and Menon 2010; Zhou et al. 2013; Franke et al. 2017; Sen et al. 2010) formulations.

Although ANN algorithms have shown some success in LES of turbulent combustion, further studies are needed to examine the predictive capabilities and robustness of such algorithms. The focus of this chapter is to discuss the application of ANN while employing one specific subgrid model using the linear eddy mixing (LEM) model in LES (referred to as LEMLES) (Menon et al. 1993; Menon and Kerstein 2011). LEMLES is a two-scale strategy, where the species transport equations are solved using a two-step procedure. In the first step, the species transport equations and FRC mechanism are solved at the subgrid level using the 1D LEM model (Kerstein 1989), where the LEM model acts as an embedded SGS model for the species equation as viewed on the LES space- and time-scales. The second step simulates the evolution of the computed subgrid scalar fields at the resolved LES level. LEMLES has been extensively used in past studies for investigation of a wide range of applications, such as gas turbine combustor (Kim et al. 1999), rocket combustor (Srinivasan et al. 2015), spray combustion (Sankaran and Menon 2002), scramjet (Menon and Jou 1991), etc. Although LEMLES allows for the handling of arbitrarily complex chemical mechanisms, its use has so far been limited to moderately complex chemical mechanisms due to the cost associated with the computation of stiff kinetics. ANN algorithm within the framework of LEMLES allows addressing this issue (Sen et al. 2010; Sen and Menon 2010), which is the main focus of this chapter.

The chapter is organized as follows. An overview of ML strategies for modeling turbulent combustion reported in the literature is presented in Sect. 2. The formulation and application of ANN within LEMLES are discussed in Sects. 3 and 4. Section 5 discusses the limitations of the past studies that employed ANN within LEMLES. Section 6 concludes with a discussion of the future of ML for subgrid modeling of turbulent combustion using LEM and their implications.

2   ML for Modeling of Turbulent Combustion

As stated in Sect. 1, ML algorithms have been used to reduce the computational cost of finite rate chemistry while using different chemistry modeling paradigms (low-dimensional manifold or FRC). So, first, a brief overview of ANN-based modeling strategy for chemistry and the constituents of ANN models are discussed. Afterward, a summary of studies focused on the use of ANN in LES of turbulent combustion is presented.

2.1   ANN Model for Chemistry

While using the FRC approach, the reaction rate terms are obtained by solving a system of first-order ordinary differential equations (ODEs) expressed as:

$$\frac{dY_k}{dt} = \mathcal{F}_k(Y_k, T, P) = \dot{\omega}_k, \qquad k = 1, 2, \ldots N_s, \tag{1}$$

where Y_k and $\dot{\omega}_k$ denote the mass fraction and the reaction rate for the kth species. Here, $\dot{\omega}_k$ can be obtained for a prescribed chemical mechanism and associated kinetic parameters, along with temperature T and pressure P. The system of ODEs given by Eq. 1 is in general stiff, particularly for detailed chemical mechanisms, due to a wide range of timescales associated with different chemical species. Therefore, to solve Eq. 1, stiff ODE solvers such as the fully implicit double-precision variable-coefficient ODE solver (DVODE) (Brown et al. 1989) are needed, which tend to be expensive. ANN can be used to approximate the ODEs with nonlinear regression, thus addressing the issue of computational cost.

ANN regression can be obtained through a MLP (Bishop 1995; Haykin and Network 2004), which involves a sum of nonlinear basis functions, also referred to as activation functions, and coefficients, which include biases and weights. A typical MLP with inputs (Y_k, T) and outputs $(\dot{\omega}_k)$ is shown in Fig. 1. ANN extracts the complex relations embedded within a given input/output training dataset through a learning procedure, and the extracted complex relations can later be used to predict the states on which the training was not performed. The learning process essentially adjusts the biases and weights for each layer of the MLP to obtain a minimal error at the output layer by using a back-propagation algorithm. These optimal weights and biases, along with the specific MLP configuration, form the ANN model. The resulting ANN model can then be used for an efficient representation of the complex dynamics of chemistry described by Eq. 1.

A typical ANN model includes parameters, hyperparameters, and training strategies. The parameters, such as the model coefficients are updated by the ANN model during the learning process, and they only require initialization. The hyperparameters such as the components of the network architecture are specified for a particular problem, which varies from one problem to the next. These include the number of hidden layers and neurons, learning rate, momentum during the back-propagation algorithm, activation function, epochs, mini-batch size, and dropout. A brief overview of the hyperparameters and training strategies is discussed next.

The two key hyperparameters are the number of hidden layers and the number of neurons, which are needed for an accurate representation of complex nonlinear input/output relationships. Although increasing them, in general, improves the accuracy, it also makes the network heavy and eventually the accuracy tends to stagnate. The activation function is through which weighted sums are passed to obtain a nonlinear output. The specification of the activation function determines the efficiency and accuracy of the ANN model. Some of the commonly used activation functions include hyperbolic tangent (tanh), rectified linear unit (ReLU), sigmoid, etc.

When dealing with big data, it is also inefficient to use the entire data for training. Therefore, batches of small-size data are typically used for efficient training, although care needs to be taken to avoid overfitting, which would face difficulties in fitting to any new data. The epochs denote the number of times the algorithm trains on the entire data, and its value is also closely associated with the accuracy of the model.

The strategies that are commonly specified while obtaining the ANN model include initialization of the parameters, data normalization, optimization algorithm, and regularization. The initialization of the parameters can be performed based on the chosen activation functions and it affects the efficiency of the ANN model. In several applications, the input data has different scales, which can affect the rate of convergence during the training of the ANN model. For example, in combustion, inputs comprise of temperature and mass fraction of species, which differs by several orders of magnitude, therefore, normalization becomes imperative for improved performance. The optimizers are algorithms used during the training to reduce the loss function, which in turn is used to update the weights. It can directly affect the convergence of the model during the training stage. Some commonly used optimizers include Adam optimizer, gradient descent, stochastic gradient descent, etc. The loss function needs to be defined during the training to compute the model error. The regularization strategy is useful to avoid the overfitting of the ANN model.

It is apparent that a robust ANN model requires a careful selection of parameters, hyperparameters, and training strategies. This becomes even more challenging for turbulent combustion, which is marked by multi-scale and highly nonlinear processes with multiple regimes and modes of combustion where complex relationships between variables representing the thermochemical space exist. Therefore, usually, a significant amount of tuning is needed to realize a robust ANN model for a particular turbulent combustion application.

2.2 LES of Turbulent Combustion Using ANN

An overview of past studies focused on the use of ANN while performing LES of turbulent combustion is summarized in Table 1. The table includes some well-established turbulent combustion models that are used with either a low-dimensional manifold or a finite-rate representation for chemistry. The FRC models include LEM-LES and transported probability density function (TPDF) approaches and the low-dimensional manifold approaches include flamelet and flame surface density (FSD) approaches. It can be observed that the ANN-based strategy has been used to study canonical as well as realistic flow configurations. In addition, both premixed and non-premixed modes of combustion have been examined. This illustrates a wide range of applicability of the use of ANN for LES of turbulent combustion.

Some key details of the ANN models employed by the past studies are also summarized in Table 1 to identify if there are any commonly used constituents of the ANN model. These constituents are labeled as 'T', 'O', 'f', and 'L' corresponding to the type of training datasets, the optimization algorithm, the activation function, and the loss function, respectively. As discussed in Sect. 2.1, these are some of the key parameters describing the ANN model.

In general, the training of the ANN model has been performed using different types of datasets such as one-dimensional (1D) laminar flamelet, 1D LEM, and DNS datasets. There are advantages and limitations of the usage of these types of datasets. For example, training solely based on a 1D laminar flamelet can not account for the effects of turbulence-chemistry interactions. While this is partly addressed in training based on the 1D LEM dataset, some key features of turbulent combustion such as large-scale curvature effects are not accounted for. The DNS datasets account for all possible states for a particular test configuration and appear to be better compared to the other two approaches. However, it has limited predictive capabilities for conditions that were not present in the DNS dataset and is computationally prohibitive.

The activation function for a neuron in the ANN model defines the output of that neuron for a given input set. Similar to other fields where ANN has been used, all

Table 1 Summary of contributions to application of ML in modeling of turbulent combustion. The ANN model components are labeled as, T: Training data, O: Optimization Algorithm, f: Activation function, L: Loss function

Method	Configuration	ANN model	Key results
LEMLES	Non-premixed syngas flame (Sen and Menon 2010; Sen et al. 2010)	T: 1D LEM, O: SGD f: tanh, L: MSE	Captured extinction/re-ignition; 5.5 times faster than DI; testing of stiff kinetics needed
	Non-premixed methane flame (Sinaei and Tabejamaat 2017)	T: 1D LEM, O: WH f: tanh, L: MSE	4.9 times faster than DI; Compared ANN-LES, DI-LES, and LUT-LES
TPDF	DLR-A methane flame (Ranade et al. 2021)	T: Flamelets, O: SGD f: tanh, L: MSE	MLP-SOP based ANN; 3 times faster and 10^3 times reduced memory requirements compared to MLP based ANN
Flamelet	Premixed methane flame (Seltz et al. 2019)	T: DNS, O: SGD f: ReLU, L: CE	CNN for subgrid source and transport terms
	Non-premixed methane flame (Kempf et al. 2005)	T: Flamelets, O: - f: –, L: –	Steady flamelet modeled with ANN 10^3 times lower memory
	Non-premixed Sydney flame (Ihme et al. 2009)	T: Flamelets, O: LM f: sigmoid, L: RMS	Network performance with respect to accuracy, data retrieval time, and storage requirements
FSD	Premixed methane flame (Lapeyre et al. 2019)	T: DNS, O: SGD f: ReLU, L: MSE	CNN based subgrid flame wrinkling factor Robust for realistic configurations
	Premixed methane flame (Ren et al. 2021)	T: DNS, O: SGD f: ReLU, L: MSE	CNN based subgrid flame wrinkling factor; robust for realistic configurations

three widely popular activation functions, namely, tanh, ReLU, and sigmoid functions (Karlik and Olgac 2011; Nwankpa et al. 2018) have been used while performing LES of turbulent combustion. For the optimizer, the stochastic gradient descent (SGD) algorithm has been typically used. However, some studies have also used Widrow-Hoff (WH) and Levenberg–Marquardt (LM) algorithms. Finally, mean-squared error (MSE) has been used commonly for the loss function in these studies.

Most of the studies summarized in Table 1 demonstrate an improved performance in terms of speedup of chemistry computation as compared to a conventional direct integration (DI) approach for handling stiff kinetics (other studies may exist and hence, this list is not considered comprehensive). In addition, these studies have also demonstrated the benefits of the use of ANN in terms of reduced computational storage requirements. Some recent studies relying on the use of CNN (Lapeyre et al. 2019; Ren et al. 2021) have shown the robustness of the approach for accurately simulating realistic flow configurations where the performance of the CNN based subgrid model was shown to be better compared to reference algebraic closures. Overall, the past and recent studies clearly demonstrate the potential of ANN-based modeling of turbulent combustion. However, further studies are also needed to identify the best practices in specifying the hyperparameters and the strategies for attaining a successful and accurate ANN model.

3   Mathematical Formulation with ANN

In this section, the mathematical formulation of LEMLES with the use of ANN for the modeling of chemistry is discussed. First, the governing equations for FRC-LES and the subgrid modeling of the scalar fields using LEM are described. Afterward, two approaches using ANN, either to model the resolved reaction rates at the subgrid level or to directly model the filtered reaction rates including the subgrid effects are discussed.

3.1   Governing Equations and Subgrid Models

3.1.1   Large-Eddy Simulation

The LES equations are obtained through Favre filtering of compressible multi-species Navier-Stokes equations, which lead to the following conservation equations for mass, momentum, energy, and species mass

$$\frac{\partial \overline{\rho}}{\partial t} + \frac{\partial \overline{\rho}\widetilde{u}_i}{\partial x_i} = 0, \tag{2}$$

$$\frac{\partial \overline{\rho}\widetilde{u}_i}{\partial t} + \frac{\partial}{\partial x_j}\left[\overline{\rho}\widetilde{u}_i\widetilde{u}_j + \overline{P}\delta_{ij} - \overline{\tau}_{ij} + \tau_{ij}^{\mathrm{sgs}}\right] = 0, \tag{3}$$

$$\frac{\partial \overline{\rho}\widetilde{E}}{\partial t} + \frac{\partial}{\partial x_i}\left[\left(\overline{\rho}\widetilde{E} + \overline{P}\right)\widetilde{u}_i + \overline{q}_i - \widetilde{u}_j\overline{\tau}_{ij} + H_i^{\mathrm{sgs}} + \sigma_i^{\mathrm{sgs}}\right] = 0, \tag{4}$$

$$\frac{\partial \overline{\rho}\widetilde{Y}_k}{\partial t} + \frac{\partial}{\partial x_i}\left[\overline{\rho}\left(\widetilde{Y}_k\widetilde{u}_i + \widetilde{Y}_k\widetilde{V}_{i,k}\right) + \mathcal{Y}_{i,k}^{\mathrm{sgs}} + \theta_{i,k}^{\mathrm{sgs}}\right] = \overline{\dot{\omega}}_k \quad k = 1, ..., N_s. \quad (5)$$

Here, $\overline{f}$ denotes a spatially filtered quantity corresponding to the variable f, and $\widetilde{f}$ is a Favre-filtered quantity, which is defined as: $\widetilde{f} = \overline{\rho f}/\overline{\rho}$. In the above equations, ρ is the density, u_i is the velocity vector, P represents the pressure, E is the total energy per unit mass, Y_k is the mass fraction of the kth species, and N_s is the total number of species. In addition, τ_{ij} is the viscous stress tensor, q_i is the heat flux vector, and $V_{i,k}$, and $\dot{\omega}_k$ are species diffusion velocity vector and the reaction-rate for the kth species, respectively. The terms with superscript 'sgs' are unclosed terms resulting from the filtering operation, which require additional closure models.

The total energy per unit mass in Eq. 4, E, is defined as the sum of the internal energy per unit mass (e) and the kinetic energy per unit mass. The corresponding Favre-filtered total energy per unit mass, i.e., $\widetilde{E}$, is given as the sum of $\widetilde{e}$, the resolved kinetic energy per unit mass $(\widetilde{u}_i\widetilde{u}_i)/2$, and the SGS kinetic energy per unit mass $k^{\mathrm{sgs}} = (\widetilde{u_i u_i} - \widetilde{u}_i\widetilde{u}_i)/2$.

The above system of conservation equations is closed by using an equation of state through: $\overline{P} = \overline{\rho}\left(\widetilde{R}\widetilde{T} + T^{\mathrm{sgs}}\right)$, and the filtered enthalpy per unit mass, which is defined as: $\widetilde{h} = \left(\Sigma_{k=1}^{N_s}\widetilde{Y}_k\widetilde{h}_k + E_k^{\mathrm{sgs}}\right) + T^{\mathrm{sgs}}$. Here, $\widetilde{h}_k$ is the specific enthalpy of the kth species, $\widetilde{R}$ is the mixture gas constant and T^{sgs} is an unclosed term resulting from the filtering of the equation of state.

The filtered viscous stress tensor, $\overline{\tau}_{ij}$, and the filtered heat-flux vector, $\overline{q}_i$, are given by

$$\overline{\tau}_{ij} = 2\overline{\mu S_{ij}} - \frac{2}{3}\overline{\mu S_{kk}}\delta_{ij} \approx 2\overline{\mu}\left(\widetilde{S}_{ij} - \frac{1}{3}\widetilde{S}_{kk}\delta_{ij}\right), \quad (6)$$

$$\overline{q}_i = -\overline{\kappa\frac{\partial T}{\partial x_i}} + \overline{\rho}\sum_{k=1}^{N_s}\widetilde{h}_k\widetilde{Y}_k\widetilde{V}_{i,k} + \sum_{k=1}^{N_s}q_{i,k}^{\mathrm{sgs}} \approx -\overline{\kappa}\frac{\partial\widetilde{T}}{\partial x_i} + \overline{\rho}\sum_{k=1}^{N_s}\widetilde{h}_k\widetilde{Y}_k\widetilde{V}_{i,k} + \sum_{k=1}^{N_s}q_{i,k}^{\mathrm{sgs}}, \quad (7)$$

where $\widetilde{S}_{ij}$ is the resolved strain-rate tensor, and $\overline{\mu}$ and $\overline{\kappa}$ are filtered viscosity and thermal diffusivity, respectively, which are approximated using the resolved quantities.

The SGS terms appearing in the above equations require further modeling. These terms are given as: $\tau_{ij}^{\mathrm{sgs}} = \overline{\rho}\left(\widetilde{u_i u_j} - \widetilde{u}_i\widetilde{u}_j\right), H_i^{\mathrm{sgs}} = \overline{\rho}\left(\widetilde{Eu_i} - \widetilde{E}\widetilde{u}_i\right) + \left(\overline{u_i P} - \widetilde{u}_i\overline{P}\right),$ $\sigma_i^{\mathrm{sgs}} = \left(\overline{u_j\tau_{ij}} - \widetilde{u}_j\overline{\tau}_{ij}\right), \mathcal{Y}_{i,k}^{\mathrm{sgs}} = \overline{\rho}\left(\widetilde{u_i Y_k} - \widetilde{u}_i\widetilde{Y}_k\right), \theta_{i,k}^{\mathrm{sgs}} = \overline{\rho}\left(\widetilde{V_{i,k}Y_k} - \widetilde{V}_{i,k}\widetilde{Y}_k\right), q_{i,k}^{\mathrm{sgs}} = \overline{\rho}\left(\widetilde{h_k Y_k V_{i,k}} - \widetilde{h}_k\widetilde{Y}_k\widetilde{V}_{i,k}\right), T^{\mathrm{sgs}} = \widetilde{RT} - \widetilde{R}\widetilde{T},$ and $E_k^{\mathrm{sgs}} = \widetilde{Y_k e_k(T)} - \widetilde{Y}_k e_k(\widetilde{T})$, which result from the application of filtering operation to the non-linear terms. In the expressions for $\theta_{i,k}^{\mathrm{sgs}}, q_{i,k}^{\mathrm{sgs}}$ and E_k^{sgs} here, the repeated index k does not imply summation. Further details about these terms, their physical relevance and terms that are typically neglected in LES studies are discussed elsewhere (Fureby and Möller 1995; Ranjan et al. 2016).

In the context of reacting flows, $\mathcal{Y}_{i,k}^{\text{sgs}}, \theta_{i,k}^{\text{sgs}}, q_{i,k}^{\text{sgs}}, T^{\text{sgs}}, E_k^{\text{sgs}}$ and $\overline{\dot{\omega}}_k$ require closure models. Typically, $q_{i,k}^{\text{sgs}}, T^{\text{sgs}}, \theta_{i,k}^{\text{sgs}}$, and E_k^{sgs} are neglected in LES (Fureby and Möller 1995), and therefore, these terms are neglected here as well. The modeling of SGS scalar flux $\mathcal{Y}_{k,i}^{\text{sgs}}$ and filtered reaction rate $\overline{\dot{\omega}}_k$, is the key focus here, and they are discussed further in the following sections.

3.1.2 Subgrid Modeling Using LEM

The linear eddy mixing (LEM) model (Kerstein 1989) is a stochastic approach to model the effects of 3D turbulent mixing in a 1D domain. It was originally a stand-alone model to account for the interactions between turbulence, molecular diffusion, and reaction kinetics. In LES, the unsteady species and temperature evolution equations are solved on a 1D subdomain embedded inside each of the LES cells, where the reaction and the diffusion processes are locally resolved, but the effects of 3D (assumed isotropic) turbulence are included via randomized stirring events. The governing equations for 1D LEM are given by

$$\rho \frac{\partial Y_k}{\partial t} = F_{k,\text{stir}} - \frac{\partial}{\partial s}\left(\rho Y_k V_{s,k}\right) + \dot{\omega}_k, \tag{8}$$

$$\rho C_{p,\text{mix}} \frac{\partial T}{\partial t} = F_{T,\text{stir}} + \frac{\partial}{\partial s}\left(\kappa \frac{\partial T}{\partial s}\right) - \frac{\partial}{\partial s}\left(\sum_{k=1}^{N_S} h_k \rho Y_k V_{s,k}\right) - \sum_{k=1}^{N_S} h_k \dot{\omega}_k, \tag{9}$$

where 's' represents the co-ordinate along the 1D LEM domain. The terms $F_{k,\text{stir}}$ and $F_{T,\text{stir}}$ represent stirring events in the above equations. The turbulent stirring is implemented as stochastic events (based on the so-called triplet maps (Kerstein 1989) that attempts to mimic the effect of vortices on the scalar field. Successive folding and compressive motions are modeled during these events, with its time/length-scale governed by the nature of turbulence. This also allows for capturing a thickened reaction zone at high turbulence intensity, as the stirring time-scales get smaller, and small-sized eddies can disturb the reactive/diffusive flame structure.

The 1D LEM domain is notionally aligned in the flame-normal direction as shown in Fig. 2a. The LEM has also been coupled with LES for subgrid closure of the terms discussed in the previous section, wherein, the 1D LEM domain is embedded within each LES cell, as shown in Fig. 2b. Two approaches, linear eddy mixing model with large eddy simulation (LEMLES) (Menon and Kerstein 2011), and reaction-rate closure for large eddy simulation (RRLES) (Ranjan et al. 2016; Panchal et al. 2019) have been used in the past, and they are briefly summarized below.

The LEMLES approach models the species evolution equation, i.e., Eq. 5 with unclosed terms $\mathcal{Y}_{i,k}^{\text{sgs}}$ and $\overline{\dot{\omega}}_k$ altogether. The species mass fractions are not evolved on the LES grid, but rather only on the 1D LEM domains embedded within the 3D LES computational cells. Since the flame is resolved on the 1D domain, the grid resolution

Fig. 2 Sketch of the 1D
LEM embedded along the
flame for its standalone
application (**a**), or within
the LES cells for
LEMLES/RRLES (**b**)

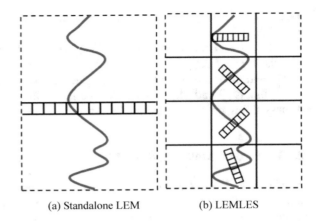

(a) Standalone LEM   (b) LEMLES

can be chosen to be fine enough to resolve the reaction and the diffusive terms, thus eliminating the need for any further closures. However, closures are needed for the subgrid turbulent mixing and the large-scale convection. While the subgrid mixing is modeled through turbulent mixing, the large-scale transport is modeled using a Lagrangian transport through the splicing algorithm (Menon and Kerstein 2011). With this approach chunks of 1D LEM domain (with Y and T) along the direction of convection across the LES cells are transported.

LEMLES has been successfully used in the past for a wide variety of problems, including premixed (Sankaran and Menon 2005), non-premixed (Sen et al. 2010; Srinivasan et al. 2015) and spray (Sankaran and Menon 2002; Patel and Menon (2008)) flames over a range of conditions. However, there are certain disadvantages of the LEMLES approach. A key limitation is that the reduction to a 1D notional dimension limits its ability in cases where the flame has to propagate in 3D as opposed to fluctuate around a statistically mean direction. At high Re, the turbulent diffusion usually dominates the molecular diffusion, which is captured by the 1D LEM model, but errors are incurred at low Re, or towards the DNS limit, where molecular diffusion, which is neglected on the large-scale, dominates.

Considering these drawbacks, the RRLES approach (Ranjan et al. 2016; Panchal et al. 2019) is a recent modification of the LEMLES approach, where only the filtered reaction-rate terms $\bar{\dot{\omega}}_k$ are modeled using a multi-scale LEM framework. Here, filtered species equations Eq. 5 are still solved using a 3D grid where a conventional gradient-diffusion closure is used for $\mathcal{Y}_{i,k}^{\text{sgs}}$, whereas, the filtered reaction rate term $\bar{\dot{\omega}}_k$ is modeled using LEM. At every time step of the evolution of the LES equations in 3D, the filtered species mass fractions ($\widetilde{Y}_k$) and the filtered temperature ($\widetilde{T}$) evolving at the resolved level are used to reconstruct SGS variation on the 1D notional LEM domain inside each LES cell, and after solving for the subgrid reaction-diffusion equation and including the effect of turbulent mixing on the LEM domain, the filtered reaction rates are computed and projected back to the 3D LES grid. The RRLES approach has an advantage over the original LEMLES approach, particularly in a well-resolved or a locally laminar condition, where it can asymptote to the DNS limit. However, this approach cannot account for counter-gradient transport of scalars, and sensitivity of results to the reconstruction procedure is another uncertainty (Ranjan et al. 2016).

3.2 ANN Based Modeling

As discussed in Sects. 1 and 2, ANNs can be considered as highly non-linear regression models, and they are used here to model the reaction rate terms $\dot{\omega}_k$ and $\overline{\dot{\omega}}_k$ described in the previous section.

3.2.1 Problem Definition: Resolved Reaction Rates

The conventional FRC allows for the inclusion of arbitrarily complex chemical kinetic mechanisms, that can range from $O(10)$ to $O(100)$ species and reactions. The individual reaction rates are computed using Arrhenius rate expressions, and these computations can get expensive with an increasing number of species/reactions. Even with reduced chemical kinetics, a stiff direct integration (DI) solver such as DVODE may have to be used, which can result in a significant computational cost, ranging 60-90% of the total computational cost of a simulation (Sen et al. 2010). As discussed in Sects. 1 and 2, a solution to this could be to tabulate these source terms over a range of conditions, instead of DI of them at each simulation step. However, this table would become very large and highly multi-dimensional as it would have $N_s + 1$ input (Y_k, T) and N_s ($\dot{\omega}_k$) output variables. Therefore, instead of tabulation, the ANN model denoted by $\mathcal{A}_k$ for the kth species is employed for estimating the reaction rates as:

$$\dot{\omega}_k = \mathcal{A}_k(Y_1, Y_2, \ldots, Y_{N_s}, T), \quad \text{for} \quad k = 1, 2, \ldots, N_s. \tag{10}$$

Considering a range of time scales associated with different chemical species, separate multi-input, and single-output MLP are used for each species. Each neuron in the ANN model $\mathcal{A}_k$ contains weights and biases, and their training is discussed in the next sections. The capabilities of the ANN model have been assessed using three chemical mechanisms in the past studies (Sen et al. 2010; Sen and Menon 2010; Sen and Menon 2009). These include, (A) 11-steps-14-species Syngas/air (Sen et al. 2010) skeletal mechanism for premixed flames, (B) 12-steps-16-species methane/air skeletal mechanism (Sung et al. 1998) for premixed flames, and (C) 21-steps-11-species Syngas (Hawkes et al. 2007) mechanism for non-premixed flame. Note that independent ANN model and training datasets are required for each chemical kinetics.

3.2.2 Training Algorithm

The training of ANN model comprise of two stages, which include, a forward propagation of the input, and a backward propagation the error. The output of a single neuron i at iteration number k is calculated as

$$y_i[k] = f\left(\sum_{m=0}^{M} W_{im}[k]y_m[k] - b_i[k]\right). \tag{11}$$

Here, $W_{im}[k]$ is the weight coefficient between neurons i and m, $y_m[k]$ is the output of the neuron m, $b_i[k]$ is the bias of the neuron i, and M is the number of neurons feeding into the neuron i. As described in Sect. 2.1, there are several options for specifying the activation function $f(\cdot)$. All the results presented in this chapter use the hyperbolic-tangent (tanh) as the activation function.

To perform tuning of the model weights and biases during the training of the ANN model, mean squared error (E) of the network are typically minimized using a gradient descent rule (GDR), i.e.,

$$W_{im}[k + 1] = W_{im}[k] - \eta \frac{\partial E[k]}{\partial W_{im}[k]}, \tag{12}$$

where k is the GDR iteration step. Standard GDR may not be able to deal with error surfaces that have local minima where it could get trapped, and therefore, a momentum modification is used as

$$W_{im}[k + 1] = W_{im}[k] - \eta \frac{\partial E[k]}{\partial W_{im}[k]} - \alpha \frac{\partial E[k]}{\partial W_{im}[k - 1]}.$$

Here, η and α are the model hyperparameters, global learning rate and momentum coefficient, respectively. Since, these model hyperparameters need to be calibrated for each new case for optimum convergence, otherwise, a modification similar to extended delta-bar-delta (EDBD) (Minai and Williams 1990) learning model has to be used. In the current approach, each neuron has their own model parameters (η_{im}, α_{im}), and they are updated at every ANN iteration based on the history of the global error as:

$$\eta_{im}[k + 1] = \eta_{im}[k] + \Delta\eta_{im}[k], \tag{13}$$

$$\Delta\eta_{im}[k] = \begin{cases} \kappa_1\lambda\eta_{im}, & \text{if } \phi_{im}[k]\overline{\phi}_{im}[k - 1] > 0 \\ -\kappa_1\lambda\eta_{im}, & \text{if } \phi_{im}[k]\overline{\phi}_{im}[k - 1] < 0 \\ 0, & \text{if } \phi_{im}[k]\overline{\phi}_{im}[k - 1] = 0. \end{cases} \tag{14}$$

Here, $\lambda = (1 - \exp(-\kappa_2\phi_{im}[k]))$, $\phi_{im}[k] = \partial E[k]/\partial W_{im}[k]$, and $\overline{\phi}_{im}[k] = (1 - \theta)$ $\phi_{im}[k - 1] + \theta\phi_{im}[k]$. Furthermore, κ_1 and κ_2 are second-order model-coefficients, which are specified to be 0.1 and 0.01, respectively, based on numerical experiments.

Some salient features of this training approach are as follows:

- Each connection has its learning coefficients.
- Changes to the model coefficients are performed based on the value of the local error gradients ($\phi_{im}[k]$ and $\phi_{im}[k-1]$), where the updates are enhanced in the regions of huge error gradient, and reduced near a minimum.
- Instead of training using the mini-batch approach, the updates are done after introducing the whole training set to establish a correlation between $\phi_{im}[k]$ and $\phi_{im}[k-1]$.
- In case the weights start to increase without bounds, the coefficients are reverted to a previously saved state.

Further details about this approach can be found elsewhere (Sen and Menon 2010, Sen and Menon 2009), however, application of more advanced approaches developed in the ML community, e.g. Adam optimizer algorithm (Kingma and Ba 2014), needs to be evaluated in the future to the problems considered here.

3.2.3 Training Dataset

For the ANN model to be able to accurately model the reaction rates $\dot{\omega}_k$, the training set has to cover a range of conditions, i.e., Y_k and T that would be encountered during the 3D simulations. Since the training set has to be generated using DI, the cost of its generation is another concern. For example, even though a DNS of the 3D application problem can generate all the states accessed during the simulation, it is not computationally feasible to do so for training, thus requiring alternate approaches. The results presented here consider the following three methods for obtaining the training dataset:

- **FANN**: The training set is generated using the tables extracted from a 2D flame-vortex interaction (FVI) simulation (Poinsot et al. 1991; Sen and Menon 2009). A premixed flame is initialized corresponding to the inflow equivalence ratio and temperature, and a coherent vortex diameter(D_C) is chosen to be of the same as the integral length scale L_F of the 3D application. Since turbulence is a superposition of multiple vortices, the maximum velocity induced by the vortex $U_{C,max}$ is varied in the range $10 < U_{C,max}/S_L < 400$, where S_L is the laminar premixed flame speed. Six cases have been considered within this range, and training samples are obtained from multiple snapshots.
- **PANN**: The training set for PANN is generated using tables obtained from 1D laminar premixed flame simulations (Sen et al. 2010). The inflow equivalence ratio and temperature are specified based on premixed flame operating conditions. A limitation of this approach is that no information about the turbulence is embedded within the training dataset.
- **LANN**: Here, the training set is generated using standalone 1D LEM simulations (Sen et al. 2010). As LEM is supposed to emulate the effects of turbulence on a flame, therefore, the resulting training dataset accounts for some effects of turbulence. This approach can be used for either premixed or non-premixed flames. The

laminar flame is initialized on the 1D LEM domain, and the reaction, diffusion, stirring equations are solved as described earlier. For premixed cases, the initial profile is a function of the equivalence ratio (ER) and inflow temperature, and for the non-premixed cases, it is also a function of the strain rate. In this approach, turbulent Reynolds number Re_t can be varied, which for the cases considered here has been varied from 10 to 180 (with 20 values in between) for LEM, and the integral length scale L corresponds to the specific 3D application.

The above strategies are computationally cheaper compared to the dataset generation using 3D simulations. The three approaches have different levels of fidelity in terms of embedding the effects of subgrid turbulence-chemistry interactions in the training datasets. For example, while PANN completely ignores the subgrid turbulence-chemistry interactions, LANN accounts for it albeit in form of stochastic stirring events. Alternate strategies need to be examined further to have an increased fidelity of the training dataset that can be generated in an efficient manner. These strategies will also need to incorporate the effects of other input variables such as pressure (and possibly heat loss) to enable applications to practical configurations.

3.2.4   Structure of ANN

Given the training dataset and the algorithm, the next step is to choose the ANN structure, e.g. number of neurons, hidden layers, etc., and normalization of input/output. A typical training dataset considered here contains approximately 5 million states. The database is first divided into nine equidistant temperature bins, and at least 100,000 data points are added to each bin to achieve proper sensitivity to temperature in reaction rate calculations. A typical flame solution would have a large number of points in the reactants and the products but not so many within the flame region, and this ensures that the ANN is not biased. The inputs and the outputs to the ANN are then normalized between ± 1 and ± 0.8, respectively, to increase the sensitivity to each parameter and remove any bias towards species with higher mass fractions. An 85/15 training/testing split has been considered to realize the ANN model. The training is stopped if there is no improvement in consecutive iterations to avoid overfitting.

The ANN can have multiple hidden layers, however, a smaller network would struggle with predicting complex reaction rate manifolds, whereas a larger network would result in a larger number of connections and a higher computational cost. To understand this, multiple ANN structures have been considered, and a few representative networks for the chemical mechanism C are summarized in Table 2. The corresponding computational speedups, with respect to DI, are plotted in Fig. 3. A significant slowdown occurs beyond 500 connections, and the ANN is even slower than DI beyond 20,000 connections. Considering this, and the testing errors in Table 2, 5/3/2 is selected as the optimal network for this particular kinetics, and it results in a 5 times speedup with testing errors below 10^{-4}. The optimal networks for mechanisms A and B are 10/5 and 10/8/4, respectively, and they result in 11 and 35 times speedup as compared to the corresponding DI. The larger speedup in mechanism B results

Table 2 Number of connections and testing errors corresponding to different ANN architectures. The table is reproduced using the data from Sen and Menon (2010)

Number of neuron in hidden layers	$\log_{10}$ (Testing error)	Connections
5	−3.521	230
10	−3.801	340
5/4	−3.889	358
10/5	−3.920	500
20/5	−4.114	770
5/3/2	−4.201	371
20/10/5	−4.870	1240

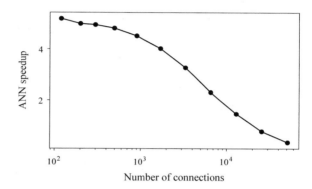

Fig. 3 Speedup of ANN against DI with various number of connections. The figure is reproduced using the digitized data from Sen and Menon (2010)

from its stiffness. The number of training samples was always specified more than 10 times the number of neurons to avoid overfitting.

Note that the errors discussed in this section are testing errors based on the dataset that was selected for training, and not the actual errors as they would result in a 3D application. These errors can occur when thermochemical states, which are accessed by the ANN model were not available in the training dataset. Further details about these errors are discussed later.

3.2.5 Modeling Filtered Reaction Rates

Prediction of $\dot{\omega}_k$ using ANN was discussed in the previous section, and these can be used instead of DI, either for a direct numerical simulation (DNS) or with the LEMLES/RRLES approach but within the LEM domain where a turbulence closure is not required for the reaction rates. Solution of LEM within each LES cell could still be costly for problems of practical interest, and therefore, a modified LES approach, referred to as TANN, where the filtered reaction rates $\overline{\dot{\omega}}_k$ are directly computed using ANN was developed (Sen 2009). This approach has similarities with the RRLES approach, for instance, subgrid species diffusion $\mathcal{Y}_{i,k}^{\text{sgs}}$ is computed using a gradient-diffusion approach, however, instead of using the LEM solver online within each cell as the simulation progresses, the filtered reaction rates are trained beforehand. The filtered reaction rates for the kth species are modeled using the ANN model $\mathcal{B}_k$ through

$$\bar{\dot{\omega}}_k = \mathcal{B}_k \left(\widetilde{Y}_1, \widetilde{Y}_2, \ldots, \widetilde{Y}_{N_s}, \widetilde{T}, Re_\Delta, \frac{\partial \widetilde{Y}_1}{\partial x}, \frac{\partial \widetilde{Y}_2}{\partial x}, \ldots, \frac{\partial \widetilde{Y}_{N_s}}{\partial x} \right). \qquad (15)$$

Here, Re_Δ corresponds to the subgrid Reynolds number $u'\Delta/\nu$, where Δ is the LES filter size, and $u' = \sqrt{2k^{sgs}/3}$. Previously described methods for ANN training and selection of optimal architecture have also been used with this approach. The ANN training database for TANN is constructed using standalone LEM solutions. Initializing with species and temperature profiles corresponding to laminar flames, a range of Re_t and L are explored corresponding to the conditions for the 3D application. The obtained 1D LEM solutions at multiple time instances are then filtered with size Δ and they are then used for ANN training.

Since, the velocity field is not available from standalone LEM, Re_Δ cannot be computed from u' or k^{sgs}. For this, an additional equation for kinetic energy $k(s)$ is solved on the LEM domain as

$$\frac{\partial k}{\partial t} = P_k - \epsilon,$$

where P_k and ϵ are turbulence production and dissipation rates, respectively. A local velocity disturbance field $u^{LEM} = \nu Re_t/L$ is computed on the segment where stirring is applied, and this is used as $P_k = 3/2(u^{LEM})^2/\Delta t$ and $\epsilon = (u^{LEM})^3/\Delta s$ to compute the production and the dissipation terms, respectively. Here, Δt and Δs are the time and space discretizations for the LEM domain, and this follows the assumption that the turbulence that is modeled by LEM is homogeneous. The evolved k over the entire domain is then filtered to compute k^{sgs} and Re_Δ.

4 Example Applications

In this section, results from the application of different types of ANN-based modeling strategies discussed in Sect. 3 are described for four test canonical configurations. These cases correspond to different modes (premixed and non-premixed) of combustion and demonstrate the application to configurations with an increasing degree of geometric complexity. The first test case is a canonical premixed flame-turbulence-vortex interaction configuration where the results are compared for LEM-LES between DI, LANN, PANN, and FANN. The second test case corresponds to a non-premixed temporally evolving jet flame that exhibits the presence of extinction and re-ignition dynamics in the presence of turbulence, and the results using LANN-LEMLES and TANN-LES are compared against available DNS data. The third test considers a stagnation point reversed flow (SPRF) premixed combustor with LANN-LEMLES and TANN-LES, and finally, the results from a cavity strut supersonic combustor obtained using TANN-LES are discussed. The third and the fourth tests illustrate application to practical configurations for which the results are compared against the available experimental data.

4.1 Premixed Flame Turbulence

The test configuration follows a previous work (Sen et al. 2010) for premixed flame-turbulence-vortex interaction for syngas/air flame. The reacting flow field is initialized using a 1D laminar steady solution for premixed flame, and a counter-rotating vortex pair is superimposed on the isotropic turbulence to induce small- and large-scale wrinkling. The chemical mechanism A is used for this test configuration and four different test conditions are considered, which include two equivalence ratios, and two values of u'/S_L. Here, u' and S_L denote turbulence intensity and laminar flame speed, respectively. The ratio of integral length scale to the laminar flame thickness $L/L_F = 5$ is selected so that the flame remains in the thin reaction zone regime. The maximum induced velocity by the vortex is chosen as $U_{C,max}/S_L = 50$. A 64^3 uniform grid is used with $\Delta/\eta = 4$, where η is the Kolmogorov length scale. The subgrid 1D LEM domain is spatially discretized using 24 cells. A 10/5 ANN model is used for this case. The use of ANN for chemistry modeling while performing LEMLES resulted in approximately $11\times$ speedup as compared to DI of the chemical kinetics.

The results for the case with ER = 0.6 and $u'/S_L = 5$ are shown in Fig. 4 at $t^* = L/U_{C,max} = 5$. For the sake of brevity, only spatially averaged profiles of a major species H_2 and two intermediate species, namely H and O are shown here, but the other species also showed a qualitatively similar trend. The model PANN shows the highest error with respect to DI even for the major species H_2, where it shows an early consumption of the fuel, which can be associated with a faster consumption speed, and the errors for PANN are even higher for the minor species.

The results with the other two models, namely, FANN and LANN are comparable to DI for this particular test case, suggesting that both the flame-vortex interaction and the standalone LEM are capable of covering a range of thermochemical states that are encountered during the 3D flame-turbulence interactions. The same conclusions are obtained for the other values of ER and u'/S_L as well. These results demonstrated both the accuracy and the efficiency of the ANN-based modeling approach for chemistry. Furthermore, the results also highlight the importance of the employed training datasets on attaining accurate results.

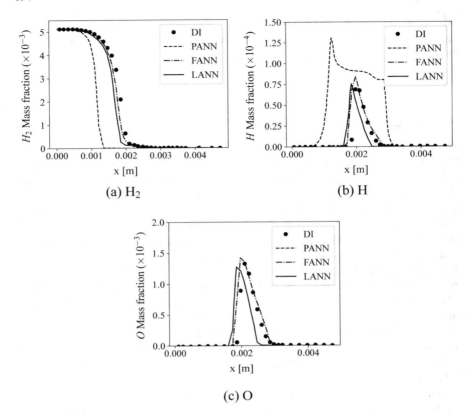

Fig. 4 Comparison of LES results for premixed flame-turbulence-vortex interaction for syngas/air at an instance for ER = 0.6 and $u'/S_L = 5$. The figures are reproduced using the digitized data from Sen and Menon (2010)

4.2 Non-premixed Temporally Evolving Jet Flame

This computational setup follows a DNS study of turbulent non-premixed syngas/air combustion in a temporally evolving jet (Hawkes et al. 2007; Sen et al. 2010). An inner fuel jet and an outer oxidizer jet flow in opposite directions, with the jet Reynolds number of $Re_{jet} = 4478$, and a Damköhler number of $Da = 0.011$. While, DNS was performed using 350 million grid points, for LES, 5.5 million ($\Delta/\eta = 8.3$) cells are used. The 1D LEM domain is discretized using 12 cells. For this test case, the chemical mechanism C has been considered. Here, the results from LANN-LEMLES and TANN-LES are discussed. In terms of the computational cost, LANN-LEMLES provided a 5.5 times speedup compared to DI-LEMLES, whereas, TANN-LES provided 18.3 times speedup, showing a significant computational gain.

The time variation of mean temperature at stoichiometric mixture fraction is shown in Fig. 5. The temperature is expected to be the maximum on the stoichiometric surface for a non-premixed flame. The initially stable non-premixed flame approaches extinction as a result of the shear-generated background turbulence, and the temperature decreases from an initial 1450 K to 1100 K at a non-dimensional time $t_j = 20$ in DNS. After this time instant, the temperature starts increasing again as a result of the re-ignition process, and finally reaches up to 1300 K at $t_j = 40$, close to its initial value. These global features are captured by both LANN-LEMLES and TANN-LES, with 5-10% error near extinction.

The contours of mass-fraction of OH species in the central $x - y$ plane are shown in Fig. 6 at time instances $t_j = 20$ and $t_j = 40$ obtained from DNS and LANN-LEMLES cases. The OH mass fraction from DNS peaks along with the shear layers, showing a broken structure due to local extinctions at $t_j = 20$, but this is followed by re-ignition at $t_j = 40$ within these pockets. Qualitatively, the features observed in the DNS case are also captured in the LANN-LEMLES case.

Mass-fractions and temperature statistics in the compositional space were also analyzed for a quantitative comparison of the flame structure by different models. The variation of OH mass fraction is shown in Fig. 7 at $t_j = 20$ and $t_j = 40$. Results with all, DNS, LANN-LEMLES and TANN-LES drop below the laminar flamelet value at extinction at $t_j = 20$, and shoot back up above it at $t_j = 40$ confirming re-ignition. Both LANN-LEMLES and TANN-LES are able to predict this behavior and match the DNS data with reasonable accuracy, with TANN-LES providing a slightly better match, particularly during the extinction phase.

Overall, the results presented here demonstrate the robustness of the ANN-based modeling of chemistry. This test case is particularly challenging because of the presence of the unsteady dynamics of turbulence-chemistry interaction, which is marked by the presence of extinction and reignition events.

Fig. 5 Evolution of mean temperature at stoichiometric mixture fraction for non-premixed extinction re-ignition test using DNS, LANN-LEMLES and TANN-LES. The figure is reproduced using the digitized data from Sen et al. (2010) and Sen (2009)

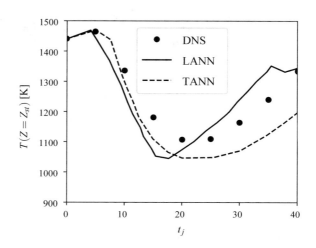

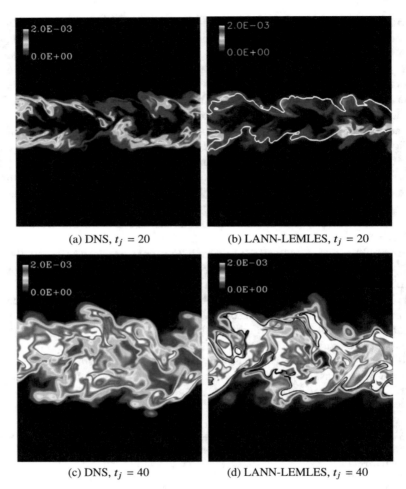

(a) DNS, $t_j = 20$ (b) LANN-LEMLES, $t_j = 20$

(c) DNS, $t_j = 40$ (d) LANN-LEMLES, $t_j = 40$

Fig. 6 Contours of OH mass fraction in the central plane at $t_j = 20$ and $t_j = 40$ obtained from DNS (**a**, **c**) and LANN-LEMLES (**b**, **d**) cases for the temporally evolving non-premixed jet configuration. The figures are borrowed from Sen et al. (2010)

4.3 SPRF Combustor

The stagnation point reversed flow (SPRF) combustor (see Fig. 8) was designed to reduce emissions (Gopalakrishnan et al. 2007; Undapalli et al. 2009). It was simulated in a premixed mode configuration for evaluating the capabilities of the LANN-LEMLES and the TANN-LES approaches (Sen 2009). Methane/air mixture is injected into the combustor at an equivalence ratio of 0.58. The flow enters and leaves the combustion chamber in the same plane, providing extensive preheating and allowing the flame to stabilize at very lean conditions. The combustion chamber

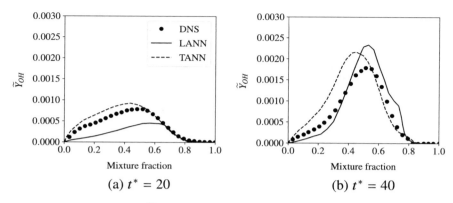

Fig. 7 Conditional average of $\widetilde{Y}_{OH}$ at $t^* = 20$ and $t^* = 40$ for non-premixed extinction re-ignition test. The symbols have the same meaning as Fig. 5. The figures are reproduced using the digitized data from Sen et al. (2010) and Sen (2009)

marked as region (5) has a wall (6) at the end. Surface (2) is closed and (3) injects the premixed mixture, with (4) as the outflow. The annular jet bulk flow velocity is 122 m/s, and it is preheated to 500 K, with $Re = 12900$. The computational domain is spatially discretized using approximately 1.2 million cells. The methane/air mechanism B is used for this test configuration. For the ANN model, Re_t varying from 10 to 400, and the integral length scale L as the radius of the whole injector assembly ($L = 8.25\,mm$) are considered. In terms of computational cost, LANN-LEMLES and TANN-LES showed 49.2 and 134.9 times speedup, respectively, as compared to DI-LEMLES for this test configuration.

The simulation results using DI-LEMLES, LANN-LEMLES and TANN-LES were time-averaged over two flow-through times and compared against experimental data along the centerline as shown in Figs. 9 and 10. Both LANN-LEMLES and

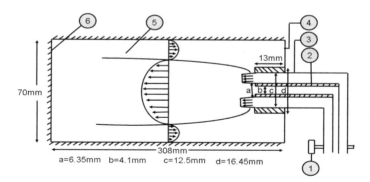

Fig. 8 Schematic of the stagnation point reversed flow combustor. This figure is borrowed from Undapalli et al. (2009)

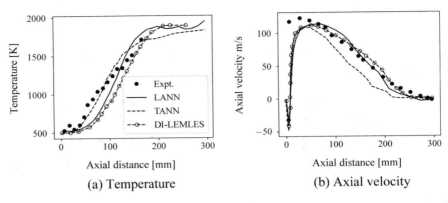

Fig. 9 Axial variations of time-averaged temperature and axial velocity for the SPRF combustor. This figures are reproduced using the digitized data from Sen (2009)

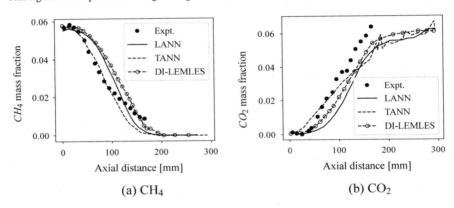

Fig. 10 Axial variations of time-averaged mass fraction of CH_4 and CO_2 for the SPRF combustor. This figures are reproduced using the digitized data from Sen (2009)

DI-LEMLES are able to capture the far-field axial velocity variation accurately. The differences near the injector could be due to differences in the boundary conditions as discussed elsewhere Sen (2009). The same holds true for temperature, CH_4, and CO_2 centerline variations, the results show approximately 10% errors with respect to the experiments, but both LANN-LEMLES and DI-LEMLES show similar results.

The centerline time-averaged variations for axial velocity are worse for TANN-LES as compared to LANN-LEMLES, whereas they are better for temperature, CH_4, and CO_2 with respect to the experiments. It was hypothesized that this could be due to differences between the use of LEM in LEMLES and TANN-LES, where, the eddy-sizes are restricted between η and Δ in the former, whereas they are between η and L in the latter, that could result in a higher wrinkling of the flame front and increased turbulence within the combustor.

The training of the ANN model using the 1D LEM dataset and subsequent use of the model while performing LES of a practical configuration again demonstrates efficiency, robustness, and generality aspects of the approach. The observed differences from the reference results, particularly with the TANN-LES need further studies so that the accuracy of the approach can be enhanced further. Some of these studies are currently underway.

4.4 Cavity Strut Flame-Holder for Supersonic Combustion

Now, the results from LES of a cavity-based flame-holder are discussed (Ghodke et al. 2011). Two configurations, as shown in Fig. 11, were considered; baseline cavity with 11 injectors on the aft ramp (no strut), and a strut positioned upstream of the cavity with 6 fuel injectors (with strut). The cavity extends 153 mm in the spanwise direction, with 90° leading edge and 22.5° ramp at the trailing edge. The cavity is 16.5 mm deep with $L/D = 2.79$, and the length of the cavity floor is 46 mm. The injected fuel mixture contains 70% methane and 30% hydrogen, whereas the mainstream contains air and water vapor at a Mach number of 2.

The computational grids for both configurations contained approximately 10 million cells, with clustering in the near-wall regions, shear layers, and near the fuel injectors. A reduced four-step methane-hydrogen kinetics was used (Peters and Kee 1987) for the simulations. The ANN model for TANN-LES was trained using the previously described method, and a 10/8/4 hidden layer structure was found to be optimal. Simulations were performed for a duration of 6 flow-through times, and the results are compared between experiments, DI-LEMLES and TANN-LES. Compared to DI-LEMLES, TANN-LES was around 50 times faster for both the no-strut and strut configurations.

Figure 12 shows instantaneous temperature field contours on a plane that is normal to the spanwise direction for both the configurations. Most of the cavity region is filled with hot products, which causes lifting of shear layer for oxidizer entrainment into the cavity. The reaction zone is even larger for the configuration with the strut due to an increased mass and heat transfer between the cavity and the main stream,

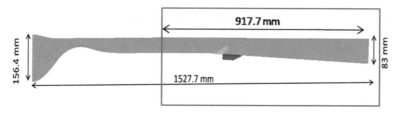

Fig. 11 Schematic of a supersonic cavity-strut flame-holder. The figure is borrowed from Ghodke et al. (2011)

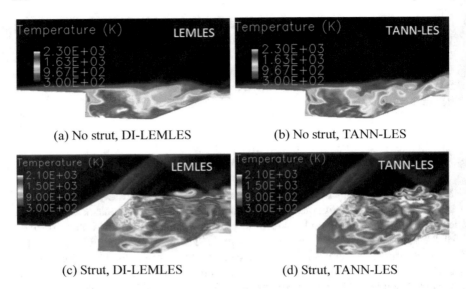

(a) No strut, DI-LEMLES   (b) No strut, TANN-LES

(c) Strut, DI-LEMLES   (d) Strut, TANN-LES

Fig. 12 Temperature contours on a center-slice at an instant for the supersonic cavity strut flame-holder. The figures are borrowed from Ghodke et al. (2011)

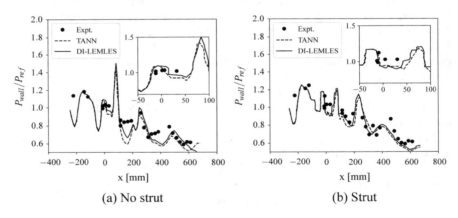

(a) No strut   (b) Strut

Fig. 13 Bottom wall pressure comparison against experimental data (Grady et al. 2010) for the supersonic cavity strut flame-holder. The strut extends from $x = -36$ mm to $x = 25$ mm, and the cavity extends from $x = 0$ mm to $x = 86$ mm. The figures are reproduced using the digitized data from Ghodke et al. (2011)

as a result of the low-pressure region behind the strut. Vortical structures behind the strut are responsible for better mixing of fuel and maintaining hot regions inside combustor by mass transfer which helps flame-holding.

Figure 13 shows the wall pressure comparison for reacting cases with available experimental data (Grady et al. 2010). For both cases, location of leading-edge shock ($x \sim -30$ mm and $x \sim 0$ mm for configurations with and without strut, respectively) and ramp expansion ($x \sim 85$ mm) are captured well, along with multiple reflections off the wall. The pressure inside the cavity is almost constant, and hence, this could be considered as a constant pressure combustion process. The peak pressures along the wall as predicted by both DI-LEMLES and TANN-LES are also in good agreement with reference experimental data, thus illustrating that the heat releases effects are accurately captured.

The use of an ANN-based strategy for modeling subgrid turbulence-chemistry interactions in this test configuration demonstrates the robustness of such an approach. This could be attributed to the efficacy of ANN to accurately represent multi-dimensional data in form of a nonlinear regression, which in turn, can account for complex input/output relations as prevalent in this particular test case where turbulence-chemistry interactions occur under supersonic flow conditions in a complex geometrical configuration. Although the approach employed here is able to capture the trends both qualitatively and quantitatively, some discrepancies with the experimental data can also be seen, which needs further investigation.

5 Limitations of Past Studies

The results discussed here used ANNs to directly represent the chemical kinetics at the subgrid level. Even though the results demonstrated various aspects of the ANN-based modeling approach for efficient computations of chemically reacting flows while utilizing FRC, there are certain challenges that need further studies. Some of the key features of ANN-based modeling that were demonstrated include a significant decrease in the computational cost and memory requirements, robustness in application to different modes and regimes of combustion, predictive ability in terms of decoupling the training dataset from the actual application, etc. Some limitations and concerns of the current work are highlighted next in order to stimulate future research:

- **Stiff kinetics with complex mechanism**: The majority of configurations in the current work explored mechanisms comprising of 11–16 species with varying levels of stiffness. The results discussed provided consistently acceptable predictions for all cases. However, scenarios relevant to practical combustion applications involve detailed kinetics with an order of 50–100 species and an order of 1000 reactions. Hence, the scalability of the current approach to such a scenario needs to be investigated further with specific attention to predictions of minor species, stiff radicals, and their interactions with turbulence.
- **ANN architecture optimization**: Even though a significant portion of the current work is focused on devising an optimal neural network for a general case, the sensitivity of the ANN to hyperparameters such as the number layers, size of the training data generated, effects of using different optimization approaches, activation functions as well as error estimation techniques need to be further exam-

ined, especially with help of open-source well-established powerful tools such as TensorFlow (Martín et al. 2015) or PyTorch (Paszke et al. 2019).

- **Training data generation**: For TANN, the filtered training data is generated by using the filter width based on the information of the actual LES grid. For canonical problems, this information is easily available as the computational grid involved is almost uniform throughout the domain. However, in complex configurations, since the computational grid varies significantly due to clustering in specific areas of interest, the filter width definition needs to be revised. Moreover, the ANN model trained on the table generated using the standalone LEM computations still suffers from the assumption made during the LEM approach used for table generation. Therefore, some assumptions involved in standalone LEM computations regarding turbulence homogeneity and isotropy, LEM solution initialization using a 1D laminar flame solution, stirring operations, etc. need to be revisited for further improvements.
- **Off line training**: The ANN training approach adopted here is based on offline training philosophy. The training dataset generated using 1D LEM, flame-vortex interaction, or 1D laminar flame are used to construct the thermochemical database, and once the ANN model is trained on this, it is used as-is in LES without any further learning. Therefore, it is expected that it may contribute to large errors at some state spaces that are far from training data. The alternate method of combining offline and online training, where ANN needs to be retrained on such states, can be adopted. A similar strategy has been employed in a recent DNS study (Chi et al. 2021).
- **Subgrid modeling of scalar fluxes**: In the TANN approach, a gradient diffusion-based eddy diffusivity model is used for the closure of the SGS scalar flux. However, in chemically reacting flows both gradient and counter-gradient subgrid turbulent transport of the scalar fields are observed (Ranjan et al. 2016). Therefore, the predictive capabilities of the TANN-LES strategy can be further improved by using ANN-based modeling of the SGS scalar flux.

6 Summary and Outlook

Rapid advancements in computational resources have led to an increased usage of ML tools, particularly supervised DL to solve challenging problems in the field of science and engineering. DL techniques relying on the ANN is a representational learning method, which transforms the representation at one level starting with the raw input to an abstract representation at a higher level, which allows learning of complex nonlinear relationships and enables the discovery of intricate structures in a high-dimensional dataset. In this chapter, different approaches relying on ANN algorithms for efficient modeling of the chemistry. Within the FRC framework have been discussed for LES of turbulent combustion.

The two major challenges associated with FRC-LES include a robust SGS closure for turbulence-chemistry interaction and efficient handling of stiffness associated with the use of detailed chemical kinetics. In the LEMLES approach, a two-scale strategy is used; LEM is used for the subgrid modeling of the reaction, diffusion, and turbulent mixing, and large-scale transport is handled in a Lagrangian manner. The approach has been demonstrated in the past for simulations of a wide variety of canonical and practical configurations. As it allows for the inclusion of arbitrarily complex chemical kinetics and resolves the flame in the 1D LEM domain, ANN-based models have been examined in terms of their ability to efficiently model the reaction-rate terms. Apart from LEMLES, a conventional LES approach has also been discussed where instead of modeling the reaction-rate term at the subgrid level as in LEMLES, a model for the filtered-reaction-rate term is devised based on ANN.

A key step in ANN-based modeling is the training database, which was generated using three approaches, namely, laminar flame solutions, flame vortex interactions (FVI), and flame turbulence interactions (FTI) using standalone 1D LEM computations. In all three approaches, the thermochemical state space is predicted using canonical configurations and with only the knowledge of large-scale parameters of the actual geometry of interest. The ANN model trained using these three approaches showed the effectiveness of FTI (LANN) and FVI (FANN) approaches over laminar flame solutions (PANN) for training data generation and predicting the behavior of canonical as well as complex (premixed and non-premixed modes) reacting flow configurations. The TANN approach utilizes a tabulation model for the filtered reaction rates, which does not employ any explicit assumption regarding the interaction of turbulence with the laminar flame front, but solves them directly on their respective time and length scales using standalone LEM computations. The ANN models considered in the example applications were based on a back-propagation algorithm with adaptive gradient descent rule (AGDR), and tanh activation function with a simple architecture using a maximum of 3 hidden layers, one input, and one output layer. Furthermore, during the learning stage of the ANN model, the training was stopped when saturation in training error was observed to ascertain the ANN generality and avoid problems of data memorizations.

The performance of ANN-based modeling strategies was examined in terms of their accuracy, robustness, and efficiency using four test cases with an increasing degree of complexity. These cases included canonical turbulent premixed and non-premixed flames where reference DNS results were used to assess the capabilities of different modeling approaches. The robustness of the use of the ANN model for FRC was demonstrated through two practical configurations corresponding to a premixed combustor and a supersonic cavity flame holder. These cases were simulated using three different chemical mechanisms. Overall, ANN-based modeling of chemistry with the LEMLES and TANN-LES framework was able to capture qualitative features of flame-turbulence interactions, and their quantitative statistics were in good agreement with direct integration approaches for chemistry. However, some discrepancies were also noted in the results, which needs further investigation for potential improvement to the employed modeling strategies.

A major challenge with modeling of chemistry using ANN is an accurate representation of detailed chemical mechanisms over a wide range of operating conditions, which usually have a higher level of stiffness due to the wide separation of timescales associated with different chemical species. So far, while using ANN with LEM only moderately complex chemical mechanisms have been considered, which need to be extended to detailed chemical mechanisms. While modeling FRC using ANN, a multi-input and single-output ANN model is needed for each chemical species, which also poses a challenging task for the training process to attain an optimal architecture. To obtain an optimal ANN model, parameters, hyperparameters, and training strategies need to be specified. While some of the hyperparameters have demonstrated their applicability to different types of problems, further usage and assessment of ANN algorithms for turbulent combustion modeling can potentially lead to some common parameters that may work for a wide range of applications.

Another key challenge for ANN-based predictive modeling is the efficient generation of reliable training data. The data generation procedure should be general enough so that it can be used with different types of geometrical configurations, and different modes and regimes of combustion. Furthermore, the procedure should be efficient to enable a faster generation of training data for a range of input conditions that can cover a large thermochemical state space. To this end, 1D LEM-based training seems to be a good strategy, however, further improvements are needed. Some improvements that should be considered are: accounting for the effects of pressure, the use of different types of energy spectra in the LEM equations, considering a range of LES filter sizes, etc. In addition, an adaptive training approach (Chi et al. 2021) can also be considered by employing a cost function associated with the accuracy and efficiency of the ANN model.

The ANN model for reaction rate discussed in this chapter relied on a different network for each species. However, reaction-rate for the species are related to each other through the constraint of conservation of mass. This aspect is not addressed in the formulation considered here, and therefore, can be considered in future studies by following the approach used by the physics-informed neural network (Raissi et al. 2019). Although turbulent combustion modeling in the context of LES has mainly focused on robust and accurate modeling of the filtered reaction-rate term, ML tools can also be used for modeling the other unclosed terms such as SGS scalar flux, temperature, equation of state, etc. Such constraints and improvements by the use of ML tools can yield improved predictions, particularly under extreme conditions when large variations in thermochemical state space can occur, and therefore, should be considered in future studies.

Acknowledgements The results reported here from the Computational Combustion Laboratory (CCL), Georgia Institute of Technology have been funded in part by NASA/GRC, AFOSR, and AFRL (Eglin AFB). Computational resources provided by NASA Advanced Supercomputing (NAS), DOD HPC and CCL (http://pace.gatech.edu) are greatly appreciated. The first author (R. Ranjan) would like to acknowledge the support of the CECASE grant from the University of Tennessee at Chattanooga.

References

Baldi P, Sadowski P, Whiteson D (2014) Searching for exotic particles in high-energy physics with deep learning. Nat Commun 5(1):1–9

Beck A, Flad D, Munz C-D (2019) Deep neural networks for data-driven LES closure models. J Comput Phys 398:108910

Bilger RW, Pope SB, Bray KNC, Driscoll JF (2005) Paradigms in turbulent combustion research. Proc Combust Inst 30(1):21–42

Bishop CM (2006) Pattern recognition. Machine Learning 128(9)

Bishop CM et al (1995) Neural networks for pattern recognition. Oxford University Press

Bradley D, Kwa LK, Lau AKC, Missaghi M, Chin SB (1988) Laminar flamelet modeling of recirculating premixed methane and propane-air combustion. Combust Flame 71(2):109–122

Brown PN, Byrne GD, Hindmarsh AC (1989) VODE: a variable-coefficient ode solver. SIAM J Sci Stat Comput 10(5):1038–1051

Chi C, Janiga G, Thévenin D (2021) On-the-fly artificial neural network for chemical kinetics in direct numerical simulations of premixed combustion. Combust Flame 226:467–477

Christo FC, Masri AR, Nebot EM, Pope SB (1996) An integrated PDF/neural network approach for simulating turbulent reacting systems. In: Symposium (International) combustion, vol 26. Elsevier, pp 43–48

Christo FC, Masri AR, Nebot EM, Turányi T (1995) Utilising artificial neural network and repromodelling in turbulent combustion. In: Proceedings of ICNN'95-international conference on neural networks, vol 2. IEEE, pp 911–916

Collobert R, Weston J (2008) A unified architecture for natural language processing: Deep neural networks with multitask learning. In: Proceedings of 25th international conference on machine learning, pp 160–167

Duraisamy K, Iaccarino G, Xiao H (2019) Turbulence modeling in the age of data. Annu Rev Fluid Mech 51:357–377

Duraisamy K, Zhang ZJ, Singh AP (2015) New approaches in turbulence and transition modeling using data-driven techniques. In: 53rd AIAA aerospace sciences meeting, p 1284

Franke LLC, Chatzopoulos AK, Rigopoulos S (2017) Tabulation of combustion chemistry via artificial neural networks (ANNs): methodology and application to LES-PDF simulation of sydney flame l. Combust Flame 185:245–260

Fureby C (2009) Large eddy simulation modelling of combustion for propulsion applications. Philos Trans R Soc A 367(1899):2957–2969

Fureby C, Möller S-I (1995) Large eddy simulation of reacting flows applied to bluff body stabilized flames. AIAA J 33(12):2339–2347

Ghodke C, Choi J, Srinivasan S, Menon S (2011) Large eddy simulation of supersonic combustion in a cavity-strut flameholder. In 49th AIAA aerospace sciences meeting including the new horizons forum and aerospace exposition, pp AIAA–2011–0323

Gonzalez-Juez ED, Kerstein AR, Ranjan R, Menon S (2017) Advances and challenges in modeling high-speed turbulent combustion in propulsion systems. Prog Energy Comb Sci 60:26–67

Goodfellow I, Bengio Y, Courville A (2016) Deep learning. MIT Press

Gopalakrishnan P, Bobba MK, Seitzman JM (2007) Controlling mechanisms for low nox emissions in a non-premixed stagnation point reverse flow combustor. Proc Combust Inst 31(2):3401–3408

Grady N, Pitz RW, Carter C, Friedlander T, Hsu K-Y (2010) Hydroxyl tagging velocimetry in a supersonic flow over a piloted cavity. In: 48th AIAA aerospace sciences meeting including the new horizons forum and aerospace exposition, pp AIAA–2010–1405

Hawkes ER, Sankaran R, Sutherland JC, Chen JH (2007) Scalar mixing in direct numerical simulations of temporally evolving plane jet flames with skeletal CO/H_2 kinetics. Proc Combust Inst 31(1):1633–1640

Haykin S, Network N (2004) A comprehensive foundation. Neural Netw 2(2004):41

Hornik K, Stinchcombe M, White H (1989) Multilayer feedforward networks are universal approximators. Neural Netw 2(5):359–366

Ihme M, Schmitt C, Pitsch H (2009) Optimal artificial neural networks and tabulation methods for chemistry representation in LES of a bluff-body swirl-stabilized flame. Proc Combust Inst 32(1):1527–1535

Karlik B, Olgac AV (2011) Performance analysis of various activation functions in generalized MLP architectures of neural networks. I. J Artif Intell Expert Syst 1(4):111–122

Kempf A, Flemming F, Janicka J (2005) Investigation of lengthscales, scalar dissipation, and flame orientation in a piloted diffusion flame by LES. Proc Combust Inst 30(1):557–565

Kerstein AR (1989) Application to shear layer mixing Linear-eddy modeling of turbulent transport. II. Combust Flame 75:397–413

Kim W-W, Menon S, Mongia HC (1999) Large-eddy simulation of a gas turbine combustor flow. Combust Sci Tech 143(1–6):25–62

Kingma DP, Ba J(2014) Adam: a method for stochastic optimization. arXiv:1412.6980

Krizhevsky A, Sutskever I, Hinton GE (2012) Imagenet classification with deep convolutional neural networks. Adv Neural Inf Proc Syst 25:1097–1105

Lapeyre CJ, Misdariis A, Cazard N, Veynante D, Poinsot TJ (2019) Training convolutional neural networks to estimate turbulent sub-grid scale reaction rates. Combust Flame 203:255–264

LeCun Y, Bengio Y, Hinton G (2015) Deep learning. Nature 521(7553):436–444

Leung MKK, Xiong HY, Lee LJ, Frey BJ (2014) Deep learning of the tissue-regulated splicing code. Bioinformatics 30(12):I121–I129

Libbrecht MW, Noble WS (2015) Machine learning applications in genetics and genomics. Nat Rev Genetics 16(6):321–332

Ling J, Kurzawski A, Templeton J (2016) Reynolds averaged turbulence modelling using deep neural networks with embedded invariance. J Fluid Mech 807:155–166

Maas U, Pope SB (1992) Simplifying chemical kinetics: intrinsic low-dimensional manifolds in composition space. Combust Flame 88(3–4):239–264

Martín A, Agarwal A, Barham P, Brevdo E, Chen Z, Citro C, Corrado CS, Davis A, Dean J, Devin M, Ghemawat S, Goodfellow I, Harp A, Irving G, Isard M, Jia Y, Jozefowicz R, Kaiser L, Kudlur M, Levenberg J, Mané D, Monga R, Moore S, Murray D, Olah C, Schuster, Shlens J, Steiner B, Sutskever I, Talwar K, Tucker P, Vanhoucke V, Vasudevan V, Viégas F, Vinyals O, Warden P, Wattenberg M, Wicke M, Yu Y, Zheng X (2015) TensorFlow: large-scale machine learning on heterogeneous systems. Software available from tensorflow.org

Maulik R, San O (2017) A neural network approach for the blind deconvolution of turbulent flows. J Fluid Mech 831:151–181

Menon PA, McMurthy, Kerstein AR (1993) A linear eddy mixing model for large eddy simulation of turbulent combustion. In Galperin B, Orszag SA (eds) Large eddy simulation of complex engineering and Geological flows. Cambridge University Press, pp 87–314

Menon S, Jou W-H (1991) Large-eddy simulations of combustion instability in an axisymmetric ramjet combustor. Combust Sci Tech 75:53–72

Menon S, Kerstein AR (2011) The linear-eddy model. In: Turbulent combustion modeling. Springer, pp 221–247

Minai AA, Williams RD (1990) Back-propagation heuristics: a study of the extended delta-bar-delta algorithm. In: 1990 IJCNN international joint conference on neural networks. IEEE, pp 595–600

Nikolaou ZM, Chrysostomou C, Vervisch L, Cant S (2019) Progress variable variance and filtered rate modelling using convolutional neural networks and flamelet methods. Flow Turbul Combust 103(2):485–501

Nwankpa C, Ijomah W, Gachagan A, Marshall S (2018) Activation functions: comparison of trends in practice and research for deep learning. arXiv:1811.03378

Panchal A, Ranjan R, Menon S (2019) A comparison of finite-rate kinetics and flamelet-generated manifold using a multiscale modeling framework for turbulent premixed combustion. Combust Sci Tech 191(5–6):921–955

Paszke A, Gross S, Massa F, Lerer A, Bradbury J, Chanan G, Killeen T, Lin Z, Gimelshein N, Antiga L, Desmaison A, Kopf A, Yang E, DeVito Z, Raison M, Tejani A, Chilamkurthy S, Steiner B, Fang L, Bai J, Chintala S (2019) PyTorch: an imperative style, high-performance deep learning library. In Wallach H, Larochelle H, Beygelzimer A, dAlché Buc F, Fox E, Garnett R (eds) Advances in neural information processing systems, vol 32, pp 8024–8035. Curran Associates, Inc

Patel N, Menon S (2008) Simulation of spray turbulence flame interactions in a lean direct injection combustor. Combust Flame 153(7):228–257

Peters N (2000) Turbulent combustion. Cambridge University Press

Peters N, Kee RJ (1987) The computation of stretched laminar methane-air diffusion flames using a reduced four-step mechanism. Combust Flame 68(1):17–29

Pilania G, Wang C, Jiang X, Rajasekaran S, Ramprasad R (2013) Accelerating materials property predictions using machine learning. Sci Rep 3(1):1–6

Pitsch H (2006) Large-eddy simulation of turbulent combustion. Annu Rev Fluid Mech 38:453–482

Poinsot TJ, Veynante D, Candel SM (1991) Quenching processes and premixed turbulent combustion diagrams. J Fluid Mech 228:561–606

Raissi M, Perdikaris P, Karniadakis GE (2019) Physics-informed neural networks: a deep learning framework for solving forward and inverse problems involving nonlinear partial differential equations. J Comput Phys 378:686–707

Ranade R, Li G, Li S, Echekki T (2021) An efficient machine-learning approach for PDF tabulation in turbulent combustion closure. Combust Sci Tech 193(7):1258–1277

Ranjan R, Muralidharan B, Nagaoka Y, Menon S (2016) Subgrid-scale modeling of reaction-diffusion and scalar transport in turbulent premixed flames. Combust Sci Tech 188(9):1496–1537

Ren J, Wang H, Luo K, Fan J (2021) A priori assessment of convolutional neural network and algebraic models for flame surface density of high Karlovitz premixed flames. Phys Fluids 33(3):036111

Ribert G, Vervisch L, Domingo P, Niu Y-S (2014) Hybrid transported-tabulated strategy to downsize detailed chemistry for numerical simulation of premixed flames. Flow Turbulence Combust 92(1–2):175–200

Sankaran V, Menon S (2002) LES of spray combustion in swirling flows. J Turbulence 3(11):1–23

Sankaran V, Menon S (2005) Subgrid combustion modeling of 3-D premixed flames in the thin-reaction-zone regime. Proc Combust Inst 30:575–582

Seltz A, Domingo P, Vervisch L, Nikolaou ZM (2019) Direct mapping from LES resolved scales to filtered-flame generated manifolds using convolutional neural networks. Combust Flame 210:71–82

Sen BA (2009) Artificial neural networks based subgrid chemistry model for turbulent reactive flow simulations. PhD thesis, Georgia Institute of Technology

Sen BA, Menon S (2009) Turbulent premixed flame modeling using artificial neural networks based chemical kinetics. Proc Combust Inst 32(1):1605–1611

Sen BA, Menon S (2010) Linear eddy mixing based tabulation and artificial neural networks for large eddy simulations of turbulent flames. Combust Flame 157(1):62–74

Sen BA, Hawkes ER, Menon S (2010) Large eddy simulation of extinction and reignition with artificial neural networks based chemical kinetics. Combust Flame 157(3):566–578

Silver D, Schrittwieser J, Simonyan K, Antonoglou I, Huang A, Guez A, Hubert T, Baker L, Lai M, Bolton A, Chen Y (2017) Mastering the game of go without human knowledge. Nature 550(7676):354–359

Sinaei P, Tabejamaat S (2017) Large eddy simulation of methane diffusion jet flame with representation of chemical kinetics using artificial neural network. Proc Inst Mech Eng, Part E: J Proc Mech Eng 231(2):147–163

Srinivasan S, Ranjan R, Menon S (2015) Flame dynamics during combustion instability in a high-pressure, shear-coaxial injector combustor. Flow Turbulence Combust. 94(1):237–262

Sung CJ, Law CK, Chen J-Y (1998) An augmented reduced mechanism for methane oxidation with comprehensive global parametric validation. In: Symposium (International) combustion, vol 27, pp 295–304

Undapalli S, Srinivasan S, Menon S (2009) LES of premixed and non-premixed combustion in a stagnation point reverse flow combustor. Proc Combust Inst 32:1537–1544

Van Oijen JA, De Goey LPH (2000) Modelling of premixed laminar flames using flamelet-generated manifolds. Combust Sci Tech 161:113–137

Vollant A, Balarac G, Corre C (2017) Subgrid-scale scalar flux modelling based on optimal estimation theory and machine-learning procedures. J Turbulence 18(9):854–878

Yang S, Ranjan R, Yang V, Menon S, Sun W (2017) Parallel on-the-fly adaptive kinetics in direct numerical simulation of turbulent premixed flame. Proc Combust Inst 36(2):2025–2032

Yellapantula S, de Frahan MTH, King R, Day M, Grout R (2020) Machine learning of combustion LES models from reacting direct numerical simulation. In: Data analysis for direct numerical simulations of turbulent combustion. Springer, pp 273–292

Zhang Y, Xu S, Zhong S, Bai X-S, Wang H, Yao M (2020) Large eddy simulation of spray combustion using flamelet generated manifolds combined with artificial neural networks. Energy AI 2:100021

Zhou ZJ, Lü Y, Wang ZH, Xu YW, Zhou JH, Cen K (2013) Systematic method of applying ANN for chemical kinetics reduction in turbulent premixed combustion modeling. Chinese Sci Bull 58(4):486–492

On the Use of Machine Learning for Subgrid Scale Filtered Density Function Modelling in Large Eddy Simulations of Combustion Systems

S. Iavarone, H. Yang, Z. Li, Z. X. Chen, and N. Swaminathan

Abstract The application of machine learning algorithms to model subgrid-scale filtered density functions (FDFs), required to estimate filtered reaction rates for Large Eddy Simulation (LES) of chemically reacting flows, is discussed in this chapter. Three test cases, i.e., a low-swirl premixed methane-air flame, a MILD combustion of methane-air mixtures, and a kerosene spray turbulent flame, are presented. The scalar statistics in these test cases may not be easily represented using the commonly used presumed shapes for modeling FDFs of mixture fraction and progress variable. Hence, the use of ML methods is explored. Particularly, deep neural network (DNN) to infer joint FDFs of mixture fraction and progress variable is reviewed here. The Direct Numerical Simulation (DNS) datasets employed to train the DNNs in each test case are described. The DNN performances are shown and compared to typical presumed probability density function (PDF) models. Finally, this chapter examines the advantages and caveats of the DNN-based approach.

S. Iavarone (✉)
Aero-Thermo-Mechanics Laboratory, Université Libre de Bruxelles, Brussels, Belgium
e-mail: si339@cam.ac.uk; salvatore.iavarone@ulb.be

S. Iavarone · H. Yang · Z. Li · N. Swaminathan
Engineering Department, University of Cambridge, Cambridge, UK
e-mail: hy345@cam.ac.uk

Z. Li
e-mail: zl443@cam.ac.uk

N. Swaminathan
e-mail: ns341@cam.ac.uk

Z. X. Chen
State Key Laboratory of Turbulence and Complex Systems, Aeronautics and Astronautics, College of Engineering, Peking University, Beijing 100871, China
e-mail: chenzhi@pku.edu.cn; zc252@cam.ac.uk

Department of Engineering, University of Cambridge, Cambridge, UK

N. Swaminathan and A. Parente (eds.), *Machine Learning and Its Application to Reacting Flows*, Lecture Notes in Energy 44, https://doi.org/10.1007/978-3-031-16248-0_8

1 Introduction

Increasingly stringent regulations on pollutants emissions from fossil fuel combustion are demanding for novel combustion technologies which can have high fuel flexibility, increased efficiency and low emissions. Moreover, a significant adoption of renewable technologies in future years is expected to reduce carbon footprint and meet the long-term objective of CO_2 neutrality. Nevertheless, combustion-based energy technologies will play a role in the future (or low-carbon) energy mix as discussed in the chapter "Introduction". Hence, combustion research is called in to provide solutions to the expected challenges arising from issues related to fuel flexibility and improving efficiency with pollutants reduction. Current combustion studies focus on aspects such as development, validation and uncertainty quantification of new models, and involve either experiments or numerical simulations, or both. A collection of these studies represents a massive amount of data that can be leveraged to achieve significant progress in combustion science. Utilising this data has thus become a new challenge and research opportunity. Data-driven techniques such as machine learning (ML) have demonstrated their abilities to extract information from massive data and assist in developing novel models which can be leveraged for technology development.

Machine learning techniques allow us to have statistical inference, for some unknown quantities of interest, with reasonably accuracy and confidence by *carefully training* the algorithms using representative data. Since the 1990s, ML has regained increasing attention and achieved outstanding results in many areas (Jordan and Mitchell 2015), including science, technology, manufacturing, finance, education, health care, and many more. Combustion science is not an exception to this trend, there are many studies demonstrating successful use of ML for combustion and some of these studies date almost 30 years back. Christo and coworkers (Christo et al. 1995, 1996b, a) first employed a machine learning algorithm, namely the Artificial Neural Network (ANN), in the 1990s to deal with chemistry tabulation for turbulent combustion simulations. These works involved training an ANN to obtain changes in the composition of several reactive scalars rather than using the conventional direct integration of the relevant equations. Satisfactory results suggested that the ANN was able to provide, with computational efficiency, the chemical kinetics information required for turbulent combustion simulations. The computational efficiency was mainly noted to come from memory saving. The subsequent studies extended this novel approach to more complex chemical systems (Blasco et al. 1998, 1999; Chen et al. 2000), where multiple ANNs were proposed for different subdomains of the large composition space. The valuable time saving achieved by ANN compared with traditional methods was presented. The recent advances on ML applied to chemical kinetics are discussed in chapters "Machine Learning Techniques in Reactive Atomistic Simulations" and "Machine Learning for Combustion Chemistry" with different perspectives.

Blasco et al. (2000) employed two different ANNs, namely the Self-Organising Map (SOM) and the Multi-layer Perceptron (MLP), to estimate the thermochemical states during a combustion simulation. The SOM was used to partition the thermo-chemical space into subdomains, while several MLPs were trained on each sub-domain to predict the evolution of the thermochemical space in time. These early explorations identify a general route to utilise the ANN for chemistry tabulation approaches, although their generality was limited due to the similarity between train-ing and testing cases. Consequently, later studies focused on developing ANNs for a wider range of combustion conditions.

Sen et al. trained ANNs using unsteady flame-turbulence-vortex interaction cases and subsequently used them for Large Eddy Simulations (LES) of syngas/air flames quite successfully (Sen and Menon 2009; Ali Sen and Menon 2010; Sen et al. 2010). Zhou et al. demonstrated successful application of the ANN to turbulent premixed flames by including 1D laminar premixed flame cases at different turbulent inten-sities while training the ANN (Zhou et al. 2013). A wider range of combustion conditions were also considered in later studies by including non-premixed laminar flamelets (Chatzopoulos and Rigopoulos 2013) to include local extinction and reig-nition (Franke et al. 2017) and non-adiabatic conditions (Wan et al. 2020, 2021) in the training data sets. Furthermore, randomising the non-premixed flamelets before using them as training data sets were shown to improve the generality of the ANN and helped to capture the behaviour of turbulent premixed flames quite well (Readshaw et al. 2021; Ding et al. 2021). Also, other techniques were explored to improve the generalisation level of ANN: Chi et al. (2021) trained the ANN on-the-fly during a simulation, whereas An et al. (2020) trained their ANN using data from Reynolds-averaged Navier–Stokes (RANS) simulations of hydrogen/carbon monoxide/kerosene/air mixture in a rocket combustion chamber and tested it for LES.

Further to the chemical kinetics use, another application of the ANN focuses on replacing the traditional flamelet look-up table, which requires a large memory. The general procedure is to set thermochemical scalars, which are the basis of the look-up table, as the input of the ANN and to infer the tabulated values. This reduces the memory requirement significantly since only the weights and bias(es) of the ANN need to be saved. A first successful application was demonstrated in Flemming et al. (2005) by building ANNs having the mixture fraction, its variance and its scalar dissipation rate as inputs and mass fractions as outputs, and using them in LES of the Sandia flame D. This was extended in Kempf et al. (2005) and Emami and Fard (2012) to estimate scalar mass fraction variations in a turbulent $CH_4/H_2/N_2$ jet diffusion flame. The optimisation of the ANN architecture, in terms of number of hidden layers and neurons per layer, was also explored to improve the predictive accuracy of LES of the Sydney bluff-body swirl-stabilised methane-hydrogen flame (Ihme et al. 2006, 2008, 2009).

The use of ANN for inferring multi-dimensional flamelet library is also explored in recent studies. Owoyele et al. proposed a grouped multi-targets ANN approach to model 4D and 5D flamelet libraries respectively for a n-dodecane spray flame, under conditions of the Spray A flame from the Engine Combustion Network (ECN), and methyl decanoate combustion in a compression ignition engine (Owoyele et al. 2020). Ranade et al. (2021) trained a SOM-MLP method on a 4D Probability Density

Function (PDF) table and used it for RANS and LES of the DLR-A turbulent jet diffusion flame. These works showed that the ANN yielded good accuracy at reduced computational costs with low storage space requirements. Similarly, Zhang et al. (2020) extended the application of the SOM-MLP algorithm to the Flamelet Generated Manifolds (FGM) model by using species mass fractions in mixture fraction-progress variable space as training data. This ANN approach was successfully used in RANS calculations and LES of ECN Spray H flame to explore the detailed spray combustion process. More comprehensive reviews of the applications of ML in combustion research can be found in Zheng et al. (2020), Zhou et al. (2022) and Ihme et al. (2022).

Presumed PDF shapes are typically used along with tabulated chemistry approaches. The PDF of relevant scalars such as mixture fraction and progress variable are used to compute averaged temperature, density, species mass fractions, and the relevant reaction rates. These quantities can be stored in a look-up table with the first two moments of the above scalars as controlling variables. Although widely employed in several past studies, presumed PDF or Filtered Density Function (FDF), in the context of LES, approaches may not accurately represent the scalar statistical behaviour under several conditions, such as extinction and reignition, combustion among multiple streams, multi-regime burners, and multi-phase reacting flows. The FDFs having shapes different to the regular distributions such as Gaussian or β-function can be also observed prominently in Moderate or Intense Low-oxygen Dilution (MILD) combustion. This combustion mode features broadly distributed reaction zones rather than conventional flamelet-like structures, with strong interactions between autoigniting and propagating fronts. Therefore, it may not be satisfactory to use conventional PDFs/FDFs models to predict reaction rates, and advanced data-driven techniques like machine learning may be a suitable alternative for improving the accuracy. De Frahan et al. (2019) compared the performance of three different machine learning techniques, *viz.,* random forests, which is a traditional ensemble methods, deep neural networks (DNNs), and conditional variational autoencoder (CVAE), multiple hidden layers between which is also know as generative learning, to infer marginal FDFs of reaction progress variable in a swirling methane/air premixed flame and showed that DNN is superior compared to the other two techniques. The DNN is an ANN with multiple hidden layers between input and output. Yao et al. (2020) built an MLP to obtain the mixture fraction marginal FDF for LES of turbulent spray flames and observed an order of magnitude improvement compared to those of the traditional presumed FDF approaches. Chen et al. (2021) employed a DNN to predict the joint FDF of mixture fraction and progress variable in MILD combustion conditions and showed that the DNN is generally able to capture the complex FDF behaviours and their variations with excellent accuracy, outperforming other presumed FDF models.

This chapter aims to provide an overview of recent studies employing deep neural networks (interchangeably referred to as DNN, ANN or MLP hereafter) to infer subgrid-scale FDFs and reaction rates needed for LES of turbulent combustion under conventional and MILD conditions. A review of the Direct Numerical Simulation (DNS) data used to train these DNNs is also given. The chapter is structured as follows. A recap of the treatment of FDFs in LES of turbulent combustion systems is provided in Sect. 2. The DNS cases used as training datasets for the DNNs are described in Sect. 3. The characteristics of the DNNs employed for the different combustion cases are illustrated in Sect. 4. The main results in terms of FDF and reaction rate predictions are discussion in Sect. 5. The conclusions are summarised in Sect. 6.

2 FDF Modelling

The filtered reaction rate appearing in the transport equation for a species filtered mass fraction or reaction progress variable needs a closure model and recent developments in various closure models are described in the book (Swaminathan et al. 2022) and review papers (Veynante and Vervisch 2002; Pitsch 2006). Earlier chapters of this book discuss the potential application of ML techniques to some of the reaction rate closures. In the presumed PDF approach, the filtered reaction rate is modelled as an integral of the product of a conditional reaction rate and a FDF (see Eq. 6). The mixture fraction and the reaction progress variable are typically used as conditioning variables to signify the role of mixing and flame propagation on reaction rate (Bradley et al. 1998; Ihme and Pitsch 2008a). The conditional reaction rate may be estimated using one of the methods developed in past studies and these methods used canonical flames for chemistry tabulation, e.g., flamelet-generated manifolds (van Oijen and de Goey 2002), flame prolongation of intrinsic low dimensional manifold (Gicquel et al. 2000), conditional source term estimation method (Jin et al. 2008), or the solution of conditionally filtered equations for species mass fractions and energy via the conditional moment closure method (Klimenko and Bilger 1999).

The subgrid variations in the conditioning variables about their filtered values are represented by the filtered density function (FDF). The FDF can generally be obtained by solving its transport equations using various approaches, e.g., Lagrangian particles (Pope 1985), Eulerian stochastic fields (Jones and Kakhi 1998), and multi-environment (Fox 2003). However, these approaches are computationally expensive and thus using a presumed FDF can be chosen (Pitsch 2006; Pope 2013) to save computational costs. This presumed FDF approach will need only the statistical moments, usually the mean and variance, of the key variables (mixture fraction, progress variable, flame stretch/straining, heat loss, etc., depending on the physical scenario of interest) to be transported and it is therefore much more economical.

The β-PDF (Cook and Riley 1994) is the most commonly used presumed FDF in LES of turbulent flames (Raman et al. 2005; Navarro-Martinez et al. 2005; Ihme and Pitsch 2008b; Chen et al. 2017), and it usually provides a good approximation of a conserved scalar distribution. The Favre-averaged FDF of the mixture fraction Z with a presumed β-distribution is calculated as

$$\widetilde{P}_\beta(\xi; \widetilde{Z}, \widetilde{\sigma_Z^2}) = \frac{\Gamma(a+b)}{\Gamma(a)\Gamma(b)} \, \xi^{a-1} (1-\xi)^{b-1}, \tag{1}$$

where ξ is the sample space variable for Z, $\widetilde{Z}$ is the filtered mixture fraction and $\widetilde{\sigma_Z^2} \equiv \widetilde{Z''} = \overline{(Z - \widetilde{Z})^2}$ is the mixture fraction subgrid variance. The parameters of the Γ function are $a = \widetilde{Z}\,(1/\widetilde{g_Z} - 1)$ and $b = \left(1 - \widetilde{Z}\right)(1/\widetilde{g_Z} - 1)$. The segregation factor is $\widetilde{g_Z} = \widetilde{\sigma_Z^2}\,/\left(\widetilde{Z}(1 - \widetilde{Z})\right)$. The Favre-filtered FDF of the progress variable, $\widetilde{P}_\beta(\eta; \widetilde{c}, \sigma_c^2)$, can also be presumed to follow a β distribution and obtained in a similar manner using $\widetilde{c}$ and $\widetilde{\sigma_c^2} \equiv \widetilde{c''} = \overline{(c - \widetilde{c})^2}$. The joint FDF of ξ and η can be modelled as

$$\widetilde{P}\,(\xi, \eta) = \widetilde{P}_\beta\left(\xi; \widetilde{Z}, \widetilde{\sigma_Z^2}\right) \widetilde{P}_\beta\left(\eta; \widetilde{c}, \widetilde{\sigma_c^2}\right), \tag{2}$$

assuming that there is a weak correlation between the subgrid fluctuations of Z and c. Such assumption has been widely accepted for LES of conventional combustion (Pitsch 2006; Veynante and Vervisch 2002). However, stronger subgrid correlations of scalars fluctuations can occur in MILD combustion (Minamoto et al. 2014) and hence the above assumption may not applicable universally. Other analytical distributions have been considered in past studies (Grout et al. 2009; Darbyshire and Swaminathan 2012; Linse et al. 2014). Darbyshire and Swaminathan (2012) proposed a correlated joint PDF model using the *Plackett copula* (Plackett 1965) to include the covariance of Z and c in RANS calculations. The covariance, $\widetilde{\sigma}_{Zc}$, written as $\widetilde{\sigma}_{Zc} = \overline{\left(Z - \widetilde{Z}\right)(c - \widetilde{c})}$ is used in the *copula* method to obtain a joint PDF from the univariate marginal distributions, $\widetilde{P}_\beta(Z)$ and $\widetilde{P}_\beta(c)$. For non-zero values of $\widetilde{\sigma}_{Zc}$, the correlated joint PDF is calculated as

$$\widetilde{P}\,(Z, c) = \frac{\theta\,\widetilde{P}_\beta(Z)\,\widetilde{P}_\beta(c)\,(\mathscr{A} - 2\mathscr{B})}{\left(\mathscr{A}^2 - 4\theta\mathscr{B}\right)^{3/2}}, \tag{3}$$

with

$$\mathscr{A} = 1 + (\theta - 1)\left[\widetilde{\mathscr{C}}_\beta(Z) + \widetilde{\mathscr{C}}_\beta(c)\right], \tag{4}$$

and

$$\mathscr{B} = (\theta - 1)\widetilde{\mathscr{C}}_\beta(Z)\widetilde{\mathscr{C}}_\beta(c), \tag{5}$$

where $\widetilde{\mathscr{C}}_\beta$ is the β cumulative distribution function (CDF) and θ is the odds ratio calculated using a Monte Carlo approach (Ruan et al. 2014). The *copula* method has been used in RANS calculations of stratified premixed and lifted jet flames (Ruan et al. 2014; Chen et al. 2015) showing improved prediction of the lift-off height with respect to the double-β PDF given in Eq. (2).

In presumed-FDF approaches, the subgrid reaction rate is obtained as

$$\overline{\dot{\omega}} = \int_0^1 \int_0^1 \langle \dot{\omega} | Z, c \rangle P\left(Z, c; \widetilde{Z}, \widetilde{\sigma_Z^2}, \widetilde{c}, \widetilde{\sigma_c^2}\right)\, dZ\, dc, \tag{6}$$

and this approach reduces the computational cost significantly for LES by using presumed FDF in the above equation. However, the presumed FDF shapes obtained using classical functions, for example bimodal delta function, may not be fully satisfactory for situations such as (i) MILD combustion conditions, (ii) when there are evaporating droplets, and (iii) when the burnt or burning mixture is inhomogeneous leading to significant statistical correlation between Z and c (Chen et al. 2018). To overcome these issues, machine learning algorithms are employed to construct predictive models for the scalar PDFs/FDFs in recent studies. A deep neural network (DNN), among other ML techniques tested, was shown to be better than a joint β-function model in inferring subgrid FDFs in a swirling methane-air premixed flame (de Frahan et al. 2019). This behaviour was also demonstrated for MILD combustion (Chen et al. 2021) and turbulent spray flames (Yao et al. 2020). These tests were conducted using respective direct numerical simulation (DNS) datasets. DNS can be seen as a *virtual* experiment resolving all the relevant length and time scales without turbulence modelling. Thus, it is a powerful tool for investigating combustion models. It is quite straightforward to obtain filtered quantities from DNS data by applying appropriate filtering operations (Pope 2000) and these can be used as input to ML algorithms such as DNN. The data extraction and its processing prior to using them for DNN training are important steps which can play a role to improve accuracy and generality of the neural networks. Details about these steps, along with the main features of the cases studied in de Frahan et al. (2019), Chen et al. (2021) and Yao et al. (2020), are discussed in the following sections. Details on the respective DNS cases can be found in those studies as the objective here is on the use of ML techniques.

3 DNS Data Extraction and Manipulation

Three combustion cases are considered in this chapter: a low-swirl premixed methane-air flame investigated in de Frahan et al. (2019), methane-air combustion under MILD conditions studied in Chen et al. (2021), and a turbulent kerosene spray flame used in Yao et al. (2020). The corresponding DNS setups and data preparation procedures are described next.

3.1 Low-Swirl Premixed Flame

The DNS dataset considered by de Frahan et al. is a snapshot of a quasi-stationary simulation of an experimental low-swirl, premixed methane-air burner (Day et al. 2012). In this setup, a nozzle imposes a low swirl to a CH_4/air mixture with fuel-air equivalence ratio $\phi = 0.7$ at the inflow. The nozzle region is surrounded by a co-flow of cold air. A lifted premixed flame with its partially burnt mixture reacting with co-flow air in downstream locations was observed in the experiments. The presence of this multi-regime burning introduces challenges for modeling the joint FDF of mixture fraction and progress variable. Training ML models with such DNS dataset has additional advantages such as using diverse subsets as training data, avoiding overfitting, and increasing the opportunities for model generalisation. The training sets were constructed by selecting different subvolumes, indicated by $\mathcal{V}$ as in Fig. 1, spanning from premixed combustion region to downstream zone containing mixing of premixed combustion products with co-flow air. de Frahan et al. (2019) used a single time snapshot at $t = 0.0626$ s from the DNS to demonstrate the capabilities of ML for FDF modelling. In the context of LES, the FDF at a given point and time can be extracted by applying fine-grained filtering to DNS or experimental data at a given instant (Pope 1990). In each subvolume, sample moments and the associated FDF were thus obtained by using a discrete box filter:

$$\overline{\psi}(x, y, z) = \frac{1}{n_f^3} \sum_{i=-n_f/2}^{n_f/2} \sum_{j=-n_f/2}^{n_f/2} \sum_{k=-n_f/2}^{n_f/2} \psi(x + i\Delta x, y + j\Delta x, z + k\Delta x), \quad (7)$$

where ψ is the quantity of interest, n_f is the number of points in the discrete box filter, $\overline{\Delta} = 32\Delta x$ is the filter size, and $\Delta x = 100\,\mu$m is the smallest spatial cell size in the DNS (six times smaller than the laminar flame thickness). Four sample moments of the joint FDF, i.e., $\widetilde{Z}$, $\widetilde{\sigma}_Z^2$, $\widetilde{c}$, $\widetilde{\sigma}_c^2$, which are Favre-filtered mixture fraction, its subgrid scale (SGS) variance, progress variable and its SGS variance, were extracted for each subvolume. The filter size was chosen to be representative of typical LES filter scale (Pitsch 2006) and to ensure adequate samples to construct FDF. These filters were spaced equidistant of $8\Delta x$, leading to 58800 FDFs for each subvolume. The mixture fraction Z was defined using nitrogen mass fraction so that it took a value of 1 in the burner stream and 0 in the co-flow air. The progress variable, varying between 0 and 0.21, was defined using mass fractions of CO_2, CO, H_2O and H_2 as $c = Y_{CO2} + Y_{CO} + Y_{H2O} + Y_{H2}$. The density-weighted FDFs of Z and c were constructed using 64 bins in Z space and 32 bins in c space, which gives a vector of 2048 values to describe a single joint FDF. The conditional means of the reaction rate $\langle \dot{\omega} | Z, c \rangle$ were also extracted for each sample with an identical discretisation. Prior to training, the sample moments were independently centered by subtracting the median and scaled by dividing the data by the range between the 25th and 75th quantiles. It is known that appropriate centring and scaling are generally beneficial for ML algorithms (Goodfellow et al. 2016). According to the authors this centring and scaling were robust to outliers. The samples from a volume $\mathcal{V}_i$ were randomly split among two distinct datasets: a training dataset, $\mathcal{D}_i^t$, and a validation dataset, $\mathcal{D}_i^v$, comprising 5% of the total samples, as illustrated in Fig. 1.

3.2   MILD Combustion

The MILD combustion DNS dataset of Doan et al. (2018) was used to study the application of DNN for inferring subgrid FDF in MILD combustion by Chen et al. (2021). A cube of size $L_x \times L_y \times L_z = 10 \times 10 \times 10$ mm was used to conduct DNS of turbulent combustion of inhomogeneous methane-air mixtures diluted with exhaust

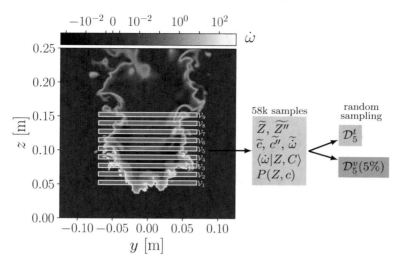

Fig. 1 Illustration of data generation procedure for $\mathcal{V}_5$

gases. A spatial resolution of $\delta x \approx 20\,\mu\mathrm{m}$ obtained using 512 points distributed uniformly in each direction was observed to be sufficient to resolve the turbulent and chemical length scales of interest as described in Doan et al. (2018). The simulation was run for 1.5 flow-through time τ_f, defined in Minamoto and Swaminathan (2015). Further detail on the DNS procedure and datasets can be found in Doan et al. (2018). Three cases, *viz.*, AZ1, AZ2 and BZ1, with different mixing length scales and dilution levels were considered for the DNN training. The conditioning variables for the FDF analyses were the Bilger mixture fraction (Bilger 1976) and a temperature-based reaction progress variable, c_T, defined as

$$c_T = \frac{T - T_u}{T_b(Z) - T_u}, \tag{8}$$

where T_u is 1500 K and the value of burnt mixture temperature T_b depends on Z and it can be obtained using MILD Flame Element (MIFE) laminar calculations (Minamoto and Swaminathan 2014). Favre-filtered fields were extracted from the DNS by applying a low-pass box filter. For example, the Favre-filtered mixture fraction $\widetilde{Z}$ was obtained as:

$$\widetilde{Z}(\boldsymbol{x}, t) = \frac{1}{\overline{\rho}(\boldsymbol{x}, t)} \int_{x-\frac{\Delta}{2}}^{x+\frac{\Delta}{2}} \rho\left(\boldsymbol{x}', t\right)\, Z\left(\boldsymbol{x}', t\right)\, d\boldsymbol{x}', \tag{9}$$

where $\overline{\cdot}$ and $\widetilde{\cdot}$ denote the Reynolds and Favre filtering respectively, ρ is the mixture density and Δ is the filter width. The position vectors are $\boldsymbol{x}$ and $\boldsymbol{x}'$. The subgrid variance was obtained as

$$\widetilde{\sigma_Z^2}(x, t) = \frac{1}{\overline{\rho}(x, t)} \int_{x-\frac{\Delta}{2}}^{x+\frac{\Delta}{2}} \rho\left(x', t\right) \left[Z\left(x', t\right) - \widetilde{Z}\left(x, t\right)\right]^2 dx'. \qquad (10)$$

Similarly, the $\widetilde{c}_T$ and $\widetilde{\sigma_{c_T}^2}$ fields were calculated as above. The Z-c_T joint FDF was then computed as

$$\widetilde{P}(\xi, \eta; x, t) = \frac{1}{\overline{\rho}(x, t)} \int_{x-\frac{\Delta}{2}}^{x+\frac{\Delta}{2}} \rho\left(x', t\right) \delta\left[\xi - Z\left(x', t\right)\right] \delta\left[\eta - c_T\left(x', t\right)\right] dx', \qquad (11)$$

where ξ and η were the sample-space variables of Z and c_T respectively, $\delta[\cdot]$ is the Dirac delta function. The discrete FDFs were obtained for a given point in a given DNS snapshot by binning the Z and c_T samples in the corresponding filtering subspace with 35 non-uniform bins in Z space (clustered around the stoichiometric value) and 31 uniform bins in c_T space. The subgrid-scale covariance, $\widetilde{\sigma}_{Zc_T}$, also used by the *copula* model, was computed as

$$\widetilde{\sigma_{Zc_T}}(x, t) = \frac{1}{\overline{\rho}(x, t)} \int_{x-\frac{\Delta}{2}}^{x+\frac{\Delta}{2}} \rho(x', t) \left[Z(x', t) - \widetilde{Z}(x, t)\right] \\ \times \left[c_T(x', t) - \widetilde{c}_T(x, t)\right] dx'. \qquad (12)$$

The filtered scalar fields $\widetilde{Z}$, $\widetilde{c}_T$, $\widetilde{\sigma_Z^2}$, $\widetilde{\sigma_{c_T}^2}$ and $\widetilde{\sigma_{Zc_T}}$ formed the DNN input matrix **X**. The unfiltered ρ, Z and c_T fields were used to obtain the Favre filtered FDFs required for the target matrix **Y**. The procedure is shown schematically in Fig. 2 for a snapshot of case AZ1. The filtered fields are presented in 2D with the thin DNS grid-lines for visual clarity. The indices i, j and k pertain to the x, y and z directions in 3D space, respectively, and are assigned to each "LES filter cube" indicated by a red box in Fig. 2. The total number of samples taken in each direction is n_{cube}. The effects of filter size were also investigated by considering a range of filter sizes relevant to typical LES. The filter sizes were normalized using the thermal thickness of the stoichiometric MIFE, $\delta_{\text{th}}^{\text{st}} = 1.6$ mm. A filter size of $\Delta = 80\delta x$ corresponded to $\Delta^+ = \Delta/\delta_{\text{th}}^{\text{st}} = 1$. The extracted matrices **X** and **Y** were flattened to be two-dimensional, with as many rows as the number of samples and as many columns as the number of features. The input matrix **X** had 5 columns, while the target matrix **Y** had 1085 columns, obtained from the discretisation step mentioned above.

Centring and scaling of the input matrix **X** were performed as follows: each column vector, having n_{cube}^3 elements, was centred by subtracting its mean and scaled by dividing by its standard deviation. Centring and scaling were not applied to the output matrix **Y**. However, to address the issue of having unbounded values of the FDFs, the discrete density function values were considered. As such, every number in **Y** varied between 0 and 1, and the sum of the elements in each target row is equal to 1.

Subsequent to the scaling procedures, a dimensionality reduction technique like Principal Component Analysis (PCA), discussed in chapter "Reduced-Order Mod-

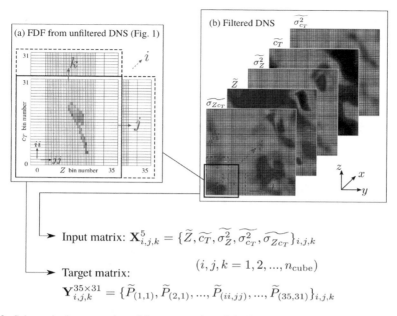

Fig. 2 Schematic demonstration of the construction of the DNN input and target matrices (Chen et al. 2021)

eling of Reacting Flows Using Data-Driven Approaches" was used to identify and remove the outliers in the training data. Two types of outliers, *viz.*, *leverage* and *orthogonal*, Verdonck et al. (2009) were determined and discarded. Details about the identification and removal step are provided in Chen et al. (2021). Once leverage and orthogonal outliers were removed from the dataset, the DNN training was then performed on the remaining observations as discussed in the following Sect. 4.2.

3.3 Spray Combustion

Carrier-phase DNS (CP-DNS) data of turbulent spray flames were used to build a deep learning training database for mixture fraction FDF predictions. In carrier-phase DNS, the flow field is resolved with a point source approximation for the droplets, thus all relevant scales of the fluid phase are resolved except the boundary layers around individual particles. The governing equations of the gas phase are solved in the Eulerian framework and coupled with a Lagrangian solver for displacement, size, and temperature of the droplets. An equilibrium state of the liquid and the vapor at the interface was assumed. A full description of the governing equations is provided in Yao et al. (2020). The computational domain is a rectangular box, discretised by a mesh with $192 \times 128 \times 128$ cells having $\delta_{\text{DNS}} = 100\,\mu\text{m}$. This grid size ensured a sufficient resolution of the small scale structures of the flow field (Pope 2000), whereas a finer resolution could compromise the point particle assumptions of the liquid phase. Kerosene droplets (treated as single-component $C_{12}H_{23}$) were randomly injected into humid air, representative of experimental (Khan et al. 2007; Wang et al. 2018) and numerical (Wright et al. 2005; Giusti et al. 2018) setups. A homogeneous isotropic turbulent velocity field, calculated by a modified von Karman spectrum (Wang et al. 2019) was imposed at the inlet. The progressive kerosene droplet evaporation led to an ignitable mixture that promoted a statistically planar turbulent partially premixed flame. Further downstream, the hot post-flame temperatures led to reduced turbulence levels due to higher viscosity and a sudden evaporation of remaining droplets that could penetrate the flame. This lack of homogeneity and the presence of a source term for the mixture fraction are prone to make the existing FDF models (O'Brien and Jiang 1991; Cook and Riley 1994) inaccurate.

Filter boxes were used for post-processing of CP-DNS data to group several DNS cells into one LES cell. A filter box example is shown in Fig. 3 along with the DNS domain and setup, and the simulated temperature contour. The mixture fraction FDF $P(\eta)$ was computed from DNS data using a mixture fraction binning, with a bin size of 0.01 for all DNS cells lying within a specific LES cell. Favre filtering was used to extract LES quantities that were employed as input variables for the ANN. According to Klimenko and Bilger (1999), the following input quantities were found to have an effect on the mixing statistics and were thus considered: mixture fraction ξ, eddy viscosity ν_t, turbulence dissipation rate ϵ_t, diffusion coefficient D, density ρ, spray evaporation rate J_m, relative velocity between the droplet and the

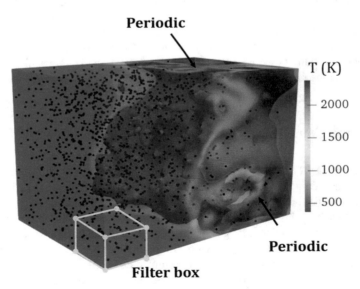

Fig. 3 Simulation setup of CP-DNS (solid points: droplets; the gas phase is colored by temperature) and an LES filter box (Yao et al. 2020)

surrounding gas U_d and droplet number density C. The turbulence dissipation rate was replaced by the more easily available strain rate $|S_{ij}|$. All the DNN inputs were filtered and Favre averaged. Therefore, the input features are commonly accessible in a typical LES of spray combustion. Moreover, Wang et al. concluded in their study that these parameters sufficiently characterize the mixture fraction FDF in turbulent spray flames. To ensure the reliability of the DNN for a reasonable range of LES meshes, the authors investigated the following LES filter sizes: $(\Delta_{LES})^3 = (8\delta_{DNS})^3$, $(\Delta_{LES})^3 = (16\delta_{DNS})^3$, $(\Delta_{LES})^3 = (32\delta_{DNS})^3$. The final database is a combination of data samples with different Δ_{LES}. The performance of the DNN for data samples using different LES filter boxes were assessed. The output target was set to be a placeholder of 60 elements covering ξ in $[0, 0.6]$, as $\xi_{max} \leq 0.6$ in the the spray flame simulations. To avoid that the binning procedure could lead to empty bins, especially for small Δ_{LES}, missing values were replaced by interpolated values computed by Stineman interpolation method, which is widely used in statistics to deal with the missing values as it preserves the monotonicity of data and prevents introducing spurious oscillations (Stineman 1980). It was found that the commonly used zero-padding operation, which fills in blank data with zeros, is not applicable as the DNN would be misled and learn erroneous patterns. A total of 18 simulation cases were run to form the full database for training and validation purposes. The validation (test) dataset consisted of five simulation cases, resulting in a test/train ratio of about 0.38. These datasets included parameter ranges that approximate conditions to be expected in real spray flames and were used for the a priori validation presented in Sect. 5.

To recap, the three studies selected several DNS cases to construct a heterogeneous training set. If only one DNS case was available then several subdomains within the DNS domain were selected. Chen et al. (2021) considered one additional DNN input feature, i.e., the scalar covariance, to the input set chosen by de Frahan et al. (2019). Yao et al. (2020) chose different DNN input features specifically for spray combustion. No scaling was adopted by Yao et al., whereas two different scaling methods were implemented in the other studies. Only Chen et al. adopted an outlier removal by using a dimensional reduction technique. Discrete density functions, bounded between 0 and 1, were the DNN target in de Frahan et al. (2019) and Chen et al. (2021) while Yao et al. (2020) considered probability density function values. The review of these studies shows that no unique algorithm needs to be adopted to prepare the input data for a ML model. The only common goal that needs to be pursued is to construct an input dataset that is as heterogeneous as possible to increase the generalisation, also known as transfer learning, of the trained ML models. The similarities and differences of the DNNs used in these three studies are discussed next.

4   Deep Neural Networks for Subgrid-Scale FDFs

A standard neural network consists of many simple connected functional units, called neurons. Each neuron receives an input which is processed through activation functions to produce an output. Multiple neurons can be combined to form fully connected networks, which are called artificial neural networks (ANNs) since they mimic the neuron arrangements in the human brain. Feed-forward networks, also called multi-layer perceptrons (MLPs), are classic ANN structures, and they are composed of layers of neurons, where a weighted output from one layer is the input to the next layer. The first layer of the MLP accepts a vector as input and the elements of this vector are known as features. The final output of the MLP is the target quantity of interest. The layer providing the final MLP output is called output layer, while the other layers in the network are called hidden layers. In a mathematical perspective (Goodfellow et al. 2016), the MLP defines a mapping from the input x to the output $y = f(x, \theta)$, where the parameters θ are the trainable network parameters. Each neuron is a functional unit that is generally described by

$$y = \phi(x^T \omega + b), \tag{13}$$

where ω and b are the weights and bias vector, and ϕ is the activation function (see Sect. 2.3.7.2, Chap. 2, this volume), which provides great flexibility to ANNs by introducing non-linearity to an otherwise linear relationship between input and output. There are several activation functions and some of these will be introduced and described later. The weight ω is a matrix of the size $k \times m$, whereas the bias b is a vector of m elements. For each layer, k is the number of inputs received from the preceding layer and m is the number of neurons in the current layer. ω and b contains the trainable parameters of the network. The training of ANNs pursues the objective of minimizing a target loss function

$$\mathcal{L}(x, \omega) = \mathcal{G}(f(x, \omega) - f^*), \tag{14}$$

where $\mathcal{G}$ is any measure of the difference between the modeled value f and the real value f^*. The most commonly used loss functions are the mean absolute error (MAE) and the mean squared error (MSE). Nonlinear optimization methods, such as backward propagation (Rumelhart et al. 1986), are used to identify the network weights that minimize the error between predictions and labeled training data. The training step gives the optimized set of weights. The MLP is a design that is suitable for regression problems, whereas other types of ANNs, such as Convolutional Neural Network (CNN) and Recurrent Neural Network (RNN), have been extensively used in processing image data and time series problems, etc., see Sect. 2.3.7.2 (Chap. 2, this volume), for further detail. A schematic of the MLP architecture with input, hidden, and output layers is shown in Fig. 4 as an example.

Fig. 4 A schematic of
3-layer MLP architecture

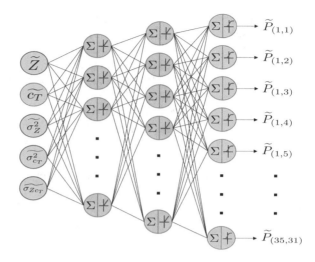

4.1 Low-Swirl Premixed Flame

A feed-forward fully connected DNN with three, two hidden and an output, layers
was trained by de Frahan et al. (2019) to predict the joint subfilter FDF of mixture
fraction and progress variable. There were 256 and 512 neurons in the two hidden
layers and neurons had a leaky rectified linear unit activation function (LeakyReLU):

$$y_i = \begin{cases} x_i & \text{if } x_i \geq 0 \\ \alpha x_i & \text{otherwise} \end{cases} \tag{15}$$

where x_i is the weighted sum of the neuron input, y_i is its output, and α, usually
equal to 0.01, is the slope. A LeakyReLU activation function avoids mapping negative input to zero values unlike its parent function ReLU having $\alpha = 0$. A large
weight update during training can yield the summed input to be always negative
regardless of the network input. A neuron featuring a ReLU function will then output a zero value leading to the *dying* ReLU case, in which the neuron neither activates
a gradient-based optimization nor adjust its weights. Furthermore, similar to the vanishing gradients problem, the learning can be slow while training ReLU networks
stumbling on constant zero gradients. The leaky rectifier allows for a small, non-zero
gradient when the unit is saturated and not active. Additionally, each hidden layer is
followed by a batch normalization layer (Ioffe and Szegedy 2015) and this technique
has been widely used to build deep networks as it leads to speed and performance
improvements. It applies the following function:

$$y_i = \gamma \frac{x_i - \mu_x}{\sqrt{\sigma_x^2 + \epsilon}} + \delta \tag{16}$$

where x_i and y_i are the i-th elements of the layer input and output vectors respectively. These vectors are of size n having a mean and variance of $\mu_x = 1/n \sum_{i=1}^{n} x_i$ and $\sigma_x^2 = 1/n \sum_{i=1}^{n} (x_i - \mu_x)^2$. A small real number ϵ is used to maintain numerical stability. Both γ and δ are learning parameter vectors of size n and they are updated iteratively during training for optimization purposes. de Frahan et al. (2019) chose $\epsilon = 10^{-5}$ and a moving average of μ_x and σ_x computed during training with a decay of 0.1 (or, equivalently, momentum of 0.9).

The DNN inputs are the four moments of the joint FDF, *viz.*, $\widetilde{Z}$, $\widetilde{\sigma_Z^2}$, $\widetilde{c}$, and $\widetilde{\sigma_c^2}$ whereas the outputs are a total of 2048 FDF values obtained from the discretisation of the joint FDF of mixture fraction Z and progress variable c as described in Sect. 3.1. Thus, an output layer having 2048 neurons, as many as the number of outputs, was considered in de Frahan et al. (2019). The output layer features a *softmax* activation function:

$$y_i = \frac{\exp(x_i)}{\sum_{i=1}^{n} \exp(x_i)} \tag{17}$$

where x_i and y_i are defined as for Eq. 16. This type of activation function ensures that $\sum_{i=1}^{n} y_i = 1$ and $y_i \in [0, 1]\ \forall\ i$. The loss function used was the binary cross entropy between the target y and the prediction $\hat{y}$ and this function is

$$\mathcal{L}(\hat{y}, y) = \frac{1}{n} \sum_{i=1}^{n} \left(y_i \log \hat{y}_i + (1 - y_i) \log \left(1 - \hat{y}_i\right) \right), \tag{18}$$

representing a proper metric for measuring the difference between two probability distributions. The total number of trainable parameters was 1.1 M. The training was performed over 500 epochs, i.e., 500 training loops through the entire training data. For each epoch, the training data is fully shuffled and divided into batches with 64 training samples per batch. All trainable parameters are updated after each epoch. A split of 95/5% between training and validation samples was applied on the entire dataset. The loss function is computed on the validation samples which are not part of the training process. Thus, the validation loss is the true indicator of the ANN's performance and provides hints regarding its generality. It is a common practice to track the losses during both training and validation steps continuously to check if the losses are decreasing over each epoch by studying learning curves (a plot of loss vs epoch number). These learning curves can be used to diagnose an underfit, overfit, or well-fit model and whether the training or validation datasets are not representative of the problem domain. A good ANN training gives loss curves that decreases continuously until a plateau is reached where the difference between the training and validation losses is small. de Frahan et al. (2019) chose Adam optimizer (Kingma and Ba 2014), which is a gradient descent algorithm, with an initial learning rate of 10^{-4}. The learning rate is a dimensionless parameter that determines the step size of the stochastic gradient descent used to adjust the weights, ω. The Adam optimizer is more sophisticated than traditional stochastic gradient descent by having a per-parameter learning rate, which can also be adapted during the training (Kingma and Ba 2014).

4.2 MILD Combustion

Chen et al. (2021) used a feed-forward fully connected DNN to infer the joint FDF of mixture fraction and progress variable. This DNN is similar to the one employed by de Frahan et al. (2019) and can be summarized as follows:

- linear hidden layer with 5 input features and bias, LeakyReLU activation function with $\alpha = 0.01$, and 256 output features;
- batch normalization layer with 256 input and output features, and momentum equal to 0.9;
- linear hidden layer with 256 input features and bias, LeakyReLU activation function with $\alpha = 0.01$, and 512 output features;
- batch normalization layer with 512 input and output features, and momentum equal to 0.9;
- linear output layer with 512 input features and bias, *softmax* activation function, and 1085 output features.

Thus, the two hidden layers had 256 and 512 fully connected neurons, where LeakyReLU activation functions were applied. Each hidden layer was followed by a batch normalization layer. The output layer contained 1085 neurons featuring a *softmax* activation function. The loss function used was the binary cross entropy given in Eq. 18 along with Adam optimizer with an initial learning rate of 10^{-4}. The model was trained for maximum 1000 epochs with batch size of 256 training samples. The ANN features were the four moments of the joint FDF and the outputs were a total of 1085 FDF values. A split of 80/20% between training and validation samples was applied on the entire dataset containing about 28000 filtered DNS boxes. An early stopping method, by using a predefined number of epochs, was used for the training to avoid overfitting. An overfitted ANN will have a validation loss that decreases for the first several epochs but increases subsequently (Goodfellow et al. 2016).

4.3 Spray Flame

Yao et al. (2020) used an MLP with four hidden layers and 500 neurons per layer to infer the Favre-filtered FDF of the mixture fraction in spray flames. As noted in Sect. 3.3, the input quantities were $\widetilde{\xi}$, $\widetilde{v}_t$, $\widetilde{|S_{ij}|}$, $\widetilde{D}$, $\overline{\rho}$, spray evaporation rate $\widetilde{J}_m$, relative velocity between the droplet and the surrounding gas $\widetilde{U}_d$, and droplet number density $\widetilde{C}$. The output was a vector with 60 elements since the FDF of the mixture fraction $P(\eta)$ (where η is the sample space variable for the mixture fraction ξ) was obtained as described in Sect. 3.3. The activation function $\phi(z) = \max(0, z)$ applied in each layer was the ReLU. A traditional stochastic gradient descent algorithm was used to minimize the mean absolute error, which was the loss function. A total of 18 DNS cases were run to form the full datasets for the training and validation steps. The validation (test) dataset consisted of five cases, resulting in a test/train ratio of ~ 0.38. An early stopping criterion was imposed for the training process. This ANN was also trained on the conditional scalar dissipation rate $\langle N|\xi = \eta \rangle$, which is another interesting application.

5 Main Results

5.1 FDF Predictions and Generalisation

An overview of the ML model performance in each of the test cases is discussed in this section. The FDF predictions provided by ML and analytical models were assessed *a priori* using the FDFs obtained from the DNS cases.

5.1.1 Premixed Flame

Three different ML models, i.e., random forest (RF), conditional variational autoencoder (CVAE), and DNN, were trained by de Frahan and coworkers using filtered DNS data from the subvolume $\mathcal{V}_3$ of the low-swirl premixed flame, i.e., the algorithms were trained on $\mathcal{D}_3^t$, and the metrics were evaluated on $\mathcal{D}_3^v$ (see Fig. 1). Figure 5 compares the marginal FDFs $P(Z)$ and $P(c)$ obtained using the three ML models, β-function model and DNS result for $\mathcal{V}_3$ for three different values (low, medium, and high) of the Jensen-Shannon divergence (JSD), which measures the similarity of two probability distributions, $Q_1 = Q^{\text{DNS}}(n)$ and $Q_2 = Q^{\text{model}}(n)$. The JSD is given by

$$J(Q_1||Q_2) = \frac{1}{2} \sum_{n=1}^{N} \left\{ Q_1(n) \ln \left[\frac{Q_1(n)}{Q_2(n)} \right] + Q_2(n) \ln \left[\frac{Q_2(n)}{Q_1(n)} \right] \right\} \quad (19)$$

The JSD divergence is symmetric, i.e., $J(Q_1||Q_2) = J(Q_2||Q_1)$, and mathematically bounded between 0 and $\ln(2)$, with 0 indicating $Q_1 = Q_2$. The JSD for the three samples shown in Fig. 5 were computed by considering the FDFs extracted from the DNS of the premixed flame and those obtained by the $\beta - \beta$ analytical model. It can be seen from Fig. 5 that the $\beta - \beta$ analytical model is unable to capture more complex FDF shapes, such as bimodal distributions, as also confirmed by high JSD values. Thus, the need for more accurate models is motivated. Accurate predictions can be expected for $J(P||P_m) < 0.3$, whereas predictions with $J(P||P_m) > 0.6$ exhibit incorrect median values and overall shapes.

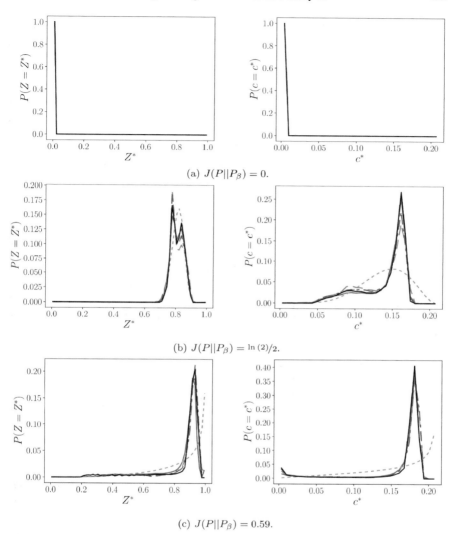

(a) $J(P||P_\beta) = 0.$

(b) $J(P||P_\beta) = \ln(2)/2.$

(c) $J(P||P_\beta) = 0.59.$

Fig. 5 Marginal FDF for low, mid-range, and high Jensen-Shannon divergence values for the $\beta - \beta$ PDF model. Red solid line is for RF model, green dashed line is for DNN model, blue dash-dotted line is for CVAE model, orange short dashed line is for $\beta - \beta$ model and black solid line is the DNS result (de Frahan et al. 2019)

The abilities of the three ML models to infer the subgrid FDF in regions other than $\mathcal{D}_3^t$ was also assessed because DNS results showed that the FDF in downstream locations were significantly different from those for $\mathcal{V}_3$. So, the ML models were trained using (1) $\mathcal{D}_3^t$ data (volume centered at z = 0.0775 m), (2) data from $\mathcal{D}_5^t$ (volume centered at z = 0.1025 m) and (3) data collected from the odd-numbered volumes $\mathcal{D}^t = \cup_{i=1,3,5,7,9} D_i^t$. The training data in the last case were representative of the entire computational domain. It was found that the models trained using data from a single volume were unable to infer the FDF in other volumes which was indicated by the high 90th percentile (J_{90}) of all the Jensen–Shannon divergences errors. The ML models trained using the odd-numbered volumes (test 3 above) gave $J_{90} < 0.2$ for the entire physical domain although only 4% of the DNS data from the entire computational domain was used for the training. Among the three ML modes, DNN yielded the lowest errors. The analytical $\beta - \beta$ model had J_{90} values which were almost twice of that for the ML models. The sample marginal FDFs of mixture fraction and progress variable for 3 different values of Jensen-Shannon divergences computed for the DNN model are shown in Fig. 6 and it is clear that the bimodal distributions are also captured quite well by the ML models.

Another generalisation test was conducted by using validation data generated from a different time snapshot of the DNS ($t = 0.059$ s). For this case, the DNN model trained on $\mathcal{D}^t = \cup_{i=1,3,5,7,9} D_i^t$ provided reasonable J_{90} values, although slightly higher than those obtained for the validation data from the same time snapshot of the training data. The $\beta - \beta$ model provided similar errors in both cases but three times higher than those of the DNN model. These generalisation tests demonstrated that the learned models are able to generalize temporally, as well as spatially. The results reported in this subsection suggest that it is important to use the training data covering the expected range of physical processes for which the ML is to be applied.

5.1.2 MILD Combustion

For the MILD combustion cases, the FDFs provided by DNN, $\beta - \beta$ and *copula* models are presented and compared to the DNS FDFs in Figs. 7, 8 and 9 for cases AZ1, AZ2 and BZ1 respectively. The DNN model significantly outperforms both analytical models and its prediction agrees very well with the DNS data for the different cases. As a general observation, the DNN captures the non-regular shapes of the marginal FDF of the progress variable quite well where the analytical models given by the β function and *copula* give *Gaussian*-like distributions. This difference has important implications for the reaction rate modelling as one shall see later in Sect. 5.2. For the mixture fraction, however, all models give good results but only the DNN is able to capture the asymmetry of the FDF which can be seen clearly in Fig. 9b and 9d for case BZ1. These results indicate promising capabilities of the DNN to predict the complex subgrid scalar statistics in MILD combustion.

It was noted by Chen et al. (2021) that the FDFs extracted directly using the instantaneous snapshots of DNS are random variables containing subgrid statistical information, as also pointed out in Pitsch (2006) and Pope (1985). The instantaneous

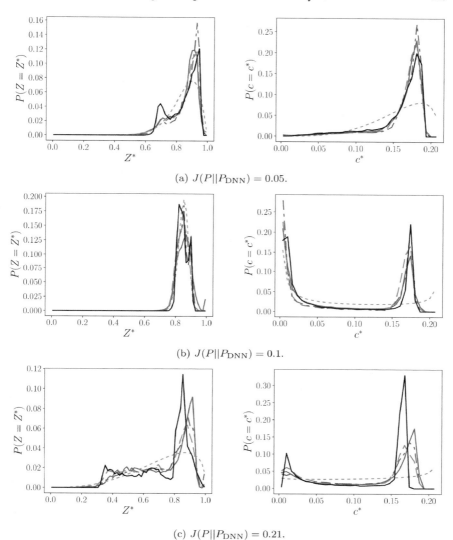

(a) $J(P||P_{\mathrm{DNN}}) = 0.05$.

(b) $J(P||P_{\mathrm{DNN}}) = 0.1$.

(c) $J(P||P_{\mathrm{DNN}}) = 0.21$.

Fig. 6 Marginal FDF for median and high Jensen-Shannon divergence values for models trained on $\mathcal{D}^t = \cup_{i=1,3,5,7,9} D_i^t$. Red solid line is for RF, green dashed line is for DNN, blue dash-dotted line is for CVAE, orange short dashed line is for $\beta - \beta$ model, and black solid line is for DNS (de Frahan et al. 2019)

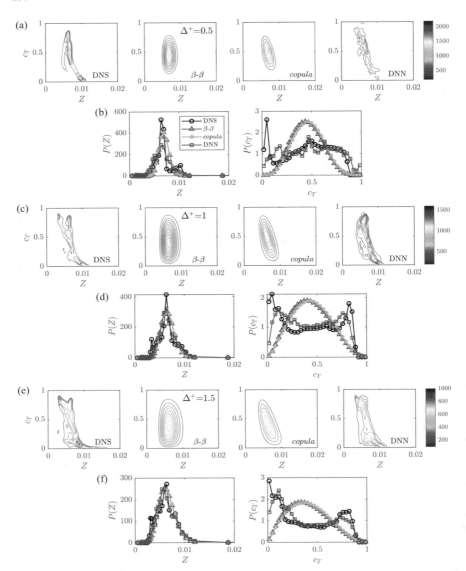

Fig. 7 Case AZ1: comparison of joint and marginal FDFs from DNS and models for filter sizes of $\Delta^+ = 0.5$ in (**a**) and (**b**), $\Delta^+ = 1$ in (**c**) and (**d**), and $\Delta^+ = 1.5$ in (**e**) and (**f**) (Chen et al. 2021)

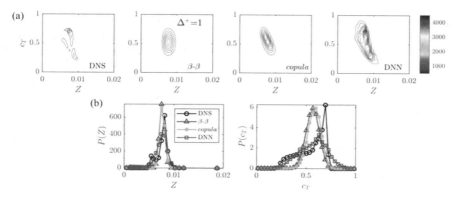

Fig. 8 Case AZ2: comparison of joint and marginal FDFs from DNS and models for filter sizes of $\Delta^+ = 0.5$ (Chen et al. 2021)

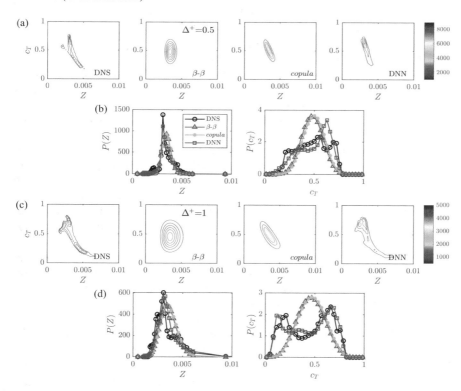

Fig. 9 Case BZ1: comparison of joint and marginal FDFs from DNS and models for filter sizes of $\Delta^+ = 0.5$ in (**a**) and (**b**), and $\Delta^+ = 1.0$ in (**c**) and (**d**) (Chen et al. 2021)

FDFs present certain levels of randomness due to the unsteady nature of single realisations. This randomness is removed to a good extent if the training data for ML are selected over many DNS realisations at a statistically stationary state. Therefore, following several experimental studies (Wang et al. 2007; Tong 2001; Cai et al. 2009), the instantaneous FDFs obtained from the DNS were conditioned on the resolved scalars, $\widetilde{Z}$ and $\widetilde{c_T}$, and then ensemble-averaged. A quantitative comparison of the conditionally averaged FDFs was then performed. Two variables, $\widetilde{Z}$ and $\widetilde{c_T}$, were considered as the number of available DNS samples was not sufficient to perform a statistically meaningful averaging on the four statistical moments used as ANN inputs. The resolved mixture fraction and progress variable were chosen so that the selected samples were located in the reaction zone ($\widetilde{c_T} \approx 0.5$). Figures 10 and 11 show the conditional FDFs, $\langle \widetilde{P}(Z, c_T) \mid \widetilde{Z}, \widetilde{c_T} \rangle$, for cases AZ1 and BZ1 respectively and the values of the conditioning variables are given in the figure captions. The DNN accurately reproduces the conditional joint and both marginal FDFs. It also captures the significant changes in the FDF shape with the varying filter size, especially for the progress variable. For case AZ1, both the β and *copula* models overpredict the peak when $\Delta^+ \leq 1$ for both Z and c_T distributions. However, for $\Delta^+ = 1.5$, the overall prediction is good for $\widetilde{P}(Z)$ and the peak of $\widetilde{P}(c_T)$ is also close to the DNS value although the shape is not captured. Similar results were reported for cases AZ2 also. For case BZ1, the mixture fraction distribution is predicted fairly well by all models for different Δ^+ values. However, both analytical models fail to predict the *bimodal-plateau* shape of $\widetilde{P}(c_T)$, which is typical of MILD combustion but seen seldom in conventional flames.

The JSD values were also calculated using Eq. (19), for the DNN and the two analytical models which confirmed the observations made using Figs. 7, 8, 9, 10 and 11. The JSD values provided by the DNN were much lower than those for the β and *copula* models. Improved predictions and lower JSD values were observed for all the models by increasing the filter size and this improvement was particularly significant for the DNN having $J_{90} < 0.05$. The DNN model performed equally well for Z and c_T.

To check for generalisation capability, the DNN was further validated using data which were not included in the learning/training step. The training and validation datasets included snapshots taken from $t = \tau_f$ to $1.2\tau_f$, where τ_f is the flow-through time, but the test data were taken using snapshots taken between $1.4\tau_f$ and $1.5\tau_f$. Substantial variations in the MILD combustion behaviour were observed among these snapshots (see Doan et al. 2018 for details). Hence, a robustly trained DNN is attractive if it can accurately infer a quantity of interest (here, FDF) for scenarios that have not been explicitly *seen* during the training process. The PDFs of the JSD values for the self-predictions (i.e., predictions performed on the training datasets) and unknown-predictions of the FDF are shown in Fig. 12. A filter size of $\Delta^+ = 1$ was used for all cases. As indicated in Fig. 12, the DNN provides a similar level of accuracy when *unseen* test data points are fed to the model. More than 80% of the JSD values are smaller than 0.05. The advantage of using DNN as FDF model is still unaffected since the majority of JSD values were larger than 0.1 for the β and *copula* FDF models. A slightly worse performance was achieved by the DNN when

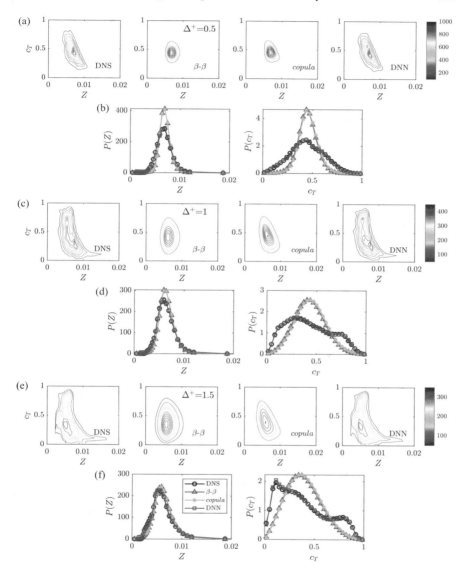

Fig. 10 Case AZ1: comparison of joint and marginal FDFs from DNS and models for **a** and **b** $\Delta^+ = 0.5, \widetilde{Z} = 0.007, \widetilde{c_T} = 0.45;$ **c** and **d** $\Delta^+ = 1, \widetilde{Z} = 0.0066, \widetilde{c_T} = 0.43;$ and **e** and **f** $\Delta^+ = 1.5,$ $\widetilde{Z} = 0.0064, \widetilde{c_T} = 0.39$ (Chen et al. 2021)

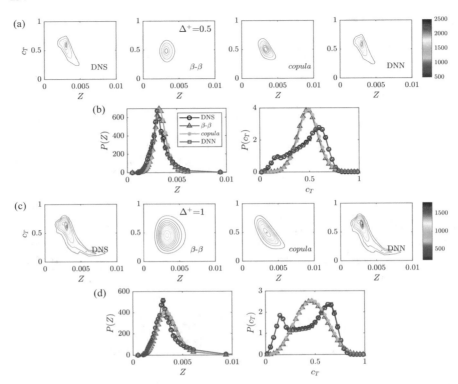

Fig. 11 Case BZ1: comparison of joint and marginal FDFs from DNS and models for **a** and **b** $\Delta^+ = 0.5$, $\widetilde{Z} = 0.00034$, $\widetilde{c_T} = 0.48$; and **c** and **d** $\Delta^+ = 1$, $\widetilde{Z} = 0.0036$, $\widetilde{c_T} = 0.46$ (Chen et al. 2021)

the training data came from cases AZ1 and BZ1, and the validation was done on case AZ2. The JSD results obtained from this new test with the self-predictions for $\Delta^+ = 0.5$ indicated that the overall performance was still good although the JSD distribution shifted towards higher JSD values. Further improvement on predictions is expected to be achieved if more datasets with different scenarios are included in the training.

5.1.3 Spray Flame

Yao et al. (2020) visually compared the FDF predicted by ANN and β-function model with the DNS values for one of the validation cases (CX1). Moreover, the data samples of this case were divided into three different groups characterized by filter size Δ_{LES}, to compute the sensitivity of the trained ANN model to LES grid sizes. The LES cells were selected randomly for a given $\widetilde{\xi}$ ranging from fuel-lean to fuel-rich conditions. The stoichiometric mixture fraction value is $\widetilde{\xi}_{st} = 0.068$.

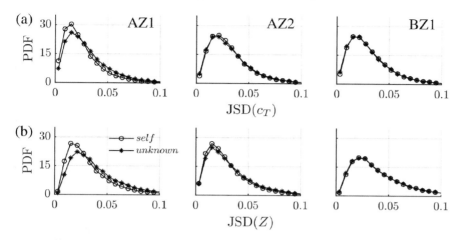

Fig. 12 Comparison of Jansen-Shannon divergence for DNN self- and unknown-predictions of FDF of **a** progress variable and **b** mixture fraction. The filter size for all cases is $\Delta^+ = 1.0$ (Chen et al. 2021)

Figure 13 compares FDF computed using ANN and β-function with DNS results for two filtered mixture fraction values and three Δ_{LES}. There is no marked differences in the ANN prediction for different Δ_{LES}. The ANN predictions of $\widetilde{P}(\eta)$ are in excellent agreement with the DNS results, including the peak value and its location. The FDF is skewed towards the lean side ($\eta < \xi_{st}$) for $\widetilde{\xi} = 0.05$ whereas it is stretched towards the rich side for $\widetilde{\xi} = 0.10$, and even a bimodal behaviour appears at larger filter sizes. The β-function does not seem to represent the FDFs well and numerical issues can arise when the mean is close to zero or unity with small SGS variance (Kronenburg et al. 2000).

5.2 Reaction Rate Predictions

The filtered reaction rate inferred by the ML models were also assessed against DNS results by de Frahan et al. (2019) for their premixed flame and by Chen et al. (2021) for the MILD combustion cases. The ML models used by de Frahan et al. inferred the unconditional filtered reaction rates $\bar{\omega}$, which are computed according to Eq. 6, and are shown in Fig. 14. Significant over predictions were observed for the $\beta - \beta$ model. The comparisons of the conditional reaction rates are also shown in Fig. 14.

The reaction rate in the transport equation for the filtered temperature-based progress variable, $\bar{\omega}_{c_T}$, can be computed using

$$\bar{\omega}_{c_T}(\boldsymbol{x}, t) = \int_0^1 \int_0^1 \langle \dot{\omega}_{c_T} \rangle \widetilde{P}(Z, c_T; \boldsymbol{x}, t) \, dZ \, dc_T, \tag{20}$$

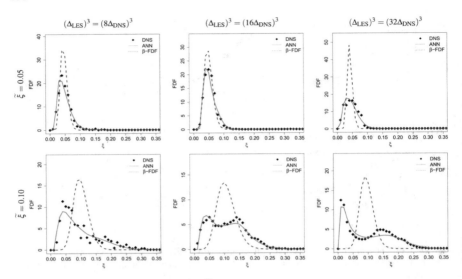

Fig. 13 Validation of ANN predictions of $\widetilde{P}(\eta)$ with DNS results for different LES grid sizes. The results are shown for $\widetilde{\xi} = 0.05$ (top) and $\widetilde{\xi} = 0.1$ (bottom) (Yao et al. 2020)

where the joint FDF $\widetilde{P}(Z, c_T)$ is obtained through the ANN in the MILD combustion cases investigated by Chen et al. (2021). The symbol $\langle \dot{\omega}_{c_T}(\boldsymbol{x}, t) \rangle = \langle \dot{\omega}_{c_T}(\boldsymbol{x}, t)/\rho(\boldsymbol{x}, t)|Z, c_T \rangle$ is defined as the doubly conditional mean reaction rate obtained from the DNS data. The instantaneous reaction rate of c_T is defined as $\dot{\omega}_{c_T} = \dot{q}/[c_p(T_b - T_u)]$, with $\dot{q}$ and c_p being the volumetric heat release rate and specific heat capacity of the mixture respectively. The conditional averages are computed using samples collected over the entire computational domain, see Sect. 3.2, and all the snapshots available (≈ 60) to achieve good statistical convergence. The authors verified that the doubly conditional mean rates have negligible variations in time and space, supporting the assumption of many turbulent combustion models (*viz.*, flamelets, see Bradley et al. 1990; Fiorina et al. 2003; Pierce and Moin 2004; van Oijen et al. 2016; and conditional moment-based methods, see Klimenko and Bilger 1999; Steiner and Bushe 2001) that the conditional means have small temporal and spatial variations if appropriate conditioning variables are used. The target filtered reaction rate $\overline{\dot{\omega}}_{c_T}^{m-DNS}$ was obtained by computing both the conditional mean reaction rate and the FDF in Eq. 20 directly from the DNS data. The scatter plots of $\overline{\dot{\omega}}_{c_T}^{m-DNS}$ and the reaction rates computed using FDFs obtained through β, *copula* and DNN models are presented in Fig. 15 for one of the DNS cases (AZ1) investigated in Chen et al. (2021). The qualitative behaviours and the trends were found to be similar for the other two cases. Although all models give reasonable predictions, the DNN outperforms the analytical models for all filter sizes. Moreover, the DNN predictions generally exhibit good symmetry about the diagonal, indicating a bias towards neither under- nor over-prediction, while the scatters for both the β

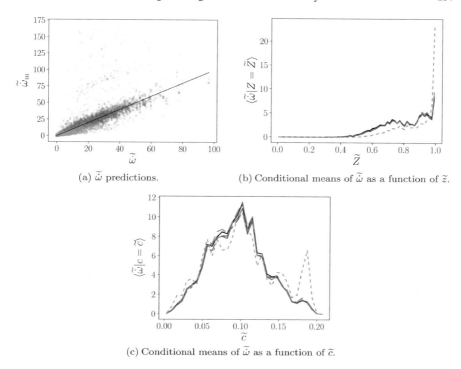

(a) $\widetilde{\dot{\omega}}$ predictions.

(b) Conditional means of $\widetilde{\dot{\omega}}$ as a function of $\widetilde{z}$.

(c) Conditional means of $\widetilde{\dot{\omega}}$ as a function of $\widetilde{c}$.

Fig. 14 Reaction rate $\overline{\dot{\omega}}$ inferred by the ML models trained on $\mathcal{D}^t = \cup_{i=1,3,5,7,9} D_i^t$. Red squares and solid line are for RF model, green diamonds and dashed line are for DNN, blue circles and dash-dotted line are for CVAE, orange pentagons and short dashed line are for $\beta - \beta$ model, and black solid line is for DNS result (de Frahan et al. 2019)

and *copula* models are asymmetric. As Δ^+ increases, the DNN prediction improves considerably whereas the performance of the analytical models does not follow this trend with the filter size. For both the β and *copula* models, a trend in the off-diagonal samples moving from under-predictions at small Δ^+ to over-predictions at larger Δ^+ can be seen.

6　Conclusions and Prospects

The application of ML algorithms to infer subgrid-scale filtered density functions (FDFs) in three test cases, i.e., swirling premixed flame, MILD and spray combustion, have been discussed in this chapter. Particularly, the promising results provided by deep neural networks (DNNs) for accurately inferring the FDFs have been shown. DNNs are generally able to capture the complex FDF behaviours and their variations with great accuracy across various combustion scenarios, turbulent and thermochem-

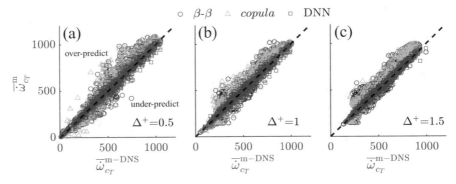

Fig. 15 Scatter plot of $\overline{\dot{\omega}}_{c_T}^{m-DNS}$ and $\overline{\dot{\omega}}_{c_T}$ (in kg/m³/s) modelled using different FDF models (denoted using different markers) for case AZ1. The results for different filter sizes are also shown (Chen et al. 2021)

ical conditions, and LES filter sizes. This can be achieved by manipulating the input data (extracted from DNS of these three cases), changing the network architecture, and tuning the network hyperparameters (e.g., learning rate, batch size). It has been shown that if the DNN training dataset is heterogeneous, i.e., it contains different possible outcomes of the quantities of interest, the DNN can handle *unknown* inputs quite well, suggesting a good model robustness. Thus, the DNN can be applied as a *black-box* model to other cases. By contrast, analytical models such as the β-function and *copula* models in most cases show their limitations quite clearly.

Although the above observations demonstrate the potential of DNN-based FDF modelling in combustion, several challenges remain and require further investigations. Searching for an optimal combination of the DNN hyperparameters can be highly time-consuming and computationally expensive. For example, an exhaustive grid search, looping through all combinations of layers and neurons to find an optimum, is not an easy task and may require cloud computing services (Yao et al. 2020). Moreover, due to the black-box nature of ML models, it is often hard to debug them to a satisfying level or improve them substantially after such a level is reached. This shifts the attention to the preprocessing of training data, which can be a daunting and time-consuming task, as mentioned in Chen et al. (2021). The lack of physical constraints in the training of ML models is yet another issue, and research is ongoing to develop physics-informed ML models that can respect physical laws and increase the interpretability and generalisation capability of ML models.

If DNNs are to replace combustion models, the overhead of retrieving predictions can also be of concern and counterbalance the observed savings in storage requirement. The overhead associated with the use of DNNs is highly machine-dependent and also network size-dependent. A posteriori LES studies need to quantify the computational times required by the DNN inference of FDFs and mean reaction rates. High inference times could hinder the development of in-situ capabilities, where the ML model is trained during the simulation, which can mitigate the risk of extrapolation. The latter can be reduced by also combining ML training and applications with uncertainty quantification or sensitivity analysis approaches that can effectively verify the performance of the ML model, provide a level of confidence in its predictions, guarantee that it does not violate physics laws and promote its more comprehensive application.

Machine Learning has induced notable advancements in combustion science. It has been effectively used for finding hidden patterns under large amounts of data, exploring and visualising high-dimensional input spaces, deriving complex mapping from inputs and outputs, and reducing computational cost and memory occupation (Zhou et al. 2022). However, many challenges and hence research opportunities are left to be addressed, and the development of physics-based ML approaches is just the starting point of a scientific paradigm shift that will bring new insights in combustion science with the help of big data. The combination of ML and combustion will provide solutions to daunting problems and enhance the understanding and deployment of novel combustion processes and technologies, which will shape a cleaner and sustainable future energy arena.

Acknowledgements Z. Li and N. Swaminathan acknowledge the support from EPSRC through the grant EP/S025650/1. Iavarone acknowledges the support of FRS-FNRS Fellowship.

References

An J, He G, Luo K, Qin F, Liu B (2020) Artificial neural network based chemical mechanisms for computationally efficient modeling of hydrogen/carbon monoxide/kerosene combustion. Int J Hydrogen Energy 45(53):29594–29605

Ali Sen B, Menon S (2010) Linear eddy mixing based tabulation and artificial neural networks for large eddy simulations of turbulent flames. Combust Flame 157(1):62–74

Bilger RW (1976) Structure of diffusion flames. Combust Sci Technol 13:155–170

Blasco JA, Fueyo N, Dopazo C, Ballester J (1998) Modelling the temporal evolution of a reduced combustion chemical system with an artificial neural network. Combust Flame 113(1–2):38–52

Blasco JA, Fueyo N, Larroya JC, Dopazo C, Chen J-Y (1999) A single-step time-integrator of a methane-air chemical system using artificial neural networks. Comp Chem Eng 23(9):1127–1133

Blasco J, Fueyo N, Dopazo C, Chen J-Y (2000) A self-organizing-map approach to chemistry representation in combustion applications. Combust Theory Model 4(1):61–76

Bradley D, Gaskell PH, Lau AKC (1990) A mixedness-reactedness flamelet model for turbulent diffusion flames. Proc Combust Inst 23(1):685–692

Bradley D, Gaskell PH, Gu XJ (1998) The mathematical modeling of liftoff and blowoff of turbulent non-premixed methane jet flames at high strain rates. Proc Combust Inst 27(1):1199–1206

Cai J, Wang D, Tong C, Barlow RS, Karpetis AN (2009) Investigation of subgrid-scale mixing of mixture fraction and temperature in turbulent partially premixed flames. Proc Combust Inst 32(1):1517–1525

Chatzopoulos AK, Rigopoulos S (2013) A chemistry tabulation approach via rate-controlled constrained equilibrium (RCCE) and artificial neural networks (ANNs), with application to turbulent non-premixed CH4/H2/N2 flames. Proc Combust Inst 34(1):1465–1473

Chen J-Y, Blasco JA, Fueyo N, Dopazo C (2000) An economical strategy for storage of chemical kinetics: fitting in situ adaptive tabulation with artificial neural networks. Proc Combust Inst 28(1):115–121

Chen ZX, Ruan S, Swaminathan N (2015) Simulation of turbulent lifted methane jet flames: effects of air-dilution and transient flame propagation. Combust Flame 162:703–716

Chen ZX, Ruan S, Swaminathan N (2017) Large eddy simulation of flame edge evolution in a spark-ignited methane-air jet. Proc Combust Inst 36:1645–1652

Chen ZX, Doan NAK, Ruan S, Langella I, Swaminathan N (2018) A priori investigation of subgrid correlation of mixture fraction and progress variable in partially premixed flames. Combust Theory Model 22:862–882

Chen ZX, Iavarone S, Ghiasi G, Kannan V, D'Alessio G, Parente A, Swaminathan N (2021) Application of machine learning for filtered density function closure in MILD combustion. Combust Flame 225:160–179

Chi C, Janiga G, Thévenin D (2021) On-the-fly artificial neural network for chemical kinetics in direct numerical simulations of premixed combustion. Combust Flame 226:467–477

Christo FC, Masri AR, Nebot EM, Pope SB (1996a) An integrated pdf/neural network approach for simulating turbulent reacting systems. Symp (International) Combust 26(1):43–48

Christo FC, Masri AR, Nebot EM, Turanyi T (1995) Utilising artificial neural network and repromodelling in turbulent combustion. In: Proceedings of ICNN'95 - international conference on neural networks, vol 2. IEEE, pp 911–916

Christo FC, Masri AR, Nebot EM (1996) Artificial neural network implementation of chemistry with pdf simulation of h2/co2 flames. Combust Flame 106(4):406–427

Cook AW, Riley JJ (1994) A subgrid model for equilibrium chemistry in turbulent flows. Phys Fluids 6:2868–2870

Darbyshire OR, Swaminathan N (2012) A presumed joint pdf model for turbulent combustion with varying equivalence ratio. Combust Sci Technol 184(12):2036–2067

Day M, Tachibana S, Bell J, Lijewski M, Beckner V, Cheng RK (2012) A combined computational and experimental characterization of lean premixed turbulent low swirl laboratory flames: I. methane flames. Combust Flame 159(1):275–290

de Frahan MTH, Yellapantula S, King R, Day MS, Grout RW (2019) Deep learning for presumed probability density function models. Combust Flame 208:436–450

Ding T, Readshaw T, Rigopoulos S, Jones WP (2021) Machine learning tabulation of thermochemistry in turbulent combustion: an approach based on hybrid flamelet/random data and multiple multilayer perceptrons. Combust Flame 231:111493

Doan NAK, Swaminathan N, Minamoto Y (2018) DNS of MILD combustion with mixture fraction variations. Combust Flame 189:173–189

Emami MD, Fard AE (2012) Laminar flamelet modeling of a turbulent ch4/h2/n2 jet diffusion flame using artificial neural networks. App Math Model 36(5):2082–2093

Fiorina B, Baron R, Gicquel O, Thevenin D, Carpentier S, Darabiha N (2003) Modelling non-adiabatic partially premixed flames using flame-prolongation of ILDM. Combust Theory Model 7:449–470

Flemming F, Sadiki A, Janicka J (2005) LES using artificial neural networks for chemistry representation. Prog Comput Fluid Dyn 5(7):375–385

Fox RO (2003) Computational models for turbulent reacting flows. Cambridge University Press, Cambridge, UK

Franke LLC, Chatzopoulos AK, Rigopoulos S (2017) Tabulation of combustion chemistry via artificial neural networks (anns): methodology and application to LES-PDF simulation of Sydney flame l. Combust Flame 185:245–260

Gicquel O, Darabiha N, Thevenin D (2000) Liminar premixed hydrogen/air counterflow flame simulations using flame prolongation of ILDM with differential diffusion. Proc Combust Inst 28(2):1901–1908

Giusti A, Sitte MP, Borghesi G, Mastorakos E (2018) Numerical investigation of kerosene single droplet ignition at high-altitude relight conditions. Fuel 225:663–670

Goodfellow I, Bengio Y, Courville A (2016) Deep learning. MIT Press

Grout RW, Swaminathan N, Cant RS (2009) Effects of compositional fluctuations on premixed flames. Combust Theory Model 13(5):823–852

Ihme M, Pitsch H (2008a) Prediction of extinction and reignition in nonpremixed turbulent flames using a flamelet/progress variable model: 1. a priori study and presumed pdf closure. Combust Flame 155(1):70–89

Ihme M, Pitsch H (2008b) Prediction of extinction and reignition in nonpremixed turbulent flames using a flamelet/progress variable model: 2. application in LES of Sandia flames D and E. Combust Flame 155(1-2):90–107

Ihme M, Marsden AL, Pitsch H (2006) On the optimization of artificial neural networks for application to the approximation of chemical systems. Center for Turbulence Research Annual Research Briefs, pp 105–118

Ihme M, Marsden AL, Pitsch H (2008) Generation of optimal artificial neural networks using a pattern search algorithm: application to approximation of chemical systems. Neural Comput 20(2):573–601

Ihme M, Schmitt C, Pitsch H (2009) Optimal artificial neural networks and tabulation methods for chemistry representation in les of a bluff-body swirl-stabilized flame. Proc Combust Inst 32(1):1527–1535

Ihme M, Chung WT, Mishra AA (2022) Combustion machine learning: principles, progress and prospects. Prog Energy Combust Sci 91:101010. https://doi.org/10.1016/j.pecs.2022.101010

Ioffe S, Szegedy C (2015) Batch normalization: accelerating deep network training by reducing internal covariate shift. arXiv:1502.03167

Jin B, Grout R, Bushe WK (2008) Conditional source-term estimation as a method for chemical closure in premixed turbulent reacting flow. Flow Turbul Combust 81(4):563–582

Jones WP, Kakhi M (1998) Pdf modeling of finite-rate chemistry effects in turbulent nonpremixed jet flames. Combust Flame 115(1–2):210–229

Jordan MI, Mitchell TM (2015) Machine learning: trends, perspectives, and prospects. Science 349(6245):255–260

Kempf A, Flemming F, Janicka J (2005) Investigation of lengthscales, scalar dissipation, and flame orientation in a piloted diffusion flame by les. Proc Combust Inst 30(1):557–565

Khan QS, Baek SW, Ghassemi H (2007) On the autoignition and combustion characteristics of kerosene droplets at elevated pressure and temperature. Combust Sci Technol 179(12):2437–2451

Kingma DP, Ba J (2014) Adam: a method for stochastic optimization. arXiv: 1412.6980

Klimenko AY, Bilger RW (1999) Conditional moment closure for turbulent combustion. Prog Energy Combust Sci 25(6):595–687

Kronenburg A, Bilger RW (2000) Kent JH (2000) Computation of conditional average scalar dissipation in turbulent jet diffusion flames. Flow Turbul Combust 64(3):145–159

Linse D, Kleemann A, Hasse C (2014) Probability density function approach coupled with detailed chemical kinetics for the prediction of knock in turbocharged direct injection spark ignition engines. Combust Flame 161(4):997–1014

Minamoto Y, Swaminathan N, Cant RS, Leung T (2014) Reaction zones and their structure in MILD combustion. Combust Sci Technol 186(8):1075–1096

Minamoto Y, Swaminathan N (2015) Subgrid scale modelling for MILD combustion. Proc Combust Inst 35(3):3529–3536

Minamoto Y, Swaminathan N (2014) Scalar gradient behaviour in MILD combustion. Combust Flame 161(4):1063–1075

Navarro-Martinez S, Kronenburg A, Di Mare F (2005) Conditional moment closure for large eddy simulations. Flow Turbul Combust 75(1):245–274

O'Brien EE, Jiang TL (1991) The conditional dissipation rate of an initially binary scalar in homogeneous turbulence. Phys Fluids A 3(12):3121–3123

Owoyele O, Kundu P, Ameen MM, Echekki T, Som S (2020) Application of deep artificial neural networks to multi-dimensional flamelet libraries and spray flames. Int J Engine Res 21(1):151–168

Pierce CD, Moin P (2004) Progress-variable approach for large-eddy simulation of non-premixed turbulent combustion. J Fluid Mech 504:73–97

Pitsch H (2006) Large-Eddy simulation of turbulent combustion. Annu Rev Fluid Mech 38:453–482

Plackett RL (1965) A class of bivariate distributions. J Amer Stat Assoc 310:516–522

Pope SB (2000) Turbulent flows. Cambridge University Press, Cambridge

Pope SB (1985) Pdf methods for turbulent reactive flows. Prog Energy Combust Sci 11:119–192

Pope SB (1990) Computations of turbulent combustion: progress and challenges. Proc Combust Inst 23:591–612

Pope SB (2013) Small scales, many species and the manifold challenges of turbulent combustion. Proc Combust Inst 34:1–31

Raman V, Pitsch H, Fox RO (2005) Hybrid large-eddy simulation/lagrangian filtered-density-function approach for simulating turbulent combustion. Combust Flame 143(1):56–78

Ranade R, Li G, Li S, Echekki T (2021) An efficient machine-learning approach for pdf tabulation in turbulent combustion closure. Combust Sci Technol 193(7):1258–1277

Readshaw T, Ding T, Rigopoulos S, Jones WP (2021) Modeling of turbulent flames with the large eddy simulation-probability density function (LES-PDF) approach, stochastic fields, and artificial neural networks. Phys. Fluids 33(3)

Ruan S, Swaminathan N, Darbyshire OR (2014) Modelling of turbulent lifted jet flames using flamelets: a priori assessment and a posteriori validation. Combust Theory Model 18(2):295–329

Rumelhart DE, Hinton GE, Williams RJ (1986) Learning representations by back-propagating errors. Nature 323:533–536

Sen BA, Menon S (2009) Turbulent premixed flame modeling using artificial neural networks based chemical kinetics. Proc Combust Inst 32(1):1605–1611

Sen BA, Hawkes ER, Menon S (2010) Large eddy simulation of extinction and reignition with artificial neural networks based chemical kinetics. Combust Flame 157(3):566–578

Steiner H, Bushe WK (2001) Large eddy simulation of a turbulent reacting jet with conditional source-term estimation. Phys Fluids 13(3):754–769

Stineman RW (1980) A consistently well-behaved method for interpolation. Creative Comput 6:54–57

Swaminathan N, Bai X-S, Haugen NEL, Fureby C, Brethouwer G (eds) (2022) Advanced turbulent combustion physics and applications. Cambridge University Press, Cambridge, UK

Tong C (2001) Measurements of conserved scalar filtered density function in a turbulent jet. Phys Fluids 13:2923

van Oijen JA, Donini A, Bastiaans RJM, ten Thije Boonkkamp JHM, de Goey LPH (2016) State-of-the-art in premixed combustion modeling using flamelet generated manifolds. Prog Energy Combust Sci 57:30–74

van Oijen JA, de Goey LPH (2002) Modelling of premixed counterflow flames using the flamelet-generated manifold method. Combust Theory Model 6:463–478

Verdonck T, Hubert M, Rousseeuw P (2009) Robust PCA for skewed data and its outlier map. Comput Stat Data Anal 53:2264–2274

Veynante D, Vervisch L (2002) Turbulent combustion modeling. Prog Energy Combust Sci 28:193–266

Wan K, Barnaud C, Vervisch L, Domingo P (2020) Chemistry reduction using machine learning trained from non-premixed micro-mixing modeling: application to dns of a syngas turbulent oxy-flame with side-wall effects. Combust Flame 220:119–129

Wan K, Barnaud C, Vervisch L, Domingo P (2021) Machine learning for detailed chemistry reduction in dns of a syngas turbulent oxy-flame with side-wall effects. Proc Combust Inst 38(2):2825–2833

Wang D, Tong C, Barlow RS, Karpetis AN (2007) Experimental study of scalar filtered mass density function in turbulent partially premixed flames. Proc Combust Inst 31(1):1533–1541

Wang F, Liu R, Li M, Yao J, Jin J (2018) Kerosene evaporation rate in high temperature air stationary and convective environment. Fuel 211:582–590

Wang B, Kronenburg A, Stein OT (2019) A new perspective on modelling passive scalar conditional mixing statistics in turbulent spray flames. Combust Flame 208:376–387

Wright YM, De Paola G, Boulouchos K, Mastorakos E (2005) Simulations of spray autoignition and flame establishment with two-dimensional cmc. Combust Flame 143(4):402–419

Yao S, Wang B, Kronenburg A, Stein OT (2020) Modeling of sub-grid conditional mixing statistics in turbulent sprays using machine learning methods. Phys Fluids 32(11)

Zhang Y, Xu S, Zhong S, Bai X-S, Wang H, Yao M (2020) Large eddy simulation of spray combustion using flamelet generated manifolds combined with artificial neural networks. Energy AI 2:100021

Zheng Z, Lin X, Yang M, He Z, Bao E, Zhang H, Tian Z (2020) Progress in the application of machine learning in combustion studies. Energy Environ 9:1–14

Zhou ZJ, Lü Y, Wang ZH, Xu YW, Zhou JH, Cen K (2013) Systematic method of applying ann for chemical kinetics reduction in turbulent premixed combustion modeling. Chinese Sci Bul 58(4):486–492

Zhou L, Song Y, Ji W, Wei H (2022) Machine learning for combustion. Energy AI 7:100128

Reduced-Order Modeling of Reacting Flows Using Data-Driven Approaches

K. Zdybał, M. R. Malik, A. Coussement, J. C. Sutherland, and A. Parente

Abstract Data-driven modeling of complex dynamical systems is becoming increasingly popular across various domains of science and engineering. This is thanks to advances in numerical computing, which provides high fidelity data, and to algorithm development in data science and machine learning. Simulations of multicomponent reacting flows can particularly profit from data-based reduced-order modeling (ROM). The original system of coupled partial differential equations that describes a reacting flow is often large due to high number of chemical species involved. While the datasets from reacting flow simulation have high state-space dimensionality, they also exhibit attracting low-dimensional manifolds (LDMs). Data-driven approaches can be used to obtain and parameterize these LDMs. Evolving the reacting system using a smaller number of parameters can yield substantial model reduction and savings in computational cost. In this chapter, we review recent advances in ROM of turbulent reacting flows. We demonstrate the entire ROM workflow with a particular focus on obtaining the training datasets and data science and machine learning techniques such as dimensionality reduction and nonlinear regression. We present recent results from ROM-based simulations of experimentally measured Sandia flames D and F. We also delineate a few remaining challenges and possible future directions to address them. This chapter is accompanied by illustrative examples using the recently developed Python software, **PCAfold**. The software can be used to obtain, analyze and improve low-dimensional data representations. The examples provided herein can be helpful to students and researchers learning to apply dimensional-

K. Zdybał · M. R. Malik · A. Coussement · A. Parente (✉)
Aero-Thermo-Mechanics Laboratory, École polytechnique de Bruxelles, Université Libre de Bruxelles, Brussels, Belgium
e-mail: alessandro.parente@ulb.be

Brussels Institute for Thermal-fluid Systems, Brussels (BRITE), Université Libre de Bruxelles and Vrije Universiteit Brussel, Brussels, Belgium

J. C. Sutherland
Department of Chemical Engineering, University of Utah, Salt Lake City, UT, USA
e-mail: james.sutherland@utah.edu

N. Swaminathan and A. Parente (eds.), *Machine Learning and Its Application to Reacting Flows*, Lecture Notes in Energy 44, https://doi.org/10.1007/978-3-031-16248-0_9

ity reduction, manifold approaches and nonlinear regression to their problems. The
Jupyter notebook with the examples shown in this chapter can be found on GitHub at
`https://github.com/kamilazdybal/ROM-of-reacting-flows-`
`Springer`.

1 Introduction

There is growing interest and numerous recent developments in reduced-order modeling (ROM) of complex dynamical systems (Kutz et al. 2016; Taira et al. 2017; Lusch
et al. 2018; Mendez et al. 2019; Raissi et al. 2019; Dalakoti et al. 2020; Ramezanian
et al. 2021; Han et al. 2022; Zhou et al. 2022). While these systems can be characterized by a large number of degrees of freedom, they often exhibit low-rank structures
(Maas and Pope 1992; Holmes et al. 1997; Pope 2013; Yang et al. 2013; Mendez et al.
2018). Describing the evolution of those structures provides a powerful modeling
approach with substantial reduction to the number of partial differential equations
(PDEs) solved in computational simulations (Sutherland and Parente 2009; Biglari
and Sutherland 2015; Echekki and Mirgolbabaei 2015; Owoyele and Echekki 2017;
Malik et al. 2018, 2020).

   Reacting flow simulations can profit from model reduction due to initially high
state-space dimensionality stemming from large chemical mechanisms. Reacting
systems can often be effectively re-parameterized with much fewer variables. Numerous physics-based parameterization techniques can be found in the combustion literature (Maas and Pope 1992; Van Oijen and De Goey 2002; Jha and Groth 2012;
Gicquel et al. 2000). An alternative to the physics-motivated parameterization is a
data-driven approach, where low-dimensional manifolds (LDMs) are constructed
directly from the training data (Sutherland and Parente 2009; Yang et al. 2013). In
particular, dimensionality reduction techniques can be used to define LDMs in the
original thermo-chemical state-space. Among many available linear and nonlinear
techniques, principal component analysis (PCA) (Jolliffe 2002) is commonly used
in combustion to obtain a linear mapping between the original variables and the
LDM (Sutherland and Parente 2009; Mirgolbabaei and Echekki 2013; Echekki and
Mirgolbabaei 2015; Isaac et al. 2015; Biglari and Sutherland 2015). In PCA, the
new parameterizing variables, called principal components (PCs), can be obtained
by projecting the training data onto a newly identified basis. A small number of the
first few PCs defines the LDM. ROMs can then be built based on this new parameterization. As one example of ROM, PDEs describing the first few PCs can be evolved
in combustion simulations (Sutherland and Parente 2009) which result in a substantial reduction of computational costs as compared to transporting the original state
variables.

   Often, ROM workflows incorporate nonlinear regression to bypass the reconstruction errors associated with an inverse basis transformation. Regression can thus
provide an effective route back from the reduced space to the original state-space
where the thermo-chemical quantities of interest such as temperature, pressure and

composition, can be retrieved. Regression models can also provide closure for any non-conserved manifold parameters. Nonlinear regression techniques such as artificial neural network (ANN) (Mirgolbabaei and Echekki 2014; Dalakoti et al. 2020), multivariate adaptive regression splines (MARS) (Biglari and Sutherland 2015) or Gaussian process regression (GPR) (Isaac et al. 2015; Malik et al. 2018, 2020) were used in the past in the context of ROM.

In this chapter, we present the complete ROM workflow for application in reacting flow simulations. We begin with a concise mathematical description of a general multicomponent reacting flow. Understanding the governing equations of the analyzed system is a crucial starting point for applying data science tools on the resulting thermo-chemical state vector. After a discussion of training datasets, we present the derivation of the ROM in the context of reacting flows. We review the combination of dimensionality reduction techniques with nonlinear regression. We discuss three popular choices for nonlinear regression: ANNs, GPR and kernel regression. Finally, we review recent results from *a priori* and *a posteriori* ROM of challenging combustion simulations.

Throughout this chapter, we delineate a few outstanding challenges that remain in ROM of combustion processes. For instance, projecting the data onto a lower-dimensional basis, as is done in many ROMs, can introduce undesired behaviors on LDMs. Observations that are distant in the original space can be collapsed into a single, overlapping region. In the overlapping region, those observations are indistinguishable and the projection can become multi-valued. When the identified manifold is used as regressor, these topological behaviors on LDMs can make the regression process more difficult. Ideally, we would like to search for such parameters defining the LDM, that the resulting regression function uniquely represents all dependent variables. Recent work by Zhang et al. (2020) has demonstrated that regressing variables that have significant spatial gradients can be challenging using ANN. Steep gradients can be particularly associated with minor species whose non-zero mass fractions can be located on small portions of the manifold. Problems with ANN reconstruction of minor species on a PCA-derived manifold have recently been reported by Dalakoti et al. (2020). Nevertheless, the attempts to link the poor regression performance with the manifold topology are still scarce in the existing literature, with only a few studies emerging recently (Malik et al. 2022a; Perry et al. 2022; Zdybał et al. 2022c). We show examples of quantitative measures to assess the quality of LDMs that can help bridge this gap. We argue that the future research efforts should focus on advancing strategies that improve regression on manifolds. This should allow to better leverage the capability of techniques such as ANNs or GPR to approximate even highly nonlinear relationships between variables (Hornik et al. 1989).

PCAfold examples

The present chapter includes illustrative examples using **PCAfold** (Zdybał et al. 2020), a Python software package for generating, analyzing and improving LDMs. It incorporates the entire ROM workflow from data preprocessing, through dimensionality reduction to novel tools for assessing the quality of LDMs. **PCAfold** is composed of three main modules: `preprocess`, `reduction` and `analysis`. In brief, the `preprocess` module allows for data preprocessing such as centering and scaling, sampling, clustering and outlier removal. The `reduction` module introduces dimensionality reduction using PCA. The available variants are global and local PCA, subset PCA and PCA on sampled datasets. Finally, the `analysis` module combines functionalities for assessing LDM quality and nonlinear regression results. Each module is accompanied by plotting functions that allow for efficient viewing of results. For instructions on installing the software and for further illustrative tutorials, the reader is referred to the documentation: `https://pcafold.readthedocs.io/`. In the **PCAfold** examples that follow, we present a complete workflow that can be adopted for a combustion dataset, using all three modules in series: `preprocess` → `reduction` → `analysis`. We begin by importing the three modules:

```
from PCAfold import preprocess
from PCAfold import reduction
from PCAfold import analysis
```

2 Governing Equations for Multicomponent Mixtures

In this section, we begin with the description of the governing equations for low-Mach multicomponent mixtures, whose solution is the starting point for obtaining training datasets for ROMs in reacting flow applications. In the discussion that follows, $\nabla \cdot \boldsymbol{\phi}$ denotes the divergence of a vector quantity $\boldsymbol{\phi}$, $\nabla \boldsymbol{\phi}$ (or $\nabla \phi$) denotes the gradient of a vector quantity $\boldsymbol{\phi}$ (or a scalar quantity ϕ) and the: symbol denotes tensor contraction. The material derivative is defined as $D/Dt := \partial/\partial t + \mathbf{v} \cdot \nabla$. We let $\mathbf{v}$ be the mass-averaged convective velocity of the mixture, defined as

$$\mathbf{v} := \sum_{i=1}^{n} Y_i \mathbf{u}_i \,, \tag{1}$$

where Y_i is the mass fraction of species i, $\mathbf{u}_i$ is the velocity of species i and n is the number of species in the mixture. At a given point in space and time, transport of physical quantities in a multicomponent mixture can be described by the following set of governing equations written in the conservative (strong) form:

- Continuity equation:

$$\frac{\partial \rho}{\partial t} = -\nabla \cdot \rho \mathbf{v}, \tag{2}$$

where ρ is the mixture density.
- Species mass conservation equation:

$$\frac{\partial \rho Y_i}{\partial t} = -\nabla \cdot \rho Y_i \mathbf{v} - \nabla \cdot \mathbf{j}_i + \omega_i \qquad \text{for } i = 1, 2, \ldots, n-1, \tag{3}$$

where $\mathbf{j}_i$ is the mass diffusive flux of species i relative to the mass-averaged velocity and ω_i is the net mass production rate of species i due to chemical reactions. Note, that summation of Eqs. (3) over all n species yields the continuity equation (Eq. (2)) since $\sum_{i=1}^{n} Y_i = 1$, $\sum_{i=1}^{n} \mathbf{j}_i = 0$ and $\sum_{i=1}^{n} \omega_i = 0$. For this reason, only $n-1$ independent species mass conservation equations are solved. Mass fraction of the nth species can be computed from the constraint $\sum_{i=1}^{n} Y_i = 1$.
- Momentum equation:

$$\frac{\partial \rho \mathbf{v}}{\partial t} = -\nabla \cdot \rho \mathbf{v} \mathbf{v} - \nabla \cdot \boldsymbol{\tau} - \nabla \cdot p\mathbf{I} + \rho \sum_{i=1}^{n} Y_i \mathbf{f}_i, \tag{4}$$

where $\boldsymbol{\tau}$ is the viscous momentum flux tensor, p is pressure, $\mathbf{I}$ is the identity tensor and $\mathbf{f}_i$ is the net acceleration from body forces applied on species i.

with one of the following forms of the energy equation:

- Total internal energy equation:

$$\frac{\partial \rho e_0}{\partial t} = -\nabla \cdot \rho e_0 \mathbf{v} - \nabla \cdot \mathbf{q} - \nabla \cdot \boldsymbol{\tau} \cdot \mathbf{v} - \nabla \cdot p\mathbf{v} + \sum_{i=1}^{n} \mathbf{f}_i \cdot \mathbf{n}_i, \tag{5}$$

where e_0 is the mixture specific total internal energy, $\mathbf{q}$ is the heat flux and $\mathbf{n}_i := \rho Y_i \mathbf{u}_i$ is the total mass flux of species i.
- Internal energy equation:

$$\frac{\partial \rho e}{\partial t} = -\nabla \cdot \rho e \mathbf{v} - \nabla \cdot \mathbf{q} - \boldsymbol{\tau} : \nabla \mathbf{v} - p\nabla \cdot \mathbf{v} + \sum_{i=1}^{n} \mathbf{f}_i \cdot \mathbf{j}_i, \tag{6}$$

where e is the mixture specific internal energy.
- Enthalpy equation:

$$\frac{\partial \rho h}{\partial t} = -\nabla \cdot \rho h \mathbf{v} - \nabla \cdot \mathbf{q} - \boldsymbol{\tau} : \nabla \mathbf{v} + \frac{Dp}{Dt} + \sum_{i=1}^{n} \mathbf{f}_i \cdot \mathbf{j}_i, \tag{7}$$

where h is the mixture specific enthalpy.

- Temperature equation:

$$\frac{\partial \rho T}{\partial t} = -\nabla \cdot \rho T \mathbf{v} - \frac{1}{c_p} \nabla \cdot \mathbf{q} + \frac{\alpha T}{c_p} \frac{Dp}{Dt} - \frac{1}{c_p} \boldsymbol{\tau} : \nabla \mathbf{v} + \frac{1}{c_p} \sum_{i=1}^{n} \left(h_i (\nabla \cdot \mathbf{j}_i - \omega_i) + \mathbf{f}_i \cdot \mathbf{j}_i \right), \quad (8)$$

where T is the temperature, α is the coefficient of thermal expansion of the mixture ($\alpha = 1/T$ for an ideal gas), c_p is the mixture isobaric specific heat capacity and h_i is the enthalpy of species i.

The governing equations can also be re-formulated using a reference velocity different from the mass-averaged velocity used here. A different mixture velocity would not only affect the terms involving $\mathbf{v}$ explicitly, but also an appropriate diffusive flux will have to be formulated.

The set of governing equations is closed by a few additional relations. The first one is an equation of state. For an ideal gas, we have

$$p = \frac{\rho R_u T}{M}, \quad (9)$$

where R_u is the universal gas constant and $M = \left(\sum_{i=1}^{n} Y_i / M_i \right)^{-1}$ is the molar mass of the mixture where M_i is the molar mass of species i. For a chemically reacting flow, we also require a chemical mechanism that relates temperature, T, pressure, p, and composition, $[Y_1, Y_2, \ldots, Y_n]$, to the chemical source terms, ω_i. The heat flux, $\mathbf{q}$, requires modeling as it in general can include all possible means of heat transfer. One encountered model for $\mathbf{q}$ can be written using the standard Fourier term and the term representing heat transfer through molecular diffusion of species:

$$\mathbf{q} = -\lambda \nabla T + \sum_{i=1}^{n} h_i \mathbf{j}_i, \quad (10)$$

where λ is the mixture thermal conductivity. We also require a model for the diffusive fluxes, $\mathbf{j}_i$. Assuming Fick's law as a model for diffusion, we can express the mass diffusive flux as

$$\mathbf{j}_i = -\rho \boldsymbol{\mathcal{D}} \nabla Y_i, \quad (11)$$

where $\boldsymbol{\mathcal{D}}$ is a matrix of Fickian diffusion coefficients that are functions of the binary diffusion coefficients and composition. Finally, we require a model for the viscous momentum flux tensor, $\boldsymbol{\tau}$. Assuming Newtonian fluids, $\boldsymbol{\tau}$ can be expressed as:

$$\boldsymbol{\tau} = -\mu \left(\nabla \mathbf{v} + (\nabla \mathbf{v})^{\top} \right) + \left(\frac{2}{3} \mu - \kappa \right) (\nabla \cdot \mathbf{v}) \mathbf{I}, \quad (12)$$

where μ is the mixture viscosity. κ is the mixture dilatational viscosity and $\top$ denotes matrix transpose. The reader is referred to numerous great resources for a deeper discussion of multicomponent mass transfer or derivation of the equations above (Taylor and Krishna 1993; Giovangigli 1999; Bird et al. 2006; Kee et al. 2005).

The governing equations given by Eqs. (2)–(8) can be written in a general matrix form:

$$\frac{\partial \mathbf{X}^\top}{\partial t} = -\nabla \cdot \mathbf{C}^\top - \nabla \cdot \mathbf{D}^\top + \mathbf{S}^\top, \tag{13}$$

where $\mathbf{X} \in \mathbb{R}^{N \times Q}$ is the thermo-chemical state vector, $\mathbf{C} \in \mathbb{R}^{d \times N \times Q}$ is the convective flux vector, $\mathbf{D} \in \mathbb{R}^{d \times N \times Q}$ is the diffusive flux vector and $\mathbf{S} \in \mathbb{R}^{N \times Q}$ is the source terms vector. Here, Q is the number of transported properties, d is the number of spatial dimensions of the problem and N is the number of observations. The observations can for instance be linked to measurements on a spatio-temporal grid of a discretized domain. Typically, $N \gg Q$, but the magnitude of Q strongly depends on the number of species in the mixture. In combustion problems, Q can easily reach the order of hundreds when large chemical mechanisms are used (Lu and Law 2009). The appropriate formulation of $\mathbf{X}$, $\mathbf{C}$, $\mathbf{D}$ and $\mathbf{S}$ will depend on a given problem and the assumed simplifications to the governing equations. In the most general case, when all transport equations are solved and no further simplifications are made to the governing equations as given by Eqs. (2)–(8), we form the columns of $\mathbf{X}$, $\mathbf{C}$, $\mathbf{D}$ and $\mathbf{S}$ as per Table 1. Note, that the order of columns in $\mathbf{X}$ does not matter, as long as the corresponding column in $\mathbf{C}$, $\mathbf{D}$ and $\mathbf{S}$ carries an appropriate term. Since the thermo-chemical state of a single-phase multicomponent system is defined by $Q = n + 1$

Table 1 Formulation of the thermo-chemical state vector, $\mathbf{X}$, the convective flux vector, $\mathbf{C}$, the diffusive flux vector, $\mathbf{D}$, and the source terms vector, $\mathbf{S}$, in the most general case, where no further assumptions are imposed to the strong form of the governing equations given by Eqs. (2)–(8)

Equation	State vector	Convective flux vector	Diffusive flux vector	Source terms vector
	(Columns of $\mathbf{X}$)	(Columns of $\mathbf{C}$)	(Columns of $\mathbf{D}$)	(Columns of $\mathbf{S}$)
Continuity	ρ	$\rho \mathbf{v}$	0	0
Species mass	ρY_i	$\rho Y_i \mathbf{v}$	$\mathbf{j}_i$	ω_i
Momentum	$\rho \mathbf{v}$	$\rho \mathbf{v} \mathbf{v}$	$\boldsymbol{\tau} + p\mathbf{I}$	$\rho \sum_{i=1}^{n} Y_i \mathbf{f}_i$
Total internal energy	ρe_0	$\rho e_0 \mathbf{v}$	$\mathbf{q} + \boldsymbol{\tau} \cdot \mathbf{v} + p\mathbf{v}$	$\sum_{i=1}^{n} \mathbf{f}_i \cdot \mathbf{n}_i$
Internal energy	ρe	$\rho e \mathbf{v}$	$\mathbf{q}$	$-\boldsymbol{\tau} : \nabla \mathbf{v} - p\nabla \cdot \mathbf{v} + \sum_{i=1}^{n} \mathbf{f}_i \cdot \mathbf{j}_i$
Enthalpy	ρh	$\rho h \mathbf{v}$	$\mathbf{q}$	$-\boldsymbol{\tau} : \nabla \mathbf{v} + \frac{Dp}{Dt} + \sum_{i=1}^{n} \mathbf{f}_i \cdot \mathbf{j}_i$
Temperature	ρT	$\rho T \mathbf{v}$	0	$-\frac{1}{c_p} \nabla \cdot \mathbf{q} + \frac{\alpha T}{c_p} \frac{Dp}{Dt} - \frac{1}{c_p} \boldsymbol{\tau} : \nabla \mathbf{v} + \frac{1}{c_p} \sum_{i=1}^{n} \left(h_i (\nabla \cdot \mathbf{j}_i - \omega_i) + \mathbf{f}_i \cdot \mathbf{j}_i \right)$

variables, an example state vector that follows from the conservative form of the governing equations can be: $\mathbf{X} = [\rho, \rho e, \rho Y_1, \rho Y_2, \ldots, \rho Y_{n-1}]$ (the conserved state vector). For the reasons explained earlier, we only include $n - 1$ independent species mass fractions. Mass fraction of the most abundant species is most often removed (Niemeyer et al. 2017). Historically, specific momentum quantity ($\rho\mathbf{v}$) has not been included in the state vector in ROM of reacting flows (Sutherland and Parente 2009). Various other definitions of the state vector, $\mathbf{X}$, can be adopted with the caveat that the system given by Eq. (13) should not be over-specified (Giovangigli 1999; Hansen and Sutherland 2018). In the next section, we review several strategies to obtain data matrices $\mathbf{X}$, $\mathbf{C}$, $\mathbf{D}$ and $\mathbf{S}$.

3 Obtaining Data Matrices for Data-Driven Approaches

High-dimensional datasets, that are typical to reacting flow applications, can come from numerical simulations or experiments. A few types of numerical datasets of varying complexity often used in the context of ROM are presented in Fig. 1. In particular, solving the governing equations presented in Sect. 2 for simple reacting systems is one computational strategy to obtain training data for ROM. Those simple systems can include zero-dimensional reactors, strained laminar flamelets (Peters 1988), one-dimensional flames or one-dimensional turbulence (ODT) (Kerstein 1999; Sutherland et al. 2010; Echekki et al. 2011). With sufficient amount of assumptions made to the governing equations, we can obtain those datasets at a relatively cheap computational cost. Relaxing some of those assumptions, on the other hand, can move us along the axis of an increasing complexity of the training data, incorporating more information about the turbulence-chemistry interaction. At the end of the complexity spectrum, we have a full direct numerical simulation (DNS), which results in high-fidelity data with all spatial and temporal scales directly resolved. Resorting to more expensive numerical simulations, such as large eddy simulation (LES) or DNS, might not be necessary for ROM purposes. For instance, ODT datasets have been shown to reproduce the DNS conditional statistics well (Punati et al. 2011; Abboud et al. 2015; Lignell et al. 2015; Punati et al. Oct 2016) and have therefore been frequently used in the context of ROM (Mirgolbabaei and Echekki 2014; Mirgolbabaei et al. 2014; Mirgolbabaei and Echekki 2015; Biglari and Sutherland 2015) since they are computationally cheaper to obtain. For an additional overview of datasets presented in Fig. 1 the reader is referred to (Zdybał et al. 2022a).

As an illustrative example, the governing equations for an adiabatic, incompressible, zero-dimensional reactor simplify to:

$$\frac{\partial T}{\partial t} = -\frac{1}{\rho c_p} \sum_{i=1}^{n} h_i \omega_i, \qquad \frac{\partial Y_i}{\partial t} = \frac{\omega_i}{\rho} \qquad \text{for } i = 1, 2, \ldots, n - 1.$$

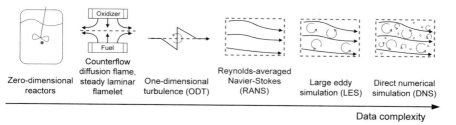

Fig. 1 Schematic overview of training datasets for ROM. As we move along the axis of an increasing complexity, more physical detail is incorporated into the reacting flow simulation

Since a zero-dimensional reactor represents combustion happening in a single point in space, all spatial derivatives present in Eqs. (2)–(8) vanish. Collecting all observations of T and Y_i into a matrix $\mathbf{X}$, and collecting all observations of $-1/\rho c_p \sum_{i=1}^{n} h_i \omega_i$ and ω_i/ρ into a matrix $\mathbf{S}$, we get

$$
\mathbf{X} = \begin{bmatrix} \vdots & \vdots & \vdots & & \vdots \\ T & Y_1 & Y_2 & \dots & Y_{n-1} \\ \vdots & \vdots & \vdots & & \vdots \end{bmatrix} \quad \text{and} \quad \mathbf{S} = \begin{bmatrix} \vdots & & \vdots & \vdots & & \vdots \\ -\frac{1}{\rho c_p} \sum_{i=1}^{n} h_i \omega_i & \frac{\omega_1}{\rho} & \frac{\omega_2}{\rho} & \dots & \frac{\omega_{n-1}}{\rho} \\ \vdots & & \vdots & \vdots & & \vdots \end{bmatrix} .
$$

Note, that even though we have removed the transport equation for the nth species, the temperature equation still couples all species through the $-\sum_{i=1}^{n} h_i \omega_i$ term, which represents the heat release rate.

4 Reduced-Order Modeling

At this point, we have learned how to construct training datasets which are the starting point for applying data-driven approaches. It has been a frequent trend in recent years to apply dimensionality reduction techniques to combustion datasets, both for ROM and for data analysis. In the context of combustion, techniques such as PCA (Sutherland and Parente 2009), local PCA (Parente et al. 2009, 2011), kernel PCA (Mirgolbabaei and Echekki 2014), t-distributed stochastic neighbor embedding (t-SNE) (Fooladgar and Duwig 2018), independent component analysis (ICA) (Gitushi et al. 2022), non-negative matrix factorization (NMF) (Zdybał et al. 2022a) or autencoders (Zhang et al. 2021) have been used. In this chapter, we focus on using dimensionality reduction techniques solely to model reduction. We use the premise that the original dataset, $\mathbf{X}$, of high rank can be efficiently approximated by a matrix of a much lower rank. The data can then be re-parameterized with the new manifold parameters (Sutherland et al. 2007). Dimensionality reduction is often coupled with nonlinear regression to provide a more robust mapping between the manifold parameters and the quantities of interest. In this section, we review ROM strategies for reacting flows that include dimensionality reduction and nonlinear regression.

4.1 Data Preprocessing

The first step towards applying dimensionality reduction is data preprocessing. The most straightforward way is data normalization (centering and scaling), which allows to equalize the importance of physical variables of different numerical ranges. Any variable ϕ in a dataset can be centered and scaled using the general formula $\widetilde{\phi} = (\phi - c)/d$, where c is the center computed as the mean value of ϕ and d is the scaling factor. Other data preprocessing means can include data sampling to tackle imbalance in sample densities, data subsetting (feature selection), or outlier removal. The effect of data preprocessing, including scaling and outlier removal, on the resulting LDMs was studied in (Parente and Sutherland 2013). In the discussion that follows, we assume that the training datasets have been appropriately preprocessed.

4.2 Reducing the Number of Governing Equations

Data-driven model reduction has emerged in recent years with applications to complex dynamical systems. Model reduction of complex systems typically starts with changing the basis to represent the original high-dimensional system. Let $\mathbf{A} \in \mathbb{R}^{Q \times Q}$ be the matrix of modes defining the new basis. The matrix $\mathbf{A}$ can be found directly from the training data using a dimensionality reduction technique, such as PCA. As long as $\mathbf{A}$ is constant in space and time, the governing equations of the form presented in Eq. (13) can be written as:

$$\frac{\partial \mathbf{A} \cdot \mathbf{X}^\top}{\partial t} = -\nabla \cdot \mathbf{A} \cdot \mathbf{C}^\top - \nabla \cdot \mathbf{A} \cdot \mathbf{D}^\top + \mathbf{A} \cdot \mathbf{S}^\top , \qquad (14)$$

where $\mathbf{X}$ can in general contain all state variables as presented in Sect. 2, or a subset of those. Equation (14) represents transformation of the original governing equations to the new basis defined by $\mathbf{A}$.

4.2.1 Principal Component Transport

PCA is one dimensionality reduction technique that can be used to obtain the basis matrix $\mathbf{A}$ by performing eigendecomposition of the data covariance matrix. PCA can provide optimal reaction variables, PCs, that are linear combinations of the original thermo-chemical state variables (Sutherland 2004; Sutherland and Parente 2009; Parente et al. 2009). We can define the matrix of PCs, $\mathbf{Z} \in \mathbb{R}^{N \times Q}$, as $\mathbf{Z} = \mathbf{XA}$, which represents the transformation of $\mathbf{X}$ to the new PCA-basis. The governing equations written in the form of Eq. (13) can be linearly transformed to this new PCA-basis as per Eq. (14). This yields a new set of transport equations for the PCs:

$$\frac{\partial \mathbf{Z}^\top}{\partial t} = -\nabla \cdot \mathbf{C_Z}^\top - \nabla \cdot \mathbf{D_Z}^\top + \mathbf{S_Z}^\top , \tag{15}$$

where $\mathbf{C_Z} = \mathbf{CA}$ are the projected convective fluxes, $\mathbf{D_Z} = \mathbf{DA}$ are the projected diffusive fluxes and $\mathbf{S_Z} = \mathbf{SA}$ are the PC source terms – the source terms of the original state-space variables transformed to the new PCA-basis. We will further refer to the jth PC (the jth column of $\mathbf{Z}$) as Z_j and to the jth PC source term (the jth column of $\mathbf{S_Z}$) as $S_{Z,j}$. By solving the transport equations for the first q PCs only, we can significantly reduce the number of PDEs in Eq. (15) as compared to Eq. (13). PCA further guarantees that the q first PCs are the most important ones in terms of the variance retained in the data. From the Eckart-Young theorem (Eckart and Young 1936), we know that approximating the dataset $\mathbf{X}$ with only q first PCs gives the closest rank-q approximation to $\mathbf{X}$. This approximation can be obtained through an inverse basis transformation: $\mathbf{X} \approx \mathbf{Z_q A_q}^{-1}$, where the subscript $\mathbf{q}$ denotes truncation to q components. With the PCA modeling approach, the first q PCs become the reaction variables that re-parameterize the original thermo-chemical state-space. They also define the q-dimensional manifold, embedded in the originally Q-dimensional state space.

Formulation of PC-transport was first proposed by Sutherland and Parente (2009). Since then, numerous *a priori* (Biglari and Sutherland 2012; Mirgolbabaei and Echekki 2013; Mirgolbabaei et al. 2014; Malik et al. 2018; Ranade and Echekki 2019; Dalakoti et al. 2020; D'Alessio G et al. 2022; Zdybał et al. 2022c) and *a posteriori* (Isaac et al. 2014; Biglari and Sutherland 2015; Echekki and Mirgolbabaei 2015; Coussement et al. 2016; Owoyele and Echekki 2017; Ranade and Echekki 2019; Malik et al. 2020, 2022a, b) studies have been conducted. The advantage of PCA-based modeling is that models can be trained on datasets coming from simpler systems that are cheap to compute (such as zero-dimensional reactors or laminar flamelets, see Sect. 3). This has been shown to be a feasible modeling strategy (Malik et al. 2018, 2020), as long as the training data covers the possible states of the reacting system that might be accessed during simulation of real systems.

There are a few additional ingredients of the PC-transport modeling approach. First, since Eq. (15) is solved for the PCs which do not have any physical relevance, we require a mapping back to the original thermo-chemical state-space, where physical quantities of interest can be retrieved. Second, we need to parameterize the source terms, $\mathbf{S_Z}$, of any non-conserved manifold parameters (Sutherland 2004; Sutherland and Parente 2009). While in the original state space we have known relations between the transported variables and their source terms, we lack such explicit relations in the space of PCs. Both these points can be handled by coupling nonlinear regression with the PC-transport model—this will be further discussed in Sect. 4.4. Finally, in the presence of diffusion, diffusive fluxes need to be represented in the new PCA-basis as well. Treatment of PC diffusive fluxes was proposed by Mirgolbabaei and Echekki (2014) and by Biglari and Sutherland (2015). A study by Echekki and Mirgolbabaei (2015) further looked into mitigating the multicomponent effects associated with diffusion of PCs. Another study by Coussement et al. (2016) looked at the influence of differential diffusion on PCA-based models. The work done in (Coussement et al. 2016) looked at how rotation of the PCs can diagonalize the PCs diffusion coefficients matrix and thus make the treatment of diffusion of PCs easier.

Computing the PCs and the PC source terms

In this example, we demonstrate how one can obtain the PCs and the PC source terms from the state vector, **X**, and the source terms vector, **S**, respectively. We use a syngas/air steady laminar flamelet dataset and generate its two-dimensional (2D) projection onto the PCA-basis. The dataset was generated using **Spitfire** Python library (Hansen et al. 2022) and the chemical mechanism by Hawkes et al. (2007).

Load the dataset, removing the nth species, N_2:

```
import numpy as np
X = np.genfromtxt('syngas-air-SLF-state-space.csv', delimiter=',')
    [:,0:-1]
S = np.genfromtxt('syngas-air-SLF-state-space-sources.csv', delimiter
    =',')[:,0:-1]
f = np.genfromtxt('syngas-air-SLF-mixture-fraction.csv', delimiter
    =',')
chi = np.genfromtxt('syngas-air-SLF-dissipation-rates.csv', delimiter
    =',')
(n_observations, n_variables) = X.shape
```

Perform PCA on the dataset:

```
pca = reduction.PCA(X, scaling='auto', n_components=2)
```

Transform the state vector, **X**, to the new PCA basis:

```
Z = pca.transform(X)
```

Transform the source terms vector, **S**, to the new PCA basis (note the nocenter=True flag):

```
S_Z = pca.transform(S, nocenter=True)
```

Visualize the 2D projection of the dataset, colored by the two PC source terms, $S_{Z,1}$ and $S_{Z,2}$ (Fig. 2):

```
plt = reduction.plot_2d_manifold(Z[:,0], Z[:,1],
                                 color=S_Z[:,0],
                                 s=15,
                                 x_label='$Z_{1}$ [$-$]', y_label='
        $Z_{2}$ [$-$]',
                                 colorbar_label='$S_{Z, 1}$\n[$-$]',
                                 color_map='inferno',
                                 grid_on=True,
                                 figure_size=(6,4))
```

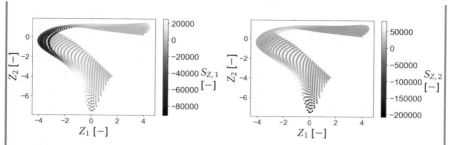

Fig. 2 Outputs of `analysis.plot_2d_manifold`

It is visible from the plot above that this 2D projection introduces significant non-uniqueness that particularly affects the dependent variable $S_{Z,1}$. At the same time, this visible overlap in the (Z_1, Z_2) space does not coincide with the region of the largest variation in the second PC source term, $S_{Z,2}$, values. We can expect that $S_{Z,1}$ will be much more strongly affected by the manifold non-uniqueness than $S_{Z,2}$.

4.3 Low-Dimensional Manifold Topology

Apart from PCA, numerous manifold learning methods can help identify LDMs in high-dimensional combustion datasets. Although the approach presented in Sect. 4.2.1 allows for substantial model reduction, several manifold challenges need to be addressed. In particular, during projection of data to a lower-dimensional basis, non-uniqueness can be introduced in the manifold topology which can hinder successful model definition. A good model should provide unique definition of all relevant dependent variables as functions of the manifold parameters (Sutherland 2004; Pope 2013). With this premise, the future research directions can be twofold. First, we require techniques to characterize the quality of LDMs. Second, we should seek strategies that provide an improved manifold topology. Both points should feed one another and can be tackled simultaneously.

Measures such as the coefficient of determination (Biglari and Sutherland 2012) or manifold nonlinearity (Isaac et al. 2014) have been used in the past to assess manifold parameterizations *a priori*. A recently proposed normalized variance derivative metric (Armstrong and Sutherland 2021) is much more informative in comparison. It can characterize manifold quality with respect to two important aspects: feature sizes and multiple scales of variation in the dependent variable space. Multiple scales of variation can often indicate non-uniqueness in manifold parameterization. A more compact metric based on the normalized variance derivative has also been proposed recently (Zdybał et al. 2022b). It reduces the manifold topology to a single number and can be used as a cost function in manifold optimization tasks.

Some topological challenges can be mitigated through appropriate data preprocessing prior to projecting to a lower-dimensional space. The most straightforward strategy is data scaling, with Pareto (Noda 2008) or VAST (Hector et al. 2003) scalings most commonly used (Biglari and Sutherland 2015; Isaac et al. 2015; Malik et al. 2018, 2020). Other authors have tackled manifold challenges by training combustion models on only a subset of the original thermo-chemical state-space variables (Chatzopoulos and Rigopoulos 2013; Mirgolbabaei and Echekki 2013, 2014; Echekki and Mirgolbabaei 2015; Isaac et al. 2015; Owoyele and Echekki 2017; Malik et al. 2020; Nguyen et al. 2021; Gitushi et al. 2022). Recent work developed a strategy for a manifold-informed state vector subset selection (Zdybał et al. 2022b). A study done by Coussement et al. (2012) suggests that tackling initial imbalance in data density can yield a more accurate low-dimensional representation of the flame region.

Another important decision that needs to be made at the modeling stage is what manifold dimensionality, q, should we select? Additional number of parameters may be required for more complex manifold topologies. While techniques such as PCA provide orthogonal manifold parameters (PCs), each bringing information about variance in another orthogonal data dimension, it is not clear how many PCs is sufficient to provide a good quality, regressible manifold topology. From the computational cost point of view, keeping low manifold dimensionality is desired. However, keeping q small should not be at the expense of the parameterization quality. Admittedly, more work is required to provide answers to those questions.

Low-dimensional manifold assessment

Below, we demonstrate how we can assess the quality of LDMs obtained from PCA using the novel normalized variance derivative metric (Armstrong and Sutherland 2021). We will assess the generated 2D projections and we take the two PC source terms as the two dependent variables. Define the bandwidth values, σ:

```
bandwidth_values = np.logspace(-5, 1, 100)
```

Specify the names of the dependent variables:

```
variable_names=['$S_{Z,1}$', '$S_{Z,2}$']
```

Compute the normalized variance derivative, $\hat{\mathcal{D}}(\sigma)$:

```
variance_data = analysis.compute_normalized_variance(Z, S_Z,
    variable_names,
                                        bandwidth_values
    =bandwidth_values)
```

Plot the $\hat{\mathcal{D}}(\sigma)$ curves for the two PC source terms (Fig. 3):

```
analysis.plot_normalized_variance_derivative(variance_data,
                            color_map='Greys',
                                figure_size=(10,2.5)
    )
```

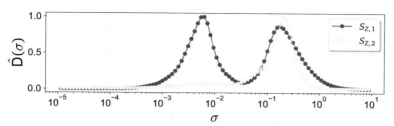

Fig. 3 Output of `analysis.plot_normalized_variance_derivative`

The normalized variance derivative, $\hat{\mathcal{D}}(\sigma)$, quantifies the information content on a manifold at various length scales specified by the bandwidth, σ. The peaks in the $\hat{\mathcal{D}}(\sigma)$ profile happening at very small length scales can often be linked to non-uniqueness in manifold topologies. In the plot above, we can observe two distinct peaks corresponding to the $\hat{\mathcal{D}}(\sigma)$ curve for the first PC source term, $S_{Z,1}$. The peak happening for smaller σ can be understood from our visualization of the manifold topology in Fig. 2. In our visualization we have seen clear overlap, where the observations corresponding to highly negative values of $S_{Z,1}$ were projected directly above observations corresponding to $S_{Z,1} \approx 0$. The information provided by $\hat{\mathcal{D}}(\sigma)$ is valuable at the modeling stage, as it allows to quantitatively assess the quality of low-dimensional data projections.

4.4 Nonlinear Regression

Nonlinear regression is often used to provide an effective mapping between the manifold parameters and the dependent variables of interest (Biglari and Sutherland 2015; Mirgolbabaei and Echekki 2015; Malik et al. 2018; Dalakoti et al. 2020). The set of dependent variables, $\boldsymbol{\phi}$, typically include the PC source terms, $\mathbf{S_Z}$, and the thermochemical state-space variables, such as temperature, density and composition. Unlike the inverse basis transformation discussed in Sect. 4.2.1, regression has the potential to yield much more accurate dependent variable reconstructions (Mirgolbabaei and Echekki 2015). Nonlinear regression techniques allow us to encode nonlinear relationships between the manifold parameters and the dependent variables. This characteristic is especially desired for modeling source terms, which are highly nonlinear functions of the independent variables. In the past research, reconstruction of the PC source terms has been shown to be much more challenging than reconstruction of the state variables (Biglari and Sutherland 2012, 2015). This is due to the fact that the state-space variables evolve nonlinearly according to the Arrhenius relations.

In this section, we are concerned with a set of n_ϕ dependent variables defined as $\boldsymbol{\phi} = [\mathbf{S_Z}, T, \rho, \mathbf{Y}_i]$, where $\mathbf{Y}_i$ is a vector of $n - 1$ species mass fractions, $\mathbf{Y}_i = [Y_1, Y_2, \ldots, Y_{n-1}]$. In mathematical terms, the goal of nonlinear regression is to find a function $\mathscr{F}$, such that:

$$\phi \approx \mathscr{F}(\mathbf{Z_q}), \tag{16}$$

where ϕ is a dependent variable and $\mathbf{Z_q}$ are the q first PCs. It is worth noting that some regression techniques allow to obtain all dependent variables at once; other require regressing dependent variables one-by-one. Three popular nonlinear regression techniques are reviewed in this section. Our main focus is in presenting how the function $\mathscr{F}$ is defined in each technique.

Nonlinear regression

In the examples that follow, we will perform and assess ANN, GPR and kernel regression of the two PC source terms defined earlier. The nonlinear regression models will be trained on 80% and tested on the remaining 20% of the data. Below, we use the sampling functionalities to randomly sample train and test data:

```
sample_random = preprocess.DataSampler(np.zeros((n_observations,)).
    astype(int),
                                    random_seed=100,
                                    verbose=True)
(idx_train, idx_test) = sample_random.random(80)
Z_train = Z[idx_train,:]; Z_test = Z[idx_test,:]
S_Z_train = S_Z[idx_train,:]; S_Z_test = S_Z[idx_test,:]
```

4.4.1 Artificial Neural Network

Artificial neural networks (ANNs) are a network of connected layers that compute the output(s) based on some convolution of the layer's input(s) (Russell and Norvig 2002). The layer's inputs and outputs are called neurons. ANNs form a parametric technique that can be used both for regression and classification and are broadly used in the context of ROM. This applies to both reacting (Mirgolbabaei and Echekki 2013, 2014, 2015; Echekki and Mirgolbabaei 2015; Ranade and Echekki 2019; Dalakoti et al. 2020; Zhang et al. 2020) and non-reacting (pure fluid) applications (Farooq et al. 2021).

For an architecture with a single neural layer (input $\rightarrow$ output), the regression function $\mathscr{F}$ at some query point P can be written as:

$$\mathscr{F}\big|_P = g_1(\mathbf{Z_q}\big|_P \mathbf{W}_1 + \mathbf{b}_1), \tag{17}$$

where $\mathbf{W}_1 \in \mathbb{R}^{q \times n_\phi}$ is the matrix of weights and $\mathbf{b}_1 \in \mathbb{R}^{1 \times n_\phi}$ is the vector of biases, and g_1 is the activation function. Both $\mathbf{W}_1$ and $\mathbf{b}_1$ are learned from the training data by solving an optimization problem. For a deep neural network (DNN) which allow for multi-layer architecture, the regression function becomes a composition of functions of the form shown in Eq. (17). Assuming m neural layers, we can write that

$$\mathscr{F}\big|_p = g_m(g_{m-1}(\cdots g_2(g_1(\mathbf{Z_q}\big|_p \mathbf{W}_1 + \mathbf{b}_1)\mathbf{W}_2 + \mathbf{b}_2)\cdots \mathbf{W}_{m-1} + \mathbf{b}_{m-1})\mathbf{W}_m + \mathbf{b}_m), \tag{18}$$

where all matrices $\mathbf{W}_l$ as well as all vectors $\mathbf{b}_l$ for layers $l = 1, 2, \ldots, m$, do not need to be of the same size, since the number of neurons can vary in different layers. Also the activation functions g_l can vary for different layers. The Eq. (18) essentially states in matrix notation that the output of one layer becomes an input of the following layer.

The advantage of using ANN regression is that predictions are relatively cheap to compute once the ANN model has been trained. As can be seen from Eqs. (17)–(18), predicting a single observation of ϕ given a set of query inputs, $\mathbf{Z_q}\big|_p$, requires vector-matrix multiplication(s), where $\mathbf{W}_l$ is typically a small matrix. This makes ANNs very appealing from the computational cost point of view. However, the optimization used to determine weights and biases is prone to reaching local minimum. The best one can hope for is that the local minimum will result in reasonable predictions. The overall performance of the trained network is dependent on many factors that the user can tune, such as the architecture or the choice of the activation function(s). The ANN predictions are also dependent on the random initial guess for the weights and biases which can greatly affect gradient descent -based algorithms. To improve the network performance, Bayesian optimization can be used to determine the ANN hyper-parameters (Mockus 2012; Bergstra et al. 2013; Barzegari and Geris 2021).

ANN regression

In this example, we create an ANN model to obtain the parameterizing function, $\mathscr{F}$. We will use a popular Python library for ANN, **Keras** (Chollet et al 2015), which is a backend of the **TensorFlow** software (Abadi et al. 2015). Below, we import the necessary libraries:

```
from keras.models import Sequential
from keras.layers import Dense
from keras import optimizers
from keras import losses
```

We use a relatively simple architecture with two hidden layers with five neurons each:

```
model = Sequential([
Dense(5, input_dim=2, activation='sigmoid'),
Dense(5, activation='sigmoid'),
Dense(2, activation='linear')])
```

Normalize the ANN outputs to the $\langle -1; 1 \rangle$ range:

```
(normalized_S_Z, centers, scales) = preprocess.center_scale(S_Z, '-1
    to1')
```

Sample the normalized train data outputs:

```
normalized_S_Z_train = normalized_S_Z[idx_train,:]
```

Compile the ANN model with the given architecture :

```
model.compile(optimizers.Adam(lr=0.001),
              loss=losses.mean_squared_error,
              metrics=['mse'])
```

Fit the compiled ANN model with the training data, specifying the hyper-parameters:

```
history = model.fit(Z_train,
                    normalized_S_Z_train,
                    batch_size=100, epochs=500,
                    validation_split=0.2, verbose=0)
```

Finally, we predict the two PC source terms, remembering to invert the $\langle -1; 1 \rangle$ normalization applied initially:

```
S_Z_ANN_predicted = model.predict(Z)
S_Z_ANN_predicted = preprocess.invert_center_scale(S_Z_ANN_predicted,
                                                   centers, scales)
```

We can visualize the regression result in 3D (Fig. 4):

```
analysis.plot_3d_regression(Z[:,0],
                            Z[:,1],
                            S_Z[:,0],
                            S_Z_ANN_predicted[:,0],
                            elev=30,
                            azim=200,
                            x_label='$Z_1$ [$-$]',
                            y_label='$Z_2$ [$-$]',
                            z_label='$S_{Z, 1}$ [$-$]',
                            figure_size=(12,6))
```

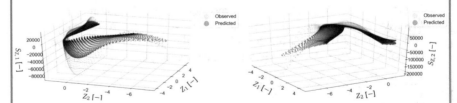

Fig. 4 Outputs of `analysis.plot_3d_regression`

The figure above demonstrates qualitatively how regression can struggle to regress dependent variables on an ill-behaved manifold. We can observe regions with large mismatch between the observed and the predicted values of the two PC source terms. In particular, highly negative values of $S_{Z,1}$ are poorly predicted. This behavior can be linked to our manifold topology assessments in the earlier examples, where we have seen non-uniqueness affecting highly negative values of $S_{Z,1}$.

4.4.2 Gaussian Process Regression

Gaussian process regression (GPR) is a kernel-based, semi-parametric regression technique (Williams and Rasmussen 2006). A powerful characteristic of GPR is that prior knowledge about the functional relationship between the independent and dependent variables can be injected at the modeling stage. For instance, if the system dynamics is known to have an oscillatory behavior, the kernel can be built using a periodic function. Another important feature of GPR is that it provides uncertainty bounds on the predicted variables, while techniques such as ANN or kernel regression only provide predictions.

In GPR, the regression function $\mathscr{F}$ is learned from the data:

$$\mathscr{F}(\mathbf{Z_q}) = \mathcal{GP}(m(\mathbf{Z_q}), \mathbf{K}(\mathbf{Z_q}, \mathbf{Z_q}')) , \tag{19}$$

where $\mathcal{GP}$ denotes a Gaussian process, m is the mean function and $\mathbf{K}$ is the covariance matrix. The covariance matrix, $\mathbf{K} \in \mathbb{R}^{n_x \times n_y}$, can be populated using any kernel of choice as long as the elements in $\mathbf{K}$ satisfy $k_{i,j} = k_{j,i}$, $\forall_{i \neq j}$. Typically, kernels are functions of the distance between data observations, x_i and x_j. Squared exponential kernel is commonly used to populate $\mathbf{K}$:

$$k_{i,j} = h^2 \exp\left(\frac{(x_i - x_j)^2}{\lambda^2}\right), \tag{20}$$

where h is the scaling factor and λ is the bandwidth of the kernel. Figure 5a visualizes the effect of increasing the kernel bandwidth, λ, on the resulting covariance matrix structure. With a larger λ, we are allowing observations that are further apart

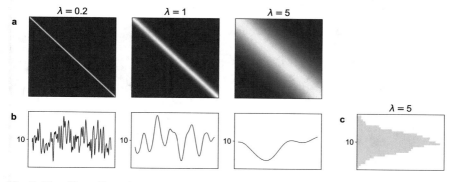

Fig. 5 The effect of kernel bandwidth on smoothing the Gaussian process regression predictions. In this example, the scaling factor $h = 0.1$. **a** Heatmaps of three covariance matrices, $\mathbf{K}$, generated using the squared exponential kernel with an increasing kernel bandwidth, λ. **b** Example regression function realizations resulting from each covariance matrix. **c** Histogram of one hundred function realizations corresponding to the $\lambda = 5$ case with the mean equal to 10. The mean dictates the most probable function value

to correlate. The structure of **K** is then reflected in possible regression function realizations (Fig. 5b). With a very narrow kernel (here $\lambda = 0.2$), the resulting realization looks very noisy—even nearby observations can have very different function values. The larger the kernel bandwidth, the smoother the realization function (Duvenaud 2014). With $\lambda = 5$ we can expect stronger correlation in function values even for observations that are further away. Figure 5c additionally shows a histogram of one hundred regression function realizations resulting from $\lambda = 5$. Since in this example we have chosen the mean equal to 10, the histogram has a Gaussian distribution centered around 10.

GPR regression

In this example, we create a GPR model to obtain the parameterizing function, $\mathscr{F}$. We will use a Python package **george** (Ambikasaran et al. 2016) to perform GPR:

```
import george
```

Create the squared exponential kernel:

```
kernel = george.kernels.ExpSquaredKernel(20, ndim=2)
```

Fit the GPR model with the training data:

```
gp = george.GP(kernel)
gp.compute(Z_train, yerr=1.25e-12,)
```

Predict the two PC source terms:

```
S_Z1_GPR_predicted, S_Z1_GPR_var = gp.predict(S_Z_train[:,0], Z,
    return_var=True)
S_Z2_GPR_predicted, S_Z2_GPR_var = gp.predict(S_Z_train[:,1], Z,
    return_var=True)
```

We visualize the predicted PC source terms (Fig. 6):

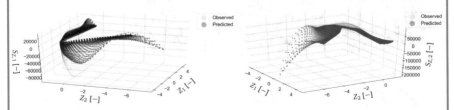

Fig. 6 Outputs of `analysis.plot_3d_regression`

In the plot above, we observe similar misprediction of the first PC source term, $S_{Z,1}$, as we have seen with ANN regression.

4.4.3   Kernel Regression

Kernel regression is a nonparametric technique that does not include the "training" step. Function $\mathscr{F}$ is inferred for each query point, P, directly from the training data samples in some vicinity of P. The regression function $\mathscr{F}$ is built from the Nadaraya-Watson estimator (Härdle 1990) as:

$$\mathscr{F}\Big|_P = \frac{\sum_{i=1}^N K_{i,P}(\mathbf{Z_q}, \sigma)\phi_i}{\sum_{i=1}^N K_{i,P}(\mathbf{Z_q}, \sigma)} , \tag{21}$$

where K is the kernel function and σ is the kernel bandwidth. The Eq. (21) essentially represents a linear combination of the weighted observations of ϕ. Similarly as in GPR, various kernels can be used in place of K. The most popular Gaussian kernel yields:

$$K_{i,P}(\mathbf{Z_q}, \sigma) = \exp\left(\frac{-||\mathbf{Z_q}|_i - \mathbf{Z_q}|_P||_2^2}{\sigma^2}\right) , \tag{22}$$

The larger the kernel bandwidth, σ, the larger the resulting coefficients K_i multiplying each data observation, ϕ_i. In other words, an increasing σ yields a stronger influence of data observations distant from P on the predicted function value at P. An implication of a larger σ on regression means that $\mathscr{F}$ becomes a smoother function – note the similarity of this concept with the covariance matrix discussion in Sect. 4.4.2.

Kernel regression

In this example, we create a kernel regression model to obtain the parameterizing function, $\mathscr{F}$. We specify the kernel bandwidth, σ, for the Nadaraya-Watson estimator:

```
bandwidth = 0.5
```

Fit the kernel regression model with the training data:

```
model = analysis.KReg(Z_train, S_Z_train)
```

Predict the two PC source terms:

```
S_Z_KReg_predicted = model.predict(Z, bandwidth=bandwidth)
```

Similarly as before, we visualize the predicted PC source terms (Fig. 7):

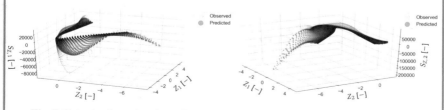

Fig. 7 Outputs of `analysis.plot_3d_regression`

Since kernel regression makes predictions by "smoothing out" function values over some neighborhood of a query point, the non-uniqueness in $S_{Z,1}$ values affected regression performance, similarly to what we have observed with ANN and GPR regression.

Nonlinear regression assessment

Here, we continue the kernel regression example and use various metrics to assess the regression performance. Two common metrics that are available are the coefficient of determination, R^2, and the normalized root mean squared error (NRMSE). For vector quantities, such as the PC source terms vector, another useful metric might be the good direction estimate (GDE) which is a measure derived from cosine similarity.

Compute the regression metrics for the two PC source terms:

```
metrics = analysis.RegressionAssessment(S_Z, S_Z_KReg_predicted,
                                        variable_names=variable_names
        ,
                                        norm='std',
                                        tolerance=0.05)
```

Display the regression metrics in a table format (Fig. 8):

```
metrics.print_metrics(table_format=['pandas'], metrics=['R2', 'NRMSE
        ', 'GDE'])
```

	R2	NRMSE	GDE
$S_{Z,1}$	0.4800	0.7211	45.2449
$S_{Z,2}$	0.8100	0.4358	45.2449

Fig. 8 Output of `analysis.RegressionAssessment print_metrics`.

The `RegressionAssessment` class also allows to compare two regression results. It can color-code the displayed table and mark the metrics that got worse red and those that got better green.

In addition to a single value of each metric for the entire dataset, we can also compute stratified metrics values, in bins (clusters) of a dependent variable. This allows us to observe how regression performed in specific regions of the manifold. Below, we compute the stratified metrics in four bins of the first PC source term, $S_{Z,1}$. We then look at the kernel regression of the first PC source term in each bin.

We first use the function from the `preprocess` module that allows to manually partition the dataset into bins of a selected variable. Compute the bins:

```
(idx, _) = preprocess.predefined_variable_bins(S_Z[:,0],
                                        split_values=[-10000,
        0, 10000],
                                        verbose=False)
```

Those data bins (clusters) are visualized below on the syngas/air flamelet dataset in the space of mixture fraction and temperature (Fig. 9):

```
preprocess.plot_2d_clustering(f, X[:,0], idx,
                              x_label='$f$ [-]', y_label='$T$ [$K$]',
                              first_cluster_index_zero=False,
                              color_map='coolwarm',
                              figure_size=(8,4))
```

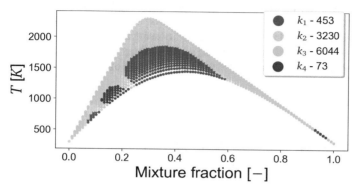

Fig. 9 Output of `preprocess.plot_2d_clustering`

Compute the stratified regression metrics:

```
metrics = analysis.RegressionAssessment(S_Z[:,0], S_Z_KReg_predicted
    [:,0],
                              idx=idx,
                              use_global_mean=True,
                              norm='std',
                              use_global_norm=True)
```

Display the stratified regression metrics in a table format (Fig. 10):

```
metrics.print_stratified_metrics(table_format=['pandas'], metrics=['
    NRMSE'])
```

	Observations	Min	Max	NRMSE
k1	453	-91,760.3782	-10,016.6359	3.0987
k2	3230	-9,989.6371	-0.0000	0.4287
k3	6044	0.0000	9,965.7716	0.1419
k4	73	10,002.6112	24,987.7263	0.6479

Fig. 10 Output of `analysis.RegressionAssessmentprint_stratified_metrics`.

The stratified metrics let us see that kernel regression performed relatively well for $S_{Z,1} > -10,000$ with NRMSE values less than 1.0 in bins k_2, k_3 and k_4. However, we see that for observations in bin k_1, corresponding to the smallest values of $S_{Z,1}$, the NRMSE is significantly higher. The results of the stratified NRMSE values are consistent with what we have seen in Fig. 7 that visualized the regression result. We have seen a significant departure from the observed and predicted data surface for highly negative values of $S_{Z,1}$. Finally, we note that the stratified regression metrics can be computed in bins obtained using any data clustering technique of choice. A good overview of data clustering algorithms can be found in (Thrun and Stier 2021). Some of those techniques are also implemented in the **scikit-learn** Python library (Pedregosa 2011).

5 Applications of the Principal Component Transport in Combustion Simulations

Using large detailed chemical mechanisms inside a numerical simulation can become a tedious task, especially when other complex phenomena are involved, such as turbulence or pollutant formation. Therefore, parameterization of the thermo-chemical state of a reacting system using a reduced set of optimally chosen variables is very appealing. In this context, the use of PCA is well-suited. PCA allows to automatically reduce dimensionality and retain most of the variance of the system. As we have seen in Sect. 4.2.1, substantial reduction in the number of governing equations of the system can be made by transporting only a subset of the PCs in a numerical simulation. In this section, we present recent applications of the PC-transport approach as reported in (Malik et al. 2018, 2020).

5.1 A Priori Validations in a Zero-Dimensional Reactor

We first show the application of the PC-transport approach in the context of zero-dimensional perfectly stirred reactor (PSR) calculations (Malik et al. 2018). The model validation was done *a priori*, meaning that the model training and validation were made using the same PSR configuration. Two different fuels were investigated: methane (CH_4) and propane (C_3H_8). For each fuel, the dataset for PCA was generated with unsteady PSR simulations, varying the residence time in the reactor from extinction to equilibrium. For each residence time inside the reactor, the entire temporal solution from initialization to steady-state was saved. The dataset for PCA generated in this way contained approximately 100,000 observations for each state variable for the methane case, and 420,000 observations for each state variable for the propane case. In methane simulations, the GRI-3.0 chemical mechanism (Smith et al. 2022) was used, with the nth species, N_2, removed, resulting in 34 species. For the propane case, the Polimi_1412 chemical mechanism (Humer et al. 2007) was used, containing 162 species. PCA-basis was computed using the species mass fractions alone ($\mathbf{X} = [Y_1, Y_2, \ldots, Y_{n-1}]$). The solution of the PC-transport model (as per Eq. (15)) without coupling with nonlinear regression was first obtained, where the predicted quantities were computed using an inverse PCA-basis transformation. Then, the PC-transport approach was coupled with GPR regression (PCA-GPR) in order to increase the dimensionality reduction potential of PCA. Both PC-transport approaches were compared with the full solution obtained by transporting the original species mass fractions (as per Eq. (3)).

5.1.1 Simulation Results for Methane/Air Combustion

Figure 11 shows the PSR solution for the temperature and the H_2O and OH mass fractions for the methane case. The results are obtained with the PC-transport model without nonlinear regression using $q = 24, q = 25$ and $q = 34$ PCs (Fig. 11a) and the PC-transport coupled with GPR regression using $q = 1$ and $q = 2$ PCs (Fig. 11b). For comparison, full solution solving governing equations for the original state variables is shown with the solid line. Using the PC-transport approach without nonlinear regression, at least $q = 25$ components out of 34 were required to obtain an accurate solution, which correspond to a model reduction of 26%. On the other hand, when the PC-transport model was coupled with GPR regression, the results show remarkable accuracy using only $q = 2$ PCs for the prediction of temperature, and both major and minor species. It can also be seen that the PCA-GPR model with $q = 1$ does not provide sufficient accuracy in the ignition region, under-estimating the ignition delay.

5.1.2 Simulation Results for the Propane/Air Combustion

Figure 12 shows the PSR solution for the temperature, and the CO_2 and O_2 mass fractions for the propane case. With the PC-transport model without regression (Fig. 12a), at least $q = 142$ components out of 162 are required in order to get an accurate description, representing a model reduction of 12%. By combining the PC-transport model with the potential offered by nonlinear regression (PCA-GPR), the number required components can be reduced down to $q = 2$. Although the reduced model performs well overall, some deviation from the full solution was observed in the ignition/extinction region. The PCA-GPR model was then further improved, by dividing the PCA manifold into two clusters and performing GPR regression locally in each cluster (PCA-L-GPR). By doing so, the level of accuracy of the model is significantly improved, leading to an almost perfect match with only $q = 2$ components instead of 162 (reduction of 98%). This improvement can be observed in Fig. 12b.

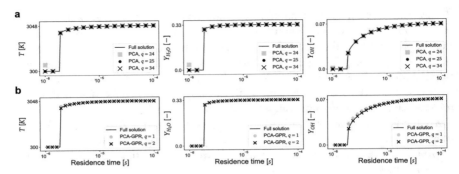

Fig. 11 Results of *a priori* PC-transport simulation of methane/air combustion in a zero-dimensional PSR reactor. Predictions of the temperature, H_2O and OH mass fractions as a function of the residence time in the reactor with the solid line representing the full solution. The results are shown for **a** the PC-transport model without regression using $q = 24$, $q = 25$ and $q = 34$ PCs and **b** the PC-transport model coupled with GPR regression using $q = 1$ and $q = 2$ PCs. Reprinted from (Malik et al. 2018) with permission from Elsevier

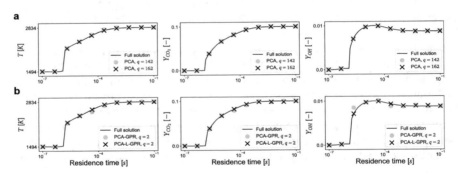

Fig. 12 Results of *a priori* PC-transport simulation of propane/air combustion in a zero-dimensional PSR reactor. Predictions of the temperature, CO_2 and OH mass fractions as a function of the residence time in the reactor with the solid line representing the full solution. The results are shown for **a** the PC-transport model without regression using $q = 142$ and $q = 162$ PCs and **b** the PC-transport model coupled with GPR regression performed globally (PCA-GPR) and locally (PCA-L-GPR) using $q = 2$ PCs. Reprinted from (Malik et al. 2018) with permission from Elsevier

5.2 A Posteriori Validations on Sandia Flame D and F

After validating the PCA-GPR approach in zero-dimensional calculations shown in the previous section, the current section shows the application of the PCA-GPR model in the framework of a non-premixed turbulent flame in a fully three-dimensional LES. The validation was done using the experimental measurements of the Sandia flames D and F (Barlow and Frank 1998). The Sandia flames D and F are piloted methane/air diffusion flames. The fuel is a mixture of CH_4 and air (25/75% by volume) at 294K. The fuel velocity is 49.6m/s for flame D and 99.2m/s for flame F,

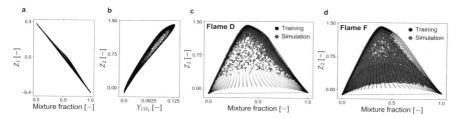

Fig. 13 The two-dimensional manifold obtained during PCA model training versus the manifold accessed during simulation of the Sandia flame D and F. With the training data preprocessing used here, **a** the first PC, Z_1, is highly correlated with mixture fraction and can be linked to the mixture stoichiometry, and **b** the second PC, Z_2, is highly correlated with the CO_2 mass fraction, Y_{CO_2}. Z_2 can thus be interpreted as a variable describing reaction progress. **c–d** Scatter plots of the PCA manifold obtained from the training dataset (black points) and the manifold accessed during simulation (pink points) of **c** the Sandia flame D, and **d** the Sandia flame F. Points on the simulation-accessed manifolds were down-sampled to 100,000 observations on each plot for clarity. Reprinted from (Malik et al. 2020) with permission from Elsevier

the latter representing the most challenging test case, being close to global extinction. The pilot jet surrounding the fuel consists of burnt gases at 1880K and a low-velocity coflow of air at 291K surrounds the flame.

The dataset for PCA model training is based on unsteady one-dimensional counter-flow diffusion methane flames. The inlet conditions for the fuel and air were set as in the experimental setup. Different counter-flow flames were generated by varying the strain rate, from equilibrium to complete extinction. The dataset generated in this way contained approximately 80,000 observations for each of the state-space variables. The GRI-3.0 chemical mechanism (Smith et al. 2022) (without N_2 species) was used. With the data preprocessing used here (including Pareto scaling and removal of temperature from the state variables), the first PC (Z_1) was highly correlated to the mixture fraction, whereas the second PC (Z_2) can be linked to a progress variable with positive weights for the products and negatives weights for the reactants. These correlations between the PCs and physical variables is shown in Fig. 13a–b. It is interesting to point out that PCA identified these controlling variables without any prior assumptions or knowledge of the system of interest. All the state-space variables, such as temperature, density, species mass fraction as well as the PCs source terms, were regressed as function of Z_1 and Z_2 using GPR (PCA-GPR). A lookup table was then generated for the simulation.

The analysis of the manifold accessed during simulation is also interesting. In Fig. 13c–d, we show the training PCA manifold (black points) overlayed with manifold accessed during simulation of flame D and F respectively (pink points). In both figures, points on the simulation-accessed manifold were down-sampled to 100,000 observations for clarity. It can be observed that both flame D and flame F simulations polled from points that stayed close to the training manifold. The highest density of points for flame D (Fig. 13c) is located near the equilibrium solution. This confirms the experimental findings that flame D does not experience significant extinction and re-ignition. On the other hand, it can be observed in Fig. 13d that flame F experiences a higher level of extinction and re-ignition phenomena, which was expected from the experimental data. For flame F, the point density is distributed more uniformly between the equilibrium solution and the extinction regions of the training manifold than for flame D. Thus, the manifold accessed during simulation of flame F covers larger region of the training manifold than for flame D.

5.2.1 Simulation Results for Methane/Air Combustion

The simulations were performed in OpenFOAM using tabulated chemistry approach. The PCs were transported, and the dependent variables $\phi = [\mathbf{S}_{Z_q}, T, \rho, \mathbf{Y}_i]$ were recovered from nonlinear regression. Details about the numerical setup can be found in (Malik et al. 2020). Figure 14 shows the temperature and the OH mass fraction profiles on the centerline (Fig. 14a), close to the burner exit (Fig. 14b) and further downstream (Fig. 14c) for flame D. It can be observed that the PCA-GPR model was able to reconstruct all variables with great accuracy. Moreover, a comparison is made between the PCA-basis calculated from the full set of 35 species and the PCA-basis computed from the reduced set of five major species only. The results are comparable for both bases, suggesting that using only the major species in order to build the PCA-basis results in no major loss of information. Figure 15 shows a comparison between the experimental and numerical profiles of temperature and selected species mass fraction on the centerline for flame F. The PCA-GPR model accurately predicts the peak and the decay in temperature and the species mass fraction profiles.

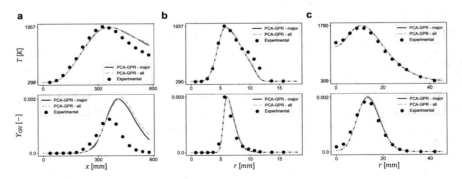

Fig. 14 Results of *a posteriori* PC-transport simulation of the Sandia flame D. Predictions of the temperature and the mass fraction of OH species **a** at the axial and **b-c** at the radial profiles. Results show a comparison between the PCA-basis calculated using the major species (PCA-GPR—major), the basis obtained using the full set of species (PCA-GPR—all) and the experimental data. Reprinted from (Malik et al. 2020) with permission from Elsevier

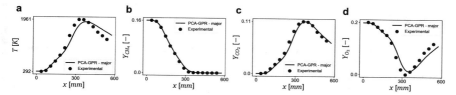

Fig. 15 Results of *a posteriori* PC-transport simulation of the Sandia flame F. Predictions of **a** the temperature and the major species mass fractions, **b** CH_4 **c** CO_2 and **d** O_2 against the experimental data at the flame centerline. The results are shown for the PC-transport model coupled with GPR regression where the PCA-basis was calculated using the major species (PCA-GPR—major). Reprinted from (Malik et al. 2020) with permission from Elsevier

6 Conclusions

In this chapter, we review the complete workflow for data-driven reduced-order modeling of reacting flows. We present strategies for model reduction using dimensionality reduction techniques and nonlinear regression. The originally high-dimensional datasets can be re-parameterized with the new manifold parameters identified directly from training data. The main focus is in the PC-transport approach, where the original system of PDEs is projected to a lower-dimensional PCA-basis. This approach allows for transporting a much smaller number of optimal manifold parameters and yields substantial model reduction. While in this chapter we review recent results from *a priori* and *a posteriori* combustion simulations using PC-transport, several important challenges still remain to be addressed in data-driven modeling of complex systems. For example, topological behaviors on manifolds, such as non-uniqueness or large spatial gradients of dependent variables, can hinder integration of model reduction with nonlinear regression. Possible future research directions that we delineate in this chapter are (1) developing tools for assessing quality of manifolds, (2) developing strategies to mitigate undesired topological behaviors on manifolds and (3) improving our understanding and performance of nonlinear regression models.

Acknowledgements The research of the first author is supported by the F.R.S.-FNRS Aspirant Research Fellow grant. Aspects of this material are based upon work supported by the National Science Foundation under Grant No. 1953350. This project has received funding from the European Research Council (ERC) under the European Union's Horizon 2020 research and innovation program under grant agreement no. 714605.

References

Abadi M et al (2015) TensorFlow: large-scale machine learning on heterogeneous systems. Software available from tensorflow.org

Abboud AW, Schroeder BB, Saad T, Smith ST, Harris DD, Lignell DO (2015) A numerical comparison of precipitating turbulent flows between large-eddy simulation and one-dimensional turbulence. AIChE J 61(10):3185–3197

Ambikasaran S, Foreman-Mackey D, Greengard L, Hogg DW, O'Neil M (2016) Fast direct methods for gaussian processes. IEEE Trans Patt Anal Mach Intell 38(2):252–265

Armstrong E, Sutherland JC (2021) A technique for characterising feature size and quality of manifolds. Combust. Theory Model. 1–23

Barlow RS, Frank JH (1998) Effects of turbulence on species mass fractions in methane/air jet flames. In: Symposium on Combustion, vol 27, pp 1087–1095. Elsevier

Barzegari M, Geris L (2021) An open source crash course on parameter estimation of computational models using a Bayesian optimization approach. J Open Source Educ 4(40):89

Bergstra J, Yamins D, Cox D (2013) Making a science of model search: hyperparameter optimization in hundreds of dimensions for vision architectures. In: Dasgupta S, McAllester D (eds) Proceedings of the 30th international conference on machine learning, vol 28, Proceedings of Machine Learning Research, Atlanta, Georgia, USA, 17–19 June 2013, pp 115–123. PMLR

Biglari A, Sutherland JC (2012) A filter-independent model identification technique for turbulent combustion modeling. Combust Flame 159(5):1960–1970

Biglari A, Sutherland JC (2015) An a-posteriori evaluation of principal component analysis-based models for turbulent combustion simulations. Combust Flame 162(10):4025–4035

Bird RB, Stewart WE, Lightfoot EN (2006) Transport phenomena. Wiley

Chatzopoulos AK, Rigopoulos S (2013) A chemistry tabulation approach via rate-controlled constrained equilibrium (RCCE) and artificial neural networks (ANNs), with application to turbulent non-premixed CH4/H2/N2 flames. Proc Combust Inst 34(1):1465–1473

Chollet F et al (2015) Keras. https://github.com/fchollet/keras

Coussement A, Gicquel O, Parente A (2012) Kernel density weighted principal component analysis of combustion processes. Combust Flame 159(9):2844–2855

Coussement A, Isaac BJ, Gicquel O, Parente A (2016) Assessment of different chemistry reduction methods based on principal component analysis: comparison of the MG-PCA and score-PCA approaches. Combust Flame 168:83–97

Dalakoti DK, Wehrfritz A, Savard B, Day MS, Bell JB, Hawkes ER (2020) An a priori evaluation of a principal component and artificial neural network based combustion model in diesel engine conditions. Proc Combust Inst

D'Alessio G, Sundaresan S, Mueller ME (2022) Automated and efficient local adaptive regression for principal component-based reduced-order modeling of turbulent reacting flows. Proc Combust Inst. https://doi.org/10.1016/j.proci.2022.07.235. https://www.sciencedirect.com/science/article/pii/S1540748922002607

Duvenaud D (2014) Automatic model construction with Gaussian processes. PhD thesis, University of Cambridge

Echekki T, Mirgolbabaei H (2015) Principal component transport in turbulent combustion: a posteriori analysis. Combust Flame 162(5):1919–1933

Echekki T, Kerstein AR, Sutherland JC (2011) The one-dimensional-turbulence model. In: Echekki T, Mastorakos E (eds) Turbulent combustion modeling, Chap. 11. Springer, pp 249–276

Eckart C, Young G (1936) The approximation of one matrix by another of lower rank. Psychometrika 1(3):211–218

Farooq H, Saeed A, Akhtar I, Bangash Z (2021) Neural network-based model reduction of hydrodynamics forces on an airfoil. Fluids 6(9):332

Fooladgar E, Duwig C (2018) A new post-processing technique for analyzing high-dimensional combustion data. Combust Flame 191:226–238

Gicquel O, Darabiha N, Thévenin D (2000) Liminar premixed hydrogen/air counterflow flame simulations using flame prolongation of ildm with differential diffusion. Proc Combust Inst 28(2):1901–1908

Giovangigli V (1999) Multicomponent flow modeling. Birkhäuser, Boston

Gitushi KM, Ranade R, Echekki T (2022) Investigation of deep learning methods for efficient high-fidelity simulations in turbulent combustion. Combust Flame 236:111814

Han X, Jia M, Chang Y, Li Y (2022) An improved approach towards more robust deep learning models for chemical kinetics. Combust Flame 238:111934

Hansen MA, Armstrong E, Sutherland JC, McConnell J, Hewson JC, Knaus, R (2022) Spitfire. https://github.com/sandialabs/Spitfire

Hansen MA, Sutherland JC (2018) On the consistency of state vectors and Jacobian matrices. Combust Flame 193:257–271

Härdle W (1990) Applied nonparametric regression. Cambridge University Press

Hawkes ER, Sankaran R, Sutherland JC, Chen JH (2007) Scalar mixing in direct numerical simulations of temporally evolving plane jet flames with skeletal CO/H2 kinetics. Proc Combust Inst 31(1):1633–1640

Holmes PJ, Lumley JL, Berkooz G, Mattingly JC, Wittenberg RW (1997) Low-dimensional models of coherent structures in turbulence. Phys Rep 287(4):337–384

Hornik K, Stinchcombe M, White H (1989) Multilayer feedforward networks are universal approximators. Neural Netw 2(5):359–366

Humer S, Frassoldati A, Granata S, Faravelli T, Ranzi E, Seiser R, Seshadri K (2007) Experimental and kinetic modeling study of combustion of JP-8, its surrogates and reference components in laminar nonpremixed flows. Proc Combust Inst 31(1):393–400

Isaac BJ, Coussement A, Gicquel O, Smith PJ, Parente A (2014) Reduced-order PCA models for chemical reacting flows. Combust Flame 161(11):2785–2800

Isaac BJ, Thornock JN, Sutherland JC, Smith PJ, Parente A (2015) Advanced regression methods for combustion modelling using principal components. Combust Flame 162(6):2592–2601

Jha PK, Groth CPT (2012) Tabulated chemistry approaches for laminar flames: evaluation of flame-prolongation of ildm and flamelet methods. Combust Theory Model 16(1):31–57

Jolliffe I (2002) Principal component analysis. Springer, New York

Kee RJ, Coltrin ME, Glarborg P (2005) Chemically reacting flow: theory and practice. Wiley

Kerstein AR (1999) One-dimensional turbulence: model formulation and application to homogeneous turbulence, shear flows, and buoyant stratified flows. J Fluid Mech 392:277–334

Keun HC, Ebbels TM, Antti H, Bollard ME, Beckonert O, Holmes E, Lindon JC, Nicholson JK (2003) Improved analysis of multivariate data by variable stability scaling: application to NMR-based metabolic profiling. Anal Chim Acta 490(1–2):265–276

Kutz JN, Brunton SL, Brunton BW, Proctor JL (2016) Dynamic mode decomposition: data-driven modeling of complex systems. SIAM

Lignell DO, Fredline GC, Lewis AD (2015) Comparison of one-dimensional turbulence and direct numerical simulations of soot formation and transport in a nonpremixed ethylene jet flame. Proc Combust Inst 35(2):1199–1206

Lu T, Law CK (2009) Toward accommodating realistic fuel chemistry in large-scale computations. Prog Energy Combust Sci 35(2):192–215

Lusch B, Kutz JN, Brunton SL (2018) Deep learning for universal linear embeddings of nonlinear dynamics. Nat Commun 9(1):1–10

Maas U, Pope SB (1992) Simplifying chemical kinetics: intrinsic low-dimensional manifolds in composition space. Combust Flame 88(3):239–264

Malik MR, Isaac BJ, Coussement A, Smith PJ, Parente A (2018) Principal component analysis coupled with nonlinear regression for chemistry reduction. Combust Flame 187:30–41

Malik MR, Vega PO, Coussement A, Parente A (2020) Combustion modeling using principal component analysis: a posteriori validation on Sandia flames D. E and F, Proc Combust Inst

Malik MR, Coussement A, Echekki T, Parente A (2022a) Principal component analysis based combustion model in the context of a lifted methane/air flame: Sensitivity to the manifold parameters and subgrid closure. Combust Flame 244:112134. https://doi.org/10.1016/j.combustflame.2022.112134. https://www.sciencedirect.com/science/article/pii/S0010218022001535

Malik MR, Khamedov R, Hernández Pérez FE, Coussement A, Parente A (2022b) Dimensionality reduction and unsupervised classification for high-fidelity reacting flow simulations. Proc Combust Inst. https://doi.org/10.1016/j.proci.2022.06.017. https://www.sciencedirect.com/science/article/pii/S1540748922000207

Mendez MA, Scelzo MT, Buchlin J-M (2018) Multiscale modal analysis of an oscillating impinging gas jet. Exp Therm Fluid Sci 91:256–276

Mendez MA, Balabane M, Buchlin J-M (2019) Multi-scale proper orthogonal decomposition of complex fluid flows. J Fluid Mech 870:988–1036

Mirgolbabaei H, Echekki T (2013) A novel principal component analysis-based acceleration scheme for LES-ODT: an a priori study. Combust Flame 160(5):898–908

Mirgolbabaei H, Echekki T (2014) Nonlinear reduction of combustion composition space with kernel principal component analysis. Combust Flame 161:118–126

Mirgolbabaei H, Echekki T (2015) The reconstruction of thermo-chemical scalars in combustion from a reduced set of their principal components. Combust Flame 162(5):1650–1652

Mirgolbabaei H, Echekki T, Smaoui N (2014) A nonlinear principal component analysis approach for turbulent combustion composition space. Int J Hydrog Energy 39(9):4622–4633

Mockus J (2012) Bayesian approach to global optimization: theory and applications, vol 37. Springer Science & Business Media

Nguyen H-T, Domingo P, Vervisch L, Nguyen P-D (2021) Machine learning for integrating combustion chemistry in numerical simulations. Energy AI 5:100082

Niemeyer KE, Curtis NJ, Sung C-J (2017) pyJac: analytical Jacobian generator for chemical kinetics. Comput Phys Commun 215:188–203

Noda I (2008) Scaling techniques to enhance two-dimensional correlation spectra. J Mol Struct 883–884:216–227

Owoyele O, Echekki T (2017) Toward computationally efficient combustion DNS with complex fuels via principal component transport. Combust Theory Model 21(4):770–798

Parente A, Sutherland JC (2013) Principal component analysis of turbulent combustion data: data pre-processing and manifold sensitivity. Combust Flame 160(2):340–350

Parente A, Sutherland JC, Tognotti L, Smith PJ (2009) Identification of low-dimensional manifolds in turbulent flames. Proc Combust Inst 32(1):1579–1586

Parente A, Sutherland JC, Dally BB, Tognotti L, Smith PJ (2011) Investigation of the MILD combustion regime via principal component analysis. Proc Combust Inst 33(2):3333–3341

Pedregosa F et al (2011) Scikit-learn: machine learning in Python. J Mach Learn Res 12(85):2825–2830

Perry BA, Henry de Frahan MT, Yellapantula S (2022) Co-optimized machine-learned manifold models for large eddy simulation of turbulent combustion. Combust Flame 244:112286. https://doi.org/10.1016/j.combustflame.2022.112286. https://www.sciencedirect.com/science/article/pii/S0010218022003017

Peters N (1988) Laminar flamelet concepts in turbulent combustion. Int Symp Combust 21(1):1231–1250. Twenty-first international symposium on combustion

Pope SB (2013) Small scales, many species and the manifold challenges of turbulent combustion. Proc Combust Inst 34(1):1–31

Punati N, Sutherland JC, Kerstein AR, Hawkes ER, Chen JH (2011) An evaluation of the one-dimensional turbulence model: comparison with direct numerical simulations of CO/H2 jets with extinction and reignition. Proc Combust Inst 33(1):1515–1522

Punati N, Wang H, Hawkes ER, Sutherland JC (2016) One-dimensional modeling of turbulent premixed jet flames—comparison to DNS. Flow Turbul Combust 97(3):913–930 Oct

Raissi M, Perdikaris P, Karniadakis GE (2019) Physics-informed neural networks: a deep learning framework for solving forward and inverse problems involving nonlinear partial differential equations. J Comput Phys 378:686–707

Ramezanian D, Nouri AG, Babaee H (2021) On-the-fly reduced order modeling of passive and reactive species via time-dependent manifolds. Comput Methods Appl Mech Eng 382:113882

Ranade R, Echekki T (2019) A framework for data-based turbulent combustion closure: a priori validation. Combust Flame 206:490–505

Ranade R, Echekki T (2019) A framework for data-based turbulent combustion closure: a posteriori validation. Combust Flame 210:279–291

Russell S, Norvig P (2022) Artificial intelligence: a modern approach. Prentice Hall

Smith GP, Golden DM, Frenklach M, Moriarty NW, Eiteneer B, Goldenberg M, Bowman CT, Hanson R, Song S, Gardiner Jr WC, Lissianski V, Qin Z (2022) GRI Mech 3.0. Available at: http://www.me.berkeley.edu/gri_mech/

Sutherland JC (2004) Evaluation of mixing and reaction models for large-eddy simulation of non-premixed combustion using direct numerical simulation. PhD thesis, Department of Chemical and Fuels Engineering, The University of Utah

Sutherland JC, Punati N, Kerstein AR (2010) A unified approach to the various formulations of the one-dimensional-turbulence model. Inst Clean Secur Energy

Sutherland JC, Parente A (2009) Combustion modeling using principal component analysis. Proc Combust Inst 32(1):1563–1570

Sutherland JC, Smith PJ, Chen JH (2007) A quantitative method for a priori evaluation of combustion reaction models. Combust Theory Model 11(2):287–303

Taira K, Brunton SL, Dawson STM, Rowley CW, Colonius T, McKeon BJ, Schmidt OT, Gordeyev S, Theofilis V, Ukeiley LS (2017) Modal analysis of fluid flows: an overview. AIAA J 55(12):4013–4041

Taylor R, Krishna R (1993) Multicomponent mass transfer. Wiley

Thrun MC, Stier Q (2021) Fundamental clustering algorithms suite. SoftwareX 13:100642

Van Oijen JA, De Goey LPH (2002) Modelling of premixed counterflow flames using the flamelet-generated manifold method. Combust Theory Model 6(3):463–478

Williams CK, Rasmussen CE (2006) Gaussian processes for machine learning. The MIT Press

Yang Y, Pope SB, Chen JH (2013) Empirical low-dimensional manifolds in composition space. Combust Flame 160(10):1967–1980

Zdybał K, Armstrong E, Parente A, Sutherland JC (2020) PCAfold: python software to generate, analyze and improve PCA-derived low-dimensional manifolds. SoftwareX 12:100630

Zdybał K, D'Alessio G, Aversano G, Malik MR, Coussement A, Sutherland JC, Parente A (2022a) Advancing reactive flow simulations with data-driven models. In: Mendez MA, Ianiro A, Noack BR, Brunton SL (eds) Data-driven fluid mechanics: combining first principles and machine learning, Chap. 15. Cambridge University Press

Zdybał K, Sutherland JC, Parente A (2022b) Manifold-informed state vector subset for reduced-order modeling. Manuscript submitted to Proc Combust Inst 39

Zdybał K, Armstrong E, Sutherland JC, Parente A (2022c) Cost function for low-dimensional manifold topology assessment. Sci Rep 12(1):1–19

Zhang Y, Xu S, Zhong S, Bai X-S, Wang H, Yao M (2020) Large eddy simulation of spray combustion using flamelet generated manifolds combined with artificial neural networks. Energy AI 2:100021

Zhang P, Liu S, Lu D, Sankaran R, Zhang G (2021) An out-of-distribution-aware autoencoder model for reduced chemical kinetics. Discrete Contin Dyn Syst - S

Zhou L, Song Y, Ji W, Wei H (2022) Machine learning for combustion. Energy AI 7:100128

AI Super-Resolution: Application to Turbulence and Combustion

M. Bode

Abstract This article summarizes and discusses recent developments with respect to artificial intelligence (AI) super-resolution as a subfilter model for large-eddy simulations. The focus is on the application of physics-informed enhanced super-resolution generative adversarial networks (PIESRGANs) for subfilter closure in turbulence and combustion applications. A priori and a posteriori results are presented for various applications, ranging from decaying turbulence to finite-rate chemistry flows. The high accuracy of AI super-resolution-based subfilter models is emphasized, and advantages and shortcoming are described.

1 Introduction

Many turbulent and reactive simulations require models to reduce the computational cost. Popular approaches include large-eddy simulation (LES) for modeling (reactive) turbulence and flamelet models for predicting chemistry. LES relies on the filtered Navier–Stokes equations. The filter operation separates the flow in larger scales above the filter width and smaller scales below the filter width, called subfilter contributions. As a result, the filtered equations can be advanced for less computational cost, however, they require modeling for subfilter contributions. Accurate modeling of these unclosed terms is one of the key challenges for predictive LES. LES has been applied successfully to many different turbulent flows including reactive turbulent flows (Smagorinsky 1963; Pope 2000; Pitsch 2006; Beck et al. 2018; Goeb et al. 2021). The flamelet concept employs asymptotic and scale arguments to motivate that flow field and chemistry are only loosely coupled by the scalar dissipation rate, a measurement for the local mixing, in combustion. Consequently, advancing chemistry is reduced to solving coupled one-dimensional (1-D) differen-

M. Bode (✉)
Jülich Supercomputing Centre, Forschungszentrum Jülich GmbH, 52425 Jülich, NRW, Germany
e-mail: m.bode@itv.rwth-aachen.de

Fakultät für Maschinenwesen, RWTH Aachen University, Templergraben 64, 52056 Aachen, NRW, Germany

N. Swaminathan and A. Parente (eds.), *Machine Learning and Its Application to Reacting Flows*, Lecture Notes in Energy 44, https://doi.org/10.1007/978-3-031-16248-0_10

tial equations, which are, for example, in mixture fraction space for non-premixed combustion. Challenges include how to tabulate the resulting flamelets efficiently and how to distribute the multiple flamelets across the domain for multiple representative interactive flamelet (MRIF) approaches (Peters 1986; Banerjee and Ierapetritou 2006; Ihme et al. 2009; Bode et al. 2019b).

Data-driven methods, such as machine learning (ML) and deep learning (DL), have gained a massive boost across almost all scientific domains, ranging from speech recognition (Hinton et al. 2012) and learning optimal complex control (Vinyals et al. 2019) to accelerating drug development (Bhati et al. 2021). Important steps towards the wider usage of ML/DL methods were the availability of more and larger (labeled) datasets as well as significant improvements with respect to graphics processing units (GPUs), which enabled high-speed GPUs and efficient execution of ML/DL operations on GPUs. One particular class of ML/DL is AI super-resolution, also called single image super-resolution (SISR), originally developed by the computer science community for increasing the resolution of 2-D images (i.e., to super-resolve images) beyond classical techniques, such as bicubic interpolation. The idea is that complex networks can extract and learn features during training with many images and are then able to add this information to images based on local information. Dong et al. (2014) introduced a super-resolution convolutional neural network (SRCNN), a deep convolutional neural network (CNN) which directly learns the end-to-end mapping between low and high resolution images. Several other works continuously improved this approach (Dong et al. 2015; Kim et al. 2016a, b; Lai et al. 2017; Simonyan and Zisserman 2014; Johnson et al. 2016; Tai et al. 2017; Zhang et al. 2018) to achieve better prediction accuracy by correcting multiple shortcomings of the original SRCNN. The switch from CNNs to generative adversarial networks (GANs) (Goodfellow et al. 2014), as proposed by Ledig et al. (2017), finally resulted in the development of enhanced super-resolution GANs (ESRGANs) by Wang et al. (2018).

The idea of AI super-resolution has been also successfully adopted for simulations of physical phenomena, from climate research (Stengel et al. 2020) to cosmology (Li et al. 2021). While many applications focus on super-resolving single time steps of simulations, Bode et al. (2019a, 2021, 2022), Bode (Bode 2022a, b, c) introduced an algorithm for employing AI super-resolution as a subfilter model for (reactive) LES. They developed the physics-informed enhanced super-resolution GAN (PIESRGAN) and demonstrated its application for various turbulent inert and reactive flows. To successfully use AI super-resolution to time-advance complex flows, accurate a priori results are necessary but not sufficient. Only if the model also gives good a posteriori results, i.e., when it is continuously used as model for multiple consecutive time steps during a simulation, it is promising for applying it to complex flows. Typically, good a posteriori results are much more difficult to achieve, as errors accumulate over time, especially if low-dissipation solvers are used. Consequently, a posteriori results are presented for all cases discussed in this article.

This work summarizes important modeling aspects of PIESRGAN in the next section. Afterward, its application to a decaying turbulence case, reactive spray setups, premixed combustion, and non-premixed combustion is described. This chapter finishes with conclusions for further developments of the AI super-resolution approach in general and the PIESRGAN in particular.

2 PIESRGAN

This section summarizes the PIESRGAN and explains the PIESRGAN-subfilter modeling approach. Details about the architecture, the time advancement algorithm, and implementation details are given. Note that the PIESRGAN modeling approach presented in this work follows a hybrid approach. AI super-resolution is only used on the smallest scales to reconstruct the subfilter contributions, while the well-known filtered equations for LES are used to advance the flow in time, i.e., the time integration is not integrated in the network. This approach is technically more complex than integrating the time integration in the network. However, it is also expected to be more general and universal. Turbulence is known to feature some universality on the smallest scales (Frisch and Kolmogorov 1995), which should be learnt by the network and be universal for many applications. The larger scales, which can be strongly affected by the geometry and setup and thus are fully case dependent, are considered by the filtered equations making PIESRGAN-subfilter models applicable for multiple cases.

2.1 Architecture

PIESRGAN is a GAN model, which is a generative model that aims to estimate the unknown probability density of observed data without an explicitly provided data likelihood function, i.e., with unsupervised learning. Technically, a GAN has two networks. The generator network is used for modeling and creates new modeled data. The discriminator network tries to distinguish whether data are generator-created or real data and provides feedback to the generator network. Thus, throughout the learning process, the generator gets better at creating data as close as possible to real data, and the discriminator learns to better identify fake data, which can be seen as two players carrying out a minimax zero-sum game to estimate the unknown data probability distribution.

The network architecture and training process are sketched in Fig. 1. Fully resolved 3-dimensional (3-D) data ("H") are filtered to get filtered data ("F"). The filtered data is used as input to the generator for creating the reconstructed data ("R"). The accuracy of the reconstructed data is evaluated by means of the fully resolved data. The discriminator tries to distinguish between reconstructed and fully resolved data. The accuracy is measured by means of the loss function, which reads

$$\mathcal{L} = \beta_1 L_{\text{adversarial}} + \beta_2 L_{\text{pixel}} + \beta_3 L_{\text{gradient}} + \beta_4 L_{\text{physics}}, \tag{1}$$

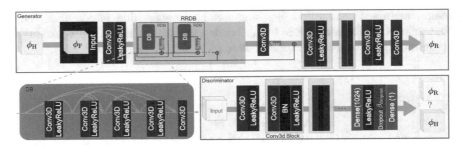

Fig. 1 Sketch of PIESRGAN. "H" denotes high-fidelity data, "F" are corresponding filtered data, and "R" are the reconstructed data. The components are: Conv3D—3-D Convolutional Layer, LeakyReLU—Activation Function, DB—Dense Block, RDB—Residual Dense Block, RRDB—Residual in Residual Dense Block, β_{RSF}—Residual Scaling Factor, BN—Batch Normalization, Dense—Fully Connected Layer, Dropout—Regularization Component, $\beta_{dropout}$—Dropout Factor. Color-modified image from Bode et al. (2021)

where β_1 to β_4 are coefficients weighting the different loss term contributions with $\sum_i \beta_i = 1$. The adversarial loss is the discriminator/generator relativistic adversarial loss (Jolicoeur-Martineau 2018), which measures both how well the generator is able to create accurate reconstructed data compared to the fully resolved data and how well the discriminator is able to identify fake data. The pixel loss and the gradient loss are defined using the mean-squared error (MSE) of the quantity and its gradient, respectively. The physics loss enforces physically motivated conditions, such as the conservation of mass, species, and elements, depending on the underlying physics of the problem. For the non-premixed temporal jet application in this work, it reads

$$L_{\text{physics}} = \beta_{41} L_{\text{mass}} + \beta_{42} L_{\text{species}} + \beta_{43} L_{\text{elements}}, \tag{2}$$

where β_{41}, β_{42}, and β_{43} are coefficients weighting the different physical loss term contributions with $\sum_i \beta_{4i} = 1$. The physically motivated loss term is very important for the application of PIESRGAN to flow problems. If the conservation laws are not fulfilled very well, the simulations tend to blow up rapidly, which is an important difference to super-resolution in the context of images. Errors which might be acceptable there can be easily too large for usage as a subfilter model (Bode et al. 2021).

The generator heavily uses 3-D CNN layers (Conv3D) (Krizhevsky et al. 2012) combined with leaky rectified linear unit (LeakyReLU) layers for activation (Maas et al. 2013). The residual in residual dense block (RRDB), which was introduced for ESRGAN, is essential for the performance of the state-of-the-art super-resolution. It replaces the residual block (RB) employed in previous architectures and contains fundamental architectural elements such as residual dense blocks (RDBs) with skip-connections. A residual scaling factor β_{RSF} helps to avoid instabilities in the forward and backward propagation. RDBs use dense connections inside. The output from each layer within the dense block (DB) is sent to all the following layers. The discriminator network is simpler. It inherits basic CNN layers (Conv3D) combined with LeakyReLU layers for activation with and without batch normalization (BN). The final layers contain a fully connected layer with LeakyReLU and dropout with dropout factor β_{dropout}. A summary of all hyperparameters is given in Table 1.

Table 1 Overview of the PIESRGAN hyperparameters. The given ranges represent the sensitivity intervals with acceptable network results. The central values were used for the decaying turbulent case in this work

β_1	$[0.2 \times 10^{-5}, 0.6 \times 10^{-4}, 0.8 \times 10^{-4}]$
β_2	$[0.79327, 0.88994, 0.91812]$
β_3	$[0.04, 0.06, 0.15]$
β_4	$[0.01, 0.05, 0.06]$
β_{RSF}	$[0.1, 0.2, 0.3]$
β_{dropout}	$[0.2, 0.4, 0.5]$
$l_{\text{generator}}$	$[1.2 \times 10^{-6}, 4.5 \times 10^{6}, 5.0 \times 10^{-6}]$
$l_{\text{discriminator}}$	$[4.4 \times 10^{-6}, 4.5 \times 10^{-6}, 8.5 \times 10^{-6}]$

2.2 Algorithm

The LES equations, which are Favre-filtered, are used to advance a PIESRGAN-LES in time. As consequence of the filter operation to the equations, unclosed terms appear, which require information from below the filter width to be evaluated. The LES subfilter algorithm aims to reconstruct this information to close the LES equations. This is done during every time step. For the cases with chemistry, the chemistry can be included in the PIESRGAN during the training process (Bode et al. 2022; Bode 2022a). As chemistry is often active locally, this can be also used to save computing time by adaptively solving only in relevant regions. The algorithm starts with the LES solution Φ_F^n at time step n, which includes the entirety of all relevant fields in the simulation, and consists of repeating the following steps:

1. Use the PIESRGAN to reconstruct Φ_R^n from Φ_{LES}^n.
2. (Only for nonuniversal quantities) Use Φ_R^n to update the scalar fields of Φ to $\Phi_R^{n;\text{update}}$ by solving the unfiltered scalar equations on the mesh of Φ_R^n.
3. Use $\Phi_R^{n;\text{update}}$ to estimate the unclosed terms Ψ_{LES}^n in the LES equations of Φ for all fields by evaluating the local terms with $\Phi_R^{n;\text{update}}$ and applying a filter operator.
4. Use Ψ_{LES}^n and Φ_{LES}^n to advance the LES equations of Φ to Φ_{LES}^{n+1}.

2.3   Implementation Details

PIESRGAN was implemented using a TensorFlow/Keras framework (Abadi et al. 2016; Keras 2019) in this work to efficiently employ GPUs. For all the examples discussed here, the data were split into training and testing sets to avoid reproduction of fully seen data. During the training and querying processes, it was found that consistent normalization of quantities is very important for highly accurate results (Bode et al. 2021). Furthermore, both operations are done based on subboxes, since reconstructing bigger boxes can become very memory intensive. Typically, each subbox is chosen large enough to cover the relevant physical scales (Bode et al. 2021). The filter width can become problematic if non-uniform meshes are employed. In these cases, training with multiple filter widths is suggested to achieve good accuracy throughout the entire domain (Bode 2022a).

The potential extrapolation capability of data-driven methods is always challenging. Many trained networks only work well in regions which were accessible during the training process. This can become very problematic for flow applications, where often data at low Reynolds numbers is abundant, while data at high Reynolds numbers is not computable at all, making transfer learning difficult. To deal with this problem, concepts such as a two-step training approaches (Bode et al. 2021) can be used relying on the further prediction width of GANs compared to single networks (Bode et al. 2022; Bode 2022a). In order to avoid this open question of extrapolation capabilities, only interpolation cases are presented in this work.

A basic version of PIESRGAN is available on GitLab (https://git.rwth-aachen. de/Mathis.Bode/PIESRGAN.git) for an interested reader.

3   Application to Turbulence

The application of PIESRGAN to non-reactive turbulence is a good starting point. Besides closing the filtered momentum equations, the evaluation of passive scalars is a key challenge toward applying PIESRGAN to turbulent reactive flows, as scalar mixing is especially important for non-premixed combustion cases. Furthermore, turbulence is assumed to be universal on the smallest scales that makes it reasonable to accurately learn the subfilter behavior by a complex network.

3.1   Case Description

A decaying turbulence case with a peak wavenumber κ_p of 15 m^{-1} and a maximum Taylor microscale-based Reynolds number Re$_\lambda$ of about 88 is used as turbulent example case here. Turbulence with an initial turbulence intensity of $u_0' = 2\langle k \rangle / 3$ with $\langle k \rangle$ as ensemble-averaged turbulent kinetic energy was initialized on a uniform mesh with 4096^3 and solved along with passive scalars. The original direct numerical simulation (DNS) was computed with the solver psOpen (Gauding et al. 2019). psOpen employs the P3DFFT library for spatial decomposition and to perform the fast Fourier transform (FFT) (Pekurovsky 2012) of the incompressible Navier–Stokes equations formulated in spectral space, but with the non-linear term computed in physical space. Over time, the turbulent intensity decays, i.e., the Reynolds number decreases, resulting in larger turbulent structures. This makes the decaying turbulence case a very good baseline application, as many practical applications also features varying Reynolds numbers.

The corresponding PIESRGAN-LES was computed with CIAO, an arbitrary order finite-difference code (Desjardins et al. 2008). The physics-informed loss function only considered a condition for enforcing mass conservation. Further details can be found in Bode et al. (2021).

3.2 A Priori Results

For evaluating the accuracy of PIESRGAN, Fig. 2 shows 2-D slices of the fully resolved velocity and scalar fields, the filtered fields, and the reconstructed fields employing PIESRGAN. The visual agreement is good, and the network seems to be able to add sufficient information to the filtered fields to reconstruct the fully resolved data. Bode et al. (2021) pointed out that high accuracy can also be achieved in scenarios in which PIESRGAN needs to "extrapolate" training data using a two-step training approach. The two-step training approach combines fully resolved data for updating generator and discriminator and underresolved training data, which further update the generator. This is an important feature of the employed GAN approach as many practical use cases feature Reynolds numbers which cannot be computed with DNS.

In addition to the visual assessment of the PIESRGAN, Fig. 3 shows the dimensionless spectra for the velocity vector field and the passive scalar, denoted as $\mathscr{S}$. The spectra are computed with the fully resolved fields, the filtered fields, and the reconstructed fields and are an important measurement for the prediction quality of PIESRGAN, as they quantify the distribution of turbulent energy and scalar among the length scales. The filter operation removes the smallest scales, and the task of the PIESRGAN model is to add the smallest scales to reconstruct the fully resolved distribution. The agreement is good for both spectra, however, not perfect for very high wavenumbers, i.e., for $\kappa/\kappa_p \approx 80$. It is important to note that the numerics have a significant impact on the results in Fig. 3. Only high order and consistent numerics avoid significant noise for high wavenumbers in the reconstructed data.

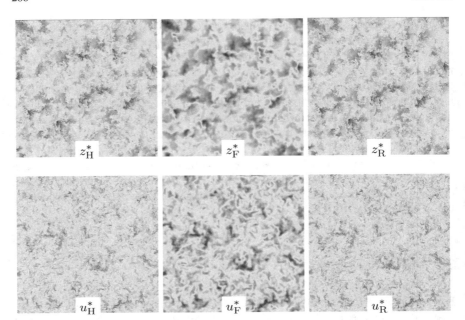

Fig. 2 Visualization of 2-D slices of the dimensionless passive scalar z^* and the dimensionless velocity component u^* for the time step with Taylor microscale-based Reynolds number of about 88. Colormaps span from blue (minimum) to red (maximum) (Bode et al. 2021)

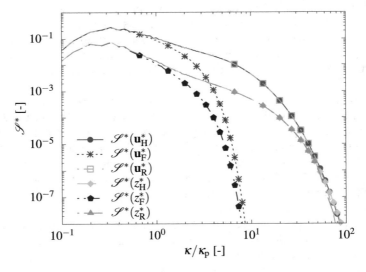

Fig. 3 Dimensionless spectra $\mathscr{S}^*$ plotted over the normalized wavenumber κ/κ_p and evaluated on DNS data, filtered data, and reconstructed data for the dimensionless velocity vector $\mathbf{u}^*$ and passive scalar z^* for the time step with Reynolds number of about 88. Note that the symbols do not represent the discretization but are only used to distinguish the different cases. Modified plot from Bode et al. (2021)

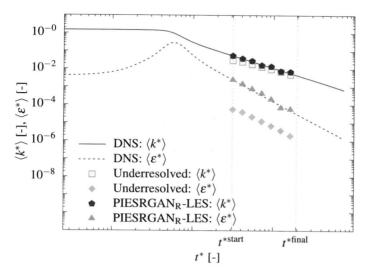

Fig. 4 Evolution over dimensionless time t^* of the ensemble-averaged dimensionless turbulent kinetic energy $\langle k^* \rangle$ and ensemble-averaged dimensionless dissipation rate $\langle \varepsilon^* \rangle$. Plot from Bode et al. (2021)

3.3   A Posteriori Results

A PIESRGAN-LES must accurately predict the decay of turbulence, usually measured by means of the ensemble-averaged turbulent kinetic energy and the ensemble-averaged dissipation rate, denoted as $\langle \varepsilon \rangle$. A uniform LES mesh of 64^3 was considered and the results are presented in Fig. 4. The prediction accuracy of PIESRGAN-LES is high. The results for a heavily underresolved simulation without LES model show that especially the ensemble-averaged dissipation rate is strongly underpredicted without model. This makes sense as the dissipation rate acts on the smallest scales which simply do not exist in the underresolved simulation due to the lack of resolution.

3.4   Discussion

The presented a posteriori results are remarkable as the trained network is able to reproduce the decay on a multiple orders of magnitude coarser mesh. One reason for this could be the universal character of turbulence on the smallest scales. From a computational point, a too drastic reduction of mesh size might not result in the fastest time-to-solution as the costs of subbox reconstruction increase with the reconstruction size. Thus, a finer LES mesh with smaller subbox reconstruction can be faster as demonstrated by the two turbulent combustion cases below. Furthermore, if the network is used as part of a multi-physics simulation, often LES meshes which are only 10–20 times coarser per direction than a DNS, which fully resolves the turbulence, are needed to accurately consider boundary conditions and other physical phenomena. In this context, it is also interesting to mention the effect of the Courant-Friedrichs-Lewy (CFL) number. Theoretically, coarser LES meshes also enable larger time steps. However, it was found that usually a time step size between the DNS and theoretical LES time step sizes is needed to accurately reproduce the DNS results. The reason might be that the CFL number is a numerical limit, however, the PIESRGAN-LES also needs to fulfil some intrinsic physical time step limitations.

Overall, PIESRGAN has many advantages for turbulent flows. It can not only be used to reduce the computing and storing cost but also to enable new workflows. For example, smaller domains can be computed first to get accurate training data. Afterward, the trained model is applied to a larger domain to achieve converged statistics. In addition to the discussed LES application, it could also be used as cheap turbulence generator for complex simulations.

4 Application to Reactive Sprays

Reactive sprays occur in many applications, such as diesel engines. Usually, the liquid fuel is injected into a combustion chamber where it finally burns. Before ignition can take place, multiple physical processes happen. The continuous liquid fuel phase splits into smaller ligaments and small droplets. These disperse droplets start evaporating and the resulting vapor mixes with the ambient gas forming a reactive mixture in which the combustion process occurs. The more these stages are spatially separated, the more similar the final combustion process becomes to classical non-premixed combustion. A measurement for this separation is the difference between lift-off length (LOL), i.e., the distance between nozzle tip and closest combustion events, and the liquid penetration length (LPL), i.e., the distance between nozzle tip and roughly furthest fuel in liquid phase. This work focuses on the Spray A and Spray C cases defined by the Engine Combustion Network (ECN) (2019).

4.1 Case Description

Spray A and Spray C are both single hole nozzles, however, while Spray A is designed to avoid cavitation, Spray C features cavitation. Additionally, Spray A has a smaller exit diameter like injectors used for diesel engines, while the exit diameter of Spray C is larger as for heavy-duty injectors. Both injectors were investigated with n-dodecane as fuel at standard reactive conditions, reading 150 MPa injection pressure, 22.8 kg/m^3 ambient density, 15 % ambient oxygen concentration, 900 K ambient temperature, and 363 K fuel temperature. Furthermore, inert conditions, i.e., without ambient oxygen, were run for Spray A, while Spray C was also simulated with 1000, 1100, and 1200 K ambient temperatures. The cases are denoted as SA900, SC900,

SC1000, SC1100, and SC1200 based on the used nozzle geometry and ambient temperature. Inert conditions are separately emphasized.

The cases were computed using CIAO with a similar setup as described by Goeb et al. (2021). More precisely, the initial droplets were generated based on a pre-computed droplet size distribution for the Spray A case (Bode et al. 2014, 2015). For the Spray C case, a blob method utilizing the effective liquid diameter at the nozzle exit was employed. Breakup and evaporation were modeled with Kelvin-Helmholtz/Rayleigh-Taylor (KH/RT) (Patterson and Reitz 1998) and Bellan's evaporation approach (Miller and Bellan 1999) for both cases. Velocity and mixing LES closure were based on PIESRGAN-subfilter modeling. Note that due to the lack of reactive spray DNS data and motivated by the separation of phenomena within the combustion process of sprays, the PIESRGAN was trained with the decaying turbulence data introduced in the previous sections.

The reaction mechanism by Yao et al. (2017) was used for all simulations. An MRIF approach was employed for chemistry modeling, which is also summarized in Fig. 5. The non-premixed flamelet approach assumes that chemistry and flow are only loosely coupled through the scalar dissipation rate. Consequently, two different sets of equations are solved in MRIF approaches. The first set are the usual flow equations solved in 3-D spatial space. The second set describes chemistry in the mixture fraction space Z which is only 1-D, and is called flamelet equations. Therefore, representing and solving the chemistry by means of the flamelet equations is much cheaper compared to solving the chemistry in full 3-D spatial space. As shown by the equations in Fig. 5, the mapping towards the flamelet space is done by weighted volume-averages, while the mapping back to physical space employs

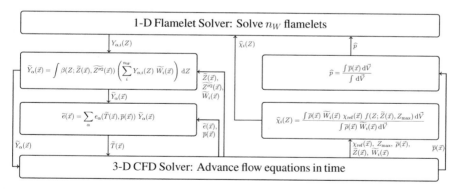

Fig. 5 Schematic representation of the MRIF approach and its coupling to 3-D computational fluid dynamics (CFD) solver. Tilde denotes Favre-filtered data. The overbar indicates Reynolds-averaging. The hat labels quantities in mixture fraction space. Z is the mixture fraction, W_i the flamelet weights, p the pressure, χ the scalar dissipation rate, ρ the density, Y_α the mass fractions, e the internal energy, and T the temperature. β denotes the presumed β-PDF, and f indicates the functional form of the scalar dissipation rate. The spatial coordinates are represented by $\mathbf{x}$, and integration over the volume of the full domain is described by $\int d\mathbf{V}$. All variables are time dependent, but t is omitted here for brevity. Image from Bode (2022c)

probability density functions (PDFs), typically constructed by means of the filtered mixture fraction and mixture fraction variance.

Thus, the MRIF approach typically requires a presumed functional form of the scalar dissipation rate in mixture fraction space f and the PDF of the mixture faction. For the functional form, often a presumed log-based profile is assumed (Pitsch et al. 1998), while a beta-PDF is often employed for the mixture fraction PDF. Both quantities are critical for LES, as they often have significant subfilter contributions. In the context of PIESRGAN modeling, both assumptions can be avoided by directly evaluating both profiles on the reconstructed fields which can improve the prediction results of the simulations. For the Spray C cases, the mixture fraction PDF was indeed evaluated based on the reconstructed data for the results presented here (Bode 2022b).

4.2   Results

The lack of DNS data makes a distinction between a priori and a posteriori results difficult. Instead LES results are compared with experimental data here (Engine Combustion Network 2019). Figures 6 and 7 compare the ignition delay time t_i and the LOL l_{LOL} for the considered spray cases. All simulations slightly underpredict the experimental results. This could be because of the chemical kinetics mechanism used which has a significant impact on the ignition delay time. Furthermore, the ignition delay time and consecutively LOL decrease with increasing ambient temperature. These trends are correctly predicted for Spray C by the PIESRGAN-LESs.

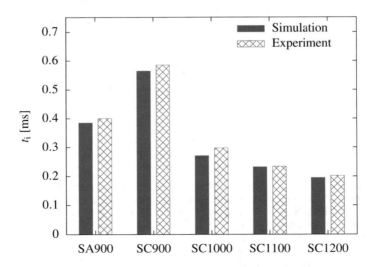

Fig. 6 Ignition delay time t_i for Spray A and Spray C cases

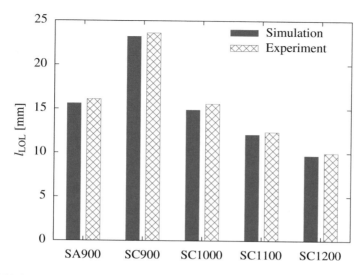

Fig. 7 LOL l_{LOL} for Spray A and Spray C cases

The near nozzle experimental data for the inert Spray A case allow a further evaluation of PIESRGAN-LES compared to classical LES with dynamic Smagorinsky (DS) model. Figure 8 compares the temporally and circumferentially averaged fuel mass fraction for an underresolved simulation without model, a DS-LES, and a PIESRGAN-LES with experimental data. The agreement is best between PIESRGAN-LES and experimental data. Note that a similar resolution is chosen for DS-LES and PIESRGAN-LES here. It seems that the PIESRGAN-LES is more robust with respect to coarser resolutions. If a finer resolution were to be used, the results for PIESRGAN-LES and DS-LES would become more similar.

4.3 Discussion

The reactive spray cases computed with PIESRGAN-subfilter model show that the PIESRGAN-based subfilter approach can be used to actually compute complex flows with high accuracy. In terms of operations needed per time step, the PIESRGAN-subfilter model is more expensive than a classical DS approach. Furthermore, the PIESRGAN approach generates additional cost for training of the network. However, the PIESRGAN approach has the advantage of naturally running on GPUs which are responsible for the majority of floating point operations per second (FLOPS) in current supercomputer systems.

As discussed, the PIESRGAN approach can be used to reduce model assumptions, such as those made for the mixture fraction PDF and functional form of the scalar dissipation rate, which is an advantage. The presented results demonstrate that

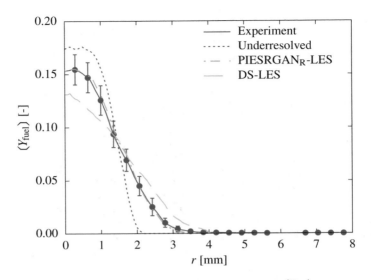

Fig. 8 Temporally and circumferentially averaged fuel mass fraction $\langle Y_{\text{fuel}} \rangle$ evaluated 18.75 mm downstream from the nozzle and plotted against the radial distance from the spray axis r. Plot from Bode et al. (2021)

simulations without the discussed presumed closures but with PIESRGAN closure are able to reasonably match experimental data. However, due to the lack of DNS data and the multiple models which are still involved, such as breakup models and the chemical mechanism, a detailed analysis of the impact of these closures on macroscopic quantities, such as LOL and ignition delay time, remains difficult. However, it can be concluded that the PIESRGAN approach is very robust even in heavily underresolved flow situations. This is an important feature for very complex simulations such as full engine simulations. In these cases, it is impossible to sufficiently resolve all parts and the robustness of closure models becomes significant.

5   Application to Premixed Combustion

In premixed combustion cases, fuel and oxidizer are completely mixed before combustion is allowed to take place. Typical examples include spark ignition engines and lean-burn gas turbines. Therefore, in contrast to non-premixed combustion, correctly predicting fuel-oxidizer mixing is less important for premixed combustion.

5.1 Case Description

Falkenstein et al. (2020a, b, c) computed a collection of premixed flame kernels with iso-octane/air mixtures under real engine conditions and with unity and constant Lewis numbers. The case with unity Lewis number, i.e., featuring the same diffusion coefficient for all scalar species, is used as demonstration case in this work. All simulations, DNS and PIESRGAN-LES, were computed with CIAO (Desjardins et al. 2008). The DNS relies on the low-Mach number limit of the Navier–Stokes equations employing the Curtiss–Hirschfelder approximation (Hirschfelder et al. 1964) for diffusive scalar transport and including the Soret effect. A mesh with 960^3 cells was used. The iso-octane reaction mechanism features 26 species (Falkenstein et al. 2020a). The setup puts one flame kernel in a homogeneous isotropic turbulence field. Consequently, the turbulence decays over time, while the flame kernel expands, wrinkles, and deforms from its originally spherical shape. As the resulting flame speed depends on the local curvature of the flame kernel, it is very important to accurately predict the flame surface density. For running PIESRGAN-LES, the training of PIESRGAN was performed with multiple filter stencil widths varying from 5 to 15 cells (Bode et al. 2022).

Often, a reaction progress variable is defined to describe the temporal state of a flame kernel. Falkenstein et al. (2020a) defined it as sum of the mass fractions of H_2, H_2O, CO, and CO_2 and introduced a simplified reaction progress variable ζ. The simplified reaction progress variable behaves according to a transport equation with the thermal diffusion coefficient as diffusion coefficient reading

$$\frac{\partial \rho \zeta}{\partial t} + \frac{\partial \rho u_j \zeta}{\partial x_j} = \frac{\partial}{\partial x_j}\left(\rho D_{\text{th}} \frac{\partial \zeta}{\partial x_j}\right) + \dot{\omega}_\zeta, \tag{3}$$

employing Einstein's summation notation, with ρ as fluid density, t as time, u_j as velocity vector, x_j as space vector, D_{th} as thermal diffusion coefficient, and $\dot{\omega}_\zeta$ as chemical source term of the simplified reaction progress variable, which is the sum of the source terms of the species used for the definition of the reaction progress variable. The evolution of one flame kernel realization is visualized in Fig. 9.

In contrast to the decaying turbulence and reactive spray cases presented in the previous sections, it is not sufficient to only train the PIESRGAN with turbulence data for finite-rate chemistry cases. Instead, the fully trained network based on decaying homogeneous isotropic turbulence was only used as starting network, which was further updated with finite-rate chemistry data. As a consequence, reconstruction is learnt for all species fields, and the optional solution step with the unfiltered transport equations on the finer mesh of the reconstructed data is employed. This combination of reconstructing and solving was found to be crucial for the accuracy of finite-rate chemistry flows (Bode et al. 2022; Bode 2022a).

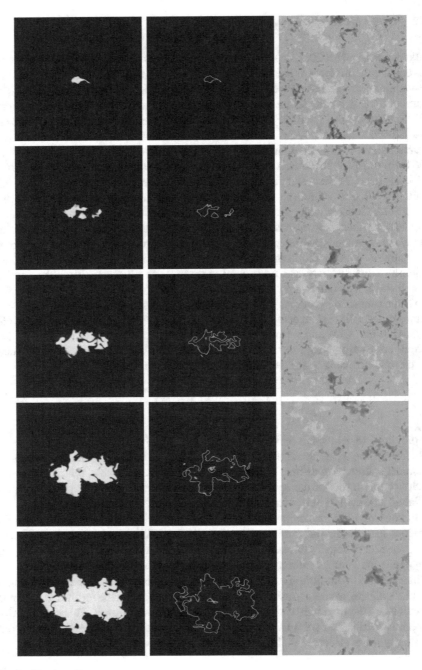

Fig. 9 (Continued)

◄**Fig. 9** (Continued) Visualization of 2-D slices of the simplified reaction progress variable ζ, the source term of the simplified reaction progress variable $\dot{\omega}_\zeta$, and the velocity component U (left to right) for five different increasing time steps (top to bottom) for the fully turbulent flame kernel with unity Lewis number. The first time shows 6.0×10^{-5} s, and the time increment is 7.5×10^{-5} s. The final time is 3.6e-4 s, which is also used for the a priori analysis in Fig. 10. Colormaps span from blue (minimum) to green to yellow (maximum). Note that the flame kernel does not break into parts at the latest time shown. A coherent flame kernel topology was maintained at all times

5.2 A Priori Results

Reconstruction results for the simplified reaction progress variable, two species mass fractions, and one velocity component are compared with fully resolved and filtered fields in Fig. 10. The agreement between fully resolved fields and reconstructed fields is good. The filtered data, which were filtered over 15 cells, are less sharp due to the smoothing of small-scale structures.

5.3 A Posteriori Results

Multiple quantities can be tracked during the evolution of the flame kernel. The flame surface density Σ can be evaluated by means of a phase indicator function $\Gamma(\mathbf{x}, t)$, defined for a reaction variable progress variable threshold value ζ_0 as $\Gamma(\mathbf{x}, t) = \mathcal{H}(\zeta(\mathbf{x}, t) - \zeta_0)$, with $\mathcal{H}$ being the Heaviside step function. The surface density is then given by

$$\Sigma = \langle |\nabla\Gamma| \rangle, \tag{4}$$

employing volume-averaging. Moreover, the corresponding characteristic length scale L_Σ can be defined as

$$L_\Sigma = \frac{4\langle\Gamma\rangle\,(1 - \langle\Gamma\rangle)}{\Sigma}. \tag{5}$$

As for the decaying turbulence case before, the averaged turbulent kinetic energy decays. In contrast to this, the flame surface density is expected to increase significantly and the characteristic length scale L_Σ should increase slightly. This is shown in Fig. 11. The agreement between DNS and PIESRGAN-LES results is good.

5.4 Discussion

The accuracy of PIESRGAN for premixed combustion cases is very promising. This enables PIESRGAN-LES to be a very useful tool for evaluation of cycle-to-cycle

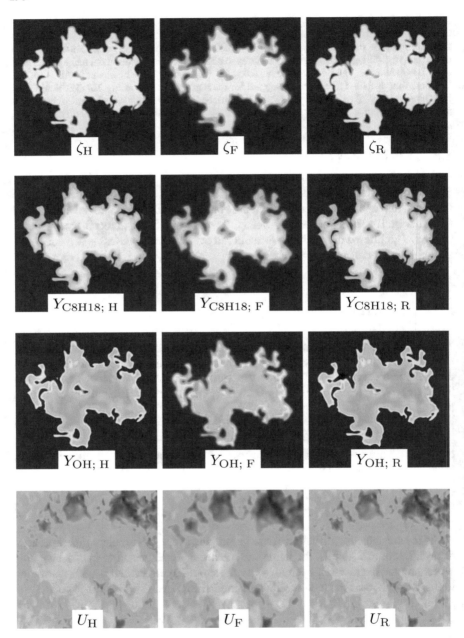

Fig. 10 Visualization of DNS, filtered, and reconstructed fields for the unity Lewis number case employing PIESRGAN. Results for the simplified reaction progress variable ζ, the C_8H_{18} mass fraction Y_{C8H18}, the OH mass fraction Y_{OH}, and the velocity component U are shown. Colormaps span from blue (minimum) to green to yellow (maximum). Note that the images are zoomed in compared to the images presented in the last row in Fig. 9

Fig. 11 Evolution over time t of the volume-averaged turbulent kinetic energy in the unburnt mixture $\langle k_u \rangle$, the surface density Σ, and the characteristic length scale L_Σ for the DNS and PIESRGAN-LES for the unity Lewis number case. Plot from Bode (2022c)

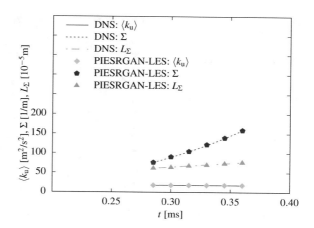

variations (CCVs) and other complex phenomena in engines. A potential workflow could first compute two DNS realizations and other complex phenomena of premixed flame kernels, which are used for on-the-fly training of the PIESRGAN. The trained network is then used to compute multiple PIESRGAN-LES realizations of the premixed flame kernel setup and enable sufficient statistics to study CCVs. Bode et al. (2022a) also showed a certain robustness of the PIESRGAN-subfilter model with respect to setup variations, which might be partly a result of the GAN approach. Consequently, PIESRGAN could also be employed to optimize geometries of turbines or devise optimal operating conditions to reduce harmful emissions.

As discussed in the context of reactive sprays, the reconstruction approach could also be used to improve conventional models, typically relying on filtered probability functions. Instead, a PIESRGAN approach allows to directly evaluate the filtered density function (FDF) increasing the model accuracy.

6 Application to Non-premixed Combustion

In non-premixed combustion cases, fuel and oxidizer are initially separated. As a consequence, mixing and continuous interdiffusion is necessary to establish a flame. Typical examples are furnaces, diesel engines, and jet engines.

6.1 Case Description

The study of non-premixed temporally evolving planar jets (Denker et al. 2020, 2021) was also performed with the CIAO code (Desjardins et al. 2008) and featured multiple nonreactive and reactive cases with a highest initial jet Reynolds number

Fig. 12 Visualization of the
turbulent non-premixed
temporal jet at a late time
step. The fuel is in the center,
two flames burn upwards and
downwards, respectively, and
the main flow direction is
from the left to the right.
Upper half: Mixture fraction
Z on a linear scale.
Colormap spans from black
(minimum) to red
(maximum). Lower half:
Scalar dissipation rate χ on a
logarithmic scale. Colormap
spans from black (minimum)
to red (yellow)

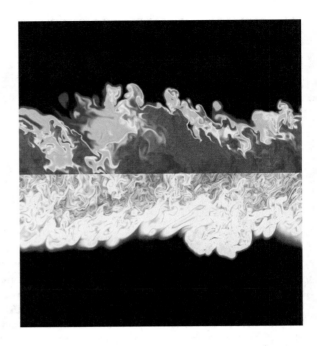

of 9850. It used methane as fuel, modeled by a reaction mechanism with 28 species.
The largest case used $1280 \times 960 \times 960$ cells and is visualized in Fig. 12 by means
of the mixture fraction Z and its scalar dissipation rate defined as

$$\chi = 2D \left(\frac{\partial Z}{\partial x_i} \right)^2 \tag{6}$$

with D as diffusivity, x_i as spatial coordinate, and utilizing Einstein's summation
notation. The temporal jet setup has two periodic directions: the flow direction (from
left to right) and the spanwise direction (perpendicular to the cut view in Fig. 12). The
moving layer of fuel is in the center and surrounded by originally quiescent air. At
the late time step shown, the central fuel stream has already experienced significant
bending due to turbulence, resulting in the lack of fuel in the upper half at about
one quarter length of the domain. Furthermore, it can be seen that the layer in which
scalar dissipation is active is broader than the fuel layer and the scalar dissipation
rate structures are much finer than the mixture fraction structures resulting from the
derivative. Only one realization per parameter combination was computed, however,
the spanwise direction was chosen in such a way that turbulent statistics evaluated in
the two periodic directions converged. The nonperiodic direction was chosen large
enough to prevent interaction of the jet with the boundary. As for the premixed case,
a PIESRGAN with learnt chemistry was employed for the results presented here.

Fig. 13 PDF $\mathscr{P}$ of the scalar dissipation rate χ evaluated with the fully resolved data, the filtered data, and the reconstructed data

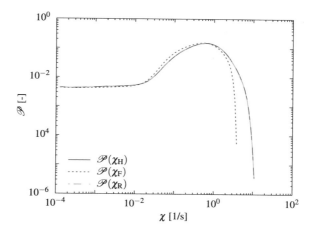

6.2 A Priori Results

The scalar dissipation rate, i.e., the measurement of local mixing, is very essential for non-premixed combustion as it requires the fuel and oxidizer streams to be mixed first, resulting in a lower limit for the scalar dissipation rate required for burning. As indicated by Fig. 12, the scalar dissipation rate is a quantity which acts on the smallest scales making it difficult for LES as it usually has significant contributions below the filter width. Furthermore, extinction (and later reignition) can occur in regions where the scalar dissipation rate becomes too large, typically estimated by the quenching scalar dissipation rate in so-called stationary flamelet solutions, denoted as χ_q. Overall, the scalar dissipation rate is a very well suited quantity to evaluate the prediction accuracy of the PIESRGAN-model. The PDF $\mathscr{P}$ of the scalar dissipation rate is shown in Fig. 13. As expected, the filtering leads to a lack of regions with very high scalar dissipation rate. These missing values are successfully reconstructed by the PIESRGAN-model via the mass fraction fields, i.e., the scalar dissipation rate shown in the figure is a post-processed quantity relying on other reconstructed quantities of the simulation data. The result in the log-log plot looks very good, however, note that the increase of probability (from about $\chi = 0.1$ to $1\ \mathrm{s}^{-1}$) is much better predicted with the reconstructed data than with filtered data alone, but far from perfect.

6.3 A Posteriori Results

Typically, a non-premixed flame is located on surfaces of roughly stoichiometric mixture fraction, which makes the scalar dissipation rate conditioned on the stoichiometric mixture fraction an interesting quantity. Furthermore, a dimensionless

Fig. 14 Evolution over dimensionless time $t^*(*)$ of the ensemble-averaged (density weighted) scalar dissipation rate conditioned on the stoichiometric mixture friction and normalized by the quenching scalar dissipation rate $\langle\chi|Z_Z\rangle\chi_q$. Modified plot from Bode (2022c)

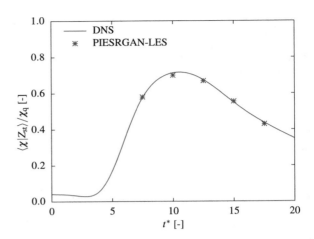

time is introduced, denoted as t^*. This time is shifted to make different cases comparable with the starting point defined as the time when the variance of the scalar dissipation rate at stoichiometric conditions is zero. The normalization is done with the jet time defined with the jet height and its bulk velocity as 32.3 mm/20.7 m/s. The time evolution of the ensemble-averaged density-weighted scalar dissipation rate conditioned on the stoichiometric mixture fraction is compared between DNS and PIESRGAN-LES in Fig. 14. The LES used training data of varying filter widths with stencil sizes of 7–15 cells per direction (Bode 2022a). The prediction of the LES is very good even though the peak is slightly underpredicted.

6.4 Discussion

The non-premixed case emphasizes two important points with respect to PIESRGAN modeling. First, as seen for the decaying turbulence case, the accuracy for predicting mixing is very high. This is crucial for many applications going far beyond combustion cases. Second, PIESRGAN is able to statistically predict a local phenomenon like quenching, which is very challenging for classical LES models. Both points make PIESRGAN very promising for predictive LES of even more complex configurations.

The non-premixed case with more than one billion grid points and 28 species, chosen as an example in this section, also highlights the capability of PIESRGAN to be used for recomputing the largest available reactive DNS. This is technically remarkable and only possible due to the rapid developments in the fields of ML/DL and supercomputers in general.

7 Conclusions

AI super-resolution is a powerful tool to improve various aspects of state-of-the-art simulations. These include the reduction of storage and input/output (I/O), a better comparability between experimental and simulation data, and highly accurate subfilter models for LES, as demonstrated by the examples discussed in this work. The remarkable progress in the fields of ML/DL and supercomputing in general, especially with respect to GPU computing, has made ML/DL-based techniques competitive and in some aspects even superior compared to classical approaches, and it is expected that the rapid developments in this field will continue in the upcoming years.

The presented applications ranging from turbulence to non-premixed combustion focused on the high accuracy of PIESRGAN-based approaches in a priori and a posteriori tests. Especially, the a posteriori accuracy is striking unveiling the potential of the PIESRGAN-subfilter approach. Compared to classical methods, the LES mesh can be often significantly reduced as the PIESRGAN technique was found to be more robust in underresolved flow situations.

From a technical point of view, PIESRGAN-based models are simple to use as they can be easily implemented in frameworks, such as Keras/TensorFlow and PyTorch, which are used by a very large community. The trained network can be coupled to any simulation code by just adapting the existing application programming interface (API) to external libraries.

PIESRGAN-based subfilter modeling is a relatively new technique and thus many questions are still open. The presented architecture resulted in good results but it is expected that it could be further improved. The approach of physics-informed loss function compared to physics-informed network layers seems to be reasonable and has the advantage of a trivial implementation while resulting in equally accurate predictions. One of the most important topics in the context of data-driven approaches is the extrapolation capability, i.e., how accurate are predictions outside of the training range. The recent publications (Bode et al. 2019a, 2021, 2022; Bode 2022a, b, c) show some promising properties in this regard for PIESRGAN, but it should be investigated in more detail in the future. Additionally, the combustion community has computed petabytes of DNS data for various combustion configuration. Given the demonstrated generality of PIESRGAN in the sense that the same architecture worked very well for multiple configurations, the combination of DNS database and PIESRGAN could be already very useful to advance combustion research. PIESRGAN was also shown to be universal enough to use the same trained network for physical parameter variations. Thus, many optimization problems could be easily accelerated.

Acknowledgements The author acknowledges computing time grants for the projects JHPC55 and TurbulenceSL by the JARA-HPC Vergabegremium provided on the JARA-HPC Partition part of the supercomputer JURECA at Jülich Supercomputing Centre, Forschungszentrum Jülich, the Gauss Centre for Supercomputing e.V. (www.gauss-centre.eu) for funding this project by providing computing time on the GCS Supercomputer JUWELS at Jülich Supercomputing Centre (JSC), and funding from the European Union's Horizon 2020 research and innovation program under the Center of Excellence in Combustion (CoEC) project, grant agreement no. 952181.

References

Abadi M, Agarwal A, Barham P, Brevdo E, Chen Z, Citro C, Corrado GS, Davis A, Dean J, Devin M, Ghemawat S, Goodfellow I, Harp A, Irving G, Isard M, Jia Y, Jozefowicz R, Kaiser L, Kudlur M, Levenberg J, Mané D, Monga R, Morre S, Murray D, Olah C, Schuster M, Shlens J, Steiner B, Sutskever I, Talwar K, Tucker P, Vanhoucke V, Vasudevan V, Viégas F, Vinyals O, Warden P, Wattengerg M, Wicke M, Yu Y, Zheng X (2016) TensorFlow: large-scale machine learning on heterogeneous systems

Banerjee I, Ierapetritou GM (2006) An adaptive reduction scheme to model reactive flow. Combust Flame 144(3):619–633

Beck AD, Flad DG, Munz C-D (2018) Neural networks for data-based turbulence models. arXiv:1806.04482

Bhati A, Wan S, Alfe D, Clyde A, Bode M, Tan L, Titov M, Merzky A, Turilli M, Jha S, Highfield RR, Rocchia W, Scafuri N, Succi S, Kranzlmüller D, Mathias G, Wifling D, Donon Y, Di Megio A, Vallecorsa S, Ma H, Trifan A, Ramathan A, Brettin T, Partin A, Xia F, Duan X, Stevens R, Coveney PV (2021) Pandemic drugs at pandemic speed: infrastructure for accelerating COVID-19 drug discovery with hybrid machine learning- and physics-based simulations on high performance computers. Interface Focus, 20210018

Bode M (2022a) Applying physics-informed enhanced super-resolution generative adversarial networks to turbulent non-premixed combustion on non-uniform meshes and demonstration of an accelerated simulation workflow. arXiv preprint arXiv:2210.16248

Bode M (2022b) Applying physics-informed enhanced super-resolution generative adversarial networks to large-eddy simulations of ECN Spray C. SAE Technical Paper 2022-01-0503

Bode M (2022c) Applying physics-informed enhanced super-resolution generative adversarial networks to finite-rate-chemistry flows and predicting lean premixed gas turbine combustors. arXiv preprig arXiv:2210.16219

Bode M, Diewald F, Broll D, Heyse J, et al (2014) Influence of the injector geometry on primary breakup in diesel injector systems. SAE Technical Paper 2014-01-1427

Bode M, Falkenstein T, Le Chenadec V, Kang S, Pitsch H, Arima T, Taniguchi H (2015) A new Euler/Lagrange approach for multiphase simulations of a multi-hole GDI injector. SAE Technical Paper 2015-01-0949

Bode M, Gauding M, Kleinheinz K, Pitsch H (2019a) Deep learning at scale for subgrid modeling in turbulent flows: regression and reconstruction. LNCS 11887:541–560

Bode M, Collier N, Bisetti F, Pitsch H (2019b) Adaptive chemistry lookup tables for combustion simulations using optimal B-spline interpolants. Combust Theory Model 23(4):674–699

Bode M, Gauding M, Lian Z, Denker D, Davidovic M, Kleinheinz K, Jitsev J, Pitsch H (2021) Using physics-informed enhanced super-resolution generative adversarial networks for subfilter modeling in turbulent reactive flows. Proc Combust Inst 38:2617–2625

Bode M, Gauding M, Goeb D, Falkenstein T, Pitsch H (2022) Applying physics-informed enhanced super-resolution generative adversarial networks to turbulent premixed combustion and engine-like flame kernel direct numerical simulation data. arXiv prepring arXiv:2210.16206

Denker D, Attili A, Gauding M, Niemietz K, Bode M, Pitsch H (2020) Dissipation element analysis of non-premixed jet flames. J Fluid Mech 905:A4

Denker D, Attili A, Boschung J, Hennig F, Gauding M, Bode M, Pitsch H (2021) A new modeling approach for mixture fraction statistics based on dissipation elements. Proc Combust Inst 38:2681–2689

Desjardins O, Blanquart G, Balarac G, Pitsch H (2008) High order conservative finite difference scheme for variable density low Mach number turbulent flows. J Comput Phys 227(15):7125–7159

Dong C, Loy CC, He K, Tang X (2014) Learning a deep convolutional network for image super-resolution. In: European conference on computer vision, pp 184–199

Dong C, Loy CC, He K, Tang X (2015) Image super-resolution using deep convolutional networks. IEEE Trans Pattern Anal Mach Intel 38(2):295–307

Engine Combustion Network (2019) https://ecn.sandia.gov

Falkenstein T, Kang S, Cai L, Bode M, Pitsch H (2020a) DNS study of the global heat release rate during early flame kernel development under engine conditions. Combust Flame 213:455–466

Falkenstein T, Rezchikova A, Langer R, Bode M, Kang S, Pitsch H (2020b) The role of differential diffusion during early flame kernel development under engine conditions - part i: analysis of the heat-release-rate response. Combust Flame 221:502–515

Falkenstein T, Chu H, Bode M, Kang S, Pitsch H (2020c) The role of differential diffusion during early flame kernel development under engine conditions - part ii: effect of flame structure and geometry. Combust Flame 221:516–529

Frisch U, Kolmogorov AN (1995) Turbulence: the legacy of AN Kolmogorov. Cambridge University Press, Cambridge

Gauding M, Wang L, Goebbert JH, Bode M, Danaila L, Varea E (2019) On the self-similarity of line segments in decaying homogeneous isotropic turbulence. Comput & Fluids 180:206–217

Goeb D, Davidovic M, Cai L, Pancharia P, Bode M, Jacobs S, Beeckmann J, Willems W, Heufer KA, Pitsch H (2021) Oxymethylene ether - n-dodecane blend spray combustion: experimental study and large-eddy simulations. Proc Combust Inst 38:3417–3425

Goodfellow IJ, Pouget-Agadie J, Mirza M, Xu B, Warde-Farley D, Ozair S, Courville A, Bengio Y (2014) Generative adversarial networks. arXiv:1406.2661

Hinton G, Deng L, Yu D, Dahl G, Mohamed A, Jaitly N, Senior A, Vanhoucke V, Nguyen P, Sainath TN, Kingsbury B (2012) Deep neural networks for acoustic modeling in speech recognition. IEEE Signal Process Mag 29

Hirschfelder JO, Curtiss CF, Bird RB, Mayer MG (1964) Molecular theory of gases and liquids

Ihme M, Schmitt C, Pitsch H (2009) Optimal artificial neural net- works and tabulation methods for chemistry representation in LES of a bluff-body swirl-stabilized flame. Proc Combust Inst 32:1527–1535

Johnson J, Alahi A, Fei-Fei L (2016) Perceptual losses for real-time style transfer and super-resolution. In: European conference on computer vision, pp 694–711

Jolicoeur-Martineau A (2018) The relativistic discriminator: a key element missing from standard GAN. arXiv:1807.00734

Keras (2019) https://keras.rstudio.com/index.html

Kim J, Lee JK, Lee KM (2016a) Accurate image super-resolution using very deep convolutional networks. In: Proceedings of the IEEE conference on computer vision and pattern recognition, pp 1646–1654

Kim J, Lee JK, Lee KM (2016b) Deeply-recursive convolutional network for image super-resolution. In: Proceedings of the IEEE conference on computer vision and pattern recognition, pp 1637–1645

Krizhevsky A, Sutskever I, Hinton GE (2012) Imagenet classification with deep convolutional neural networks. In: Advances in neural information processing systems, pp 1097–1105

Lai W-S, Huang J-B, Ahuja N, Yang M-H (2017) Deep laplacian pyramid networks for fast and accurate super-resolution. In: Proceedings of the IEEE conference on computer vision and pattern recognition, pp 624–632

Ledig C, Theis L, Huszár F, Caballero J, Cunningham A, Acosta A, Aitken A, Tejani A, Totz J, Wang Z, Shi W (2017) Photo-realistic single image super-resolution using a generative adversarial network. In: Proceedings of the IEEE conference on computer vision and pattern recognition, pp 4681–4690

Li Y, Ni Y, Croft RAC, Di Matteo T, Bird S, Feng Y (2021) AI-assisted superresolution cosmological simulations. Proc Natl Acad Sci 118:e2022038118

Maas AL, Hannun AY, Ng AY (2013) Rectifier nonlinearities improve neural network acoustic models. In: Proceedings of the 30th international conference on machine learning, p 30

Miller RS, Bellan J (1999) Direct numerical simulation of a confined three-dimensional gas mixing layer with one evaporating hydrocarbon-droplet-laden stream. J Fluid Mech 384:293–338

Patterson MA, Reitz RD (1998) Modeling the effects of fuel spray characteristics on diesel engine combustion and emission. SAE Technical Paper

Pekurovsky D (2012) P3DFFT: a framework for parallel computations of Fourier transforms in three dimensions. SIAM J Sci Comput 34:192–209

Peters N (1986) Laminar flamelet concepts in turbulent combustion. In: Twenty-First symposium (International) combustion, pp 1231–1250

Pitsch H, Chen M, Peters N (1998) Unsteady flamelet modeling of turbulent hydrogen-air diffusion flames. In: Twenty-Seventh symposium (International) combustion, vol 27, pp 1057–1064

Pitsch H (2006) Large-eddy simulation of turbulent combustion. Ann Rev Fluid Mech 38:453–482

Pope SB (2000) Turbulent flows. Cambridge University Press, Cambridge

Simonyan K, Zisserman A (2014) Very deep convolutional networks for large-scale image recognition. arXiv:1409.1556

Smagorinsky J (1963) General circulation experiments with the primitive equations: I. The basic experiment. Mon Weather Rev 91(3):99–164

Stengel K, Glaws A, Hettinger D, King RN (2020) Adversarial super-resolution of climatological wind and solar data. Proc Natl Acad Sci 117:16805–16815

Tai Y, Yang J, Liu X, Xu C (2017) Memnet: a persistent memory network for image restoration. In: Proceedings of the IEEE international conference on computer vision, pp 4539–4547

Vinyals O, Babuschkin I, Czarnecki WM, Mathieu M, Dudzik A, Chung J, Choi DH, Powell R, Ewalds T, Georgiev P, Oh J, Horgan D, Kroiss M, Danihelka I, Huang A, Sifre L, Cai T, Agapiou JP, Jaderberg M, Vezhnevets AS, Leblond R, Pohlen T, Dalibard V, Budden D, Sulsky Y, Molloy J, Paine TL, Gulcehre C, Wang Z, Pfaff T, Wu Y, Ring R, Yogatama D, Wünsch D, McKinney K, Smith O, Schaul T, Lillicrap T, Kavukcuoglu K, Hassabis D, Apps C, Silver D (2019) Grandmaster level in StarCraft II using multi-agent reinforcement learning. Nature 575:350–354

Wang X, Yu K, Wu S, Gu J, Liu Y, Dong C, Qiao Y, Loy CC (2018) ESRGAN: enhanced super-resolution generative adversarial networks. In: Proceedings of the European conference on computer vision (ECCV)

Yao T, Pei Y, Zhong B-J, Som S, Lu T, Luo KH (2017) A compact skeletal mechanism for n-dodecane with optimized semi-global low-temperature chemistry for diesel engine simulations. Fuel 191:339–349

Zhang Y, Li K, Li K, Wang L, Zhong B, Fu Y (2018) Image super-resolution using very deep residual channel attention networks. In: Proceedings of the European conference on computer vision (ECCV), pp 286–301

Machine Learning for Thermoacoustics

Matthew P. Juniper

Abstract This chapter demonstrates three promising ways to combine machine learning with physics-based modelling in order to model, forecast, and avoid thermoacoustic instability. The first method assimilates experimental data into candidate physics-based models and is demonstrated on a Rijke tube. This uses Bayesian inference to select the most likely model. This turns qualitatively-accurate models into quantitatively-accurate models that can extrapolate, which can be combined powerfully with automated design. The second method assimilates experimental data into level set numerical simulations of a premixed bunsen flame and a bluff-body stabilized flame. This uses either an Ensemble Kalman filter, which requires no prior simulation but is slow, or a Bayesian Neural Network Ensemble, which is fast but requires prior simulation. This method deduces the simulations' parameters that best reproduce the data and quantifies their uncertainties. The third method recognises precursors of thermoacoustic instability from pressure measurements. It is demonstrated on a turbulent bunsen flame, an industrial fuel spray nozzle, and full scale aeroplane engines. With this method, Bayesian Neural Network Ensembles determine how far each system is from instability. The trained BayNNEs out-perform physics-based methods on a given system. This method will be useful for practical avoidance of thermoacoustic instability.

1 Introduction

At present there is no realistic alternative to combustion engines for long distance aircraft and rockets. These engines have unrivalled power to weight ratios and their fuels have unrivalled energy to weight ratios. If we continue to fly long distances or send rockets into space, we will continue to combust fuels in increasingly high-performance gas turbines and rockets. Despite decades of research and the development of sophisticated physics-based models, thermoacoustic instability in these

M. P. Juniper (✉)
Engineering Department, University of Cambridge, Cambridge CB2 1PZ, UK
e-mail: mpj1001@cam.ac.uk

© The Author(s) 2023
N. Swaminathan and A. Parente (eds.), *Machine Learning and Its Application to Reacting Flows*, Lecture Notes in Energy 44, https://doi.org/10.1007/978-3-031-16248-0_11

engines remains difficult to predict and eliminate. The aim of this chapter is to introduce some promising avenues in which machine learning methods could be used to model, forecast, and avoid thermoacoustic instability.

1.1 The Physical Mechanism Driving Thermoacoustic Instability

The combustion chambers in aircraft and rocket engines have extraordinarily high power densities: from 100 MW/m^3 in aircraft gas turbines to 50 GW/m^3 in liquid-fuelled rocket engines (Culick 2006). They contain flames that are typically anchored by a recirculation zone (aircaft engines) or by fuel injector lips (rockets). Acoustic velocity fluctuations perturb the base of the flame, creating ripples that convect downstream and cause heat release rate fluctuations some time later, which in turn create acoustic fluctuations either directly or via entropy spots (Lieuwen 2012). If moments of higher (lower) heat release rate coincide sufficiently with moments of higher (lower) pressure around the flame, then more work is done by the heated gas during the expansion phase of the acoustic cycle than was done on it during the compression phase. If the work done by thermoacoustic driving exceeds the work dissipated through damping or acoustic radiation over a cycle, then the acoustic amplitude grows and the system is thermoacoustically unstable. This is also known as combustion instability. In high performance rocket and aircraft engines, the heat release rate is so high and the natural dissipation so low that these engines can become thermoacoustically unstable even if the thermodynamic efficiency of the cycle is as little as 0.1% (Huang and Yang 2009).

Thermoacoustic oscillations were first noticed over 200 years ago (Higgins 1802) and their physical mechanism was correctly identified nearly 150 years ago (Rayleigh 1878). They were recognized as a significant problem in rocket engines 80 years ago and have been investigated seriously for 70 years (Crocco and Cheng 1956). Nevertheless, they remain a problem for the design of gas turbine and rocket engines because engineers are rarely able to predict, at the design stage, whether a particular engine will suffer from them (Lieuwen and McManus 2003; Mongia et al. 2003). This chapter explains why thermoacoustic instability is so difficult to predict accurately and explores various data-driven approaches that could develop into alternatives or additions to current physics-based approaches.

1.2 The Extreme Sensitivity of Thermoacoustic Systems

Thermoacoustic instability is difficult to predict for two main reasons. Firstly, if the time lag between velocity fluctuations at the base of the flame and subsequent heat release rate fluctuations is similar to or greater than the acoustic period, which is usually the case, then the ratio of time lag to acoustic period strongly affects the efficiency of the thermoacoustic mechanism (Juniper and Sujith 2018). Secondly, this time lag often depends on factors that are difficult to simulate or model accurately, such as jet break-up, droplet evaporation, flame kinematics, and high Reynolds number combustion.

Rocket and aircraft engines are usually developed through component tests, sector tests, combustor tests, and full engine tests. The response of the flame to acoustic fluctuations, for example, might be measured in a well-characterized rig and then included in a model of the full engine. If, however, the flame's behaviour were to change slightly when placed in the full engine then the model would contain unknown model error in a critical component. The model would remain qualitatively accurate but become quantitatively inaccurate and therefore misleading. Indeed, it is quite common for thermoacoustic instability to recur in the later stages of engine development, even though models compiled from component tests predicted it to be stable (Mongia et al. 2003).

Encouragingly, this sensitivity also explains why thermoacoustic oscillations can usually be suppressed by making small design changes (Mongia et al. 2003; Oefelein and Yang 1993; Dowling and Morgans 2005). The challenge, of course, is to devise these small design changes from a quantitatively-accurate model rather than by trial and error. Adjoint methods combined with gradient-based optimization provide an excellent mechanism for this (Juniper and Sujith 2018; Magri and Juniper 2013; Juniper 2018; Aguilar and Juniper 2020). They rely, however, on a quantitatively accurate model. This chapter explores how experimental or numerical data could be assimilated in order to create these quantitatively-accurate models from qualitatively-accurate physics-based models or from physics-agnostic models.

1.3 The Opportunity for Data-Driven Methods in Thermoacoustics

All models contain parameters that are tuned to fit data. These range from qualitatively-accurate physics-based models with $O(10^1)$ parameters to Gaussian Process surrogate models with $O(10^3)$ parameters, and to physics-agnostic neural networks with $O(10^6)$ parameters. The challenge is to create models that are quantitatively accurate with quantified uncertainties and are sufficiently constrained to be informative.[1] To this end, all the approaches in this chapter take a Bayesian perspective and, where possible, employ rigorous statistical inference[2] (MacKay 2003).

---

[1] Freemon Dyson (2004) quoted Fermi quoting von Neumann saying: "With four parameters I can fit an elephant, and with five I can make him wiggle his trunk." Fermi was referring to arbitrary parameters rather than physics-based parameters but the general point remains that models can become un-informative if they contain too many parameters.

[2] As stated in the introduction to this book: "Machine learning is statistical inference using data collected or knowledge gained through past targeted studies or real-life experience".

The first example is a canonical thermoacoustic system: the hot wire Rijke tube (Rijke 1859; Saito 1965). Although simple and cheap to operate, it is difficult to model accurately firstly because the heat release rate is small, meaning that many visco-thermal dissipation mechanisms are sufficiently large, in comparison, that they must be included in the model, and secondly because the heat release rate fluctuations at the wire cannot be measured directly. A hot wire Rijke tube is, however, easy to automate, meaning that millions of datapoints can be obtained cheaply and elements of the system can be moved easily (Rigas et al. 2016). Physics-based models of the Rijke tube can therefore be constructed sequentially, mirroring data assimilation from component tests, sector tests, combustor tests, and full engine tests in industry. The process (MacKay 2003; Juniper and Yoko 2022) is to:

1. choose various plausible physics-based models that could explain the data;
2. tune model parameters by assimilating data from experiments;
3. quantify the uncertainties in the parameters of each model;
4. calculate the marginal likelihood (also known as the *evidence*) for each model, and thereby penalise overly-complex models;
5. compare the models against each other and select the best model;
6. add the next component and assimilate more data, allowing the parameters describing the previous components to float within constrained priors.

The second example is the assimilation of DNS and/or experimental data into a simplified combustion model, the G-equation (Williams 1985) with around 4000 degrees of freedom (Hemchandra 2009). Two approaches are demonstrated. The first approach assimilates snapshots of the data sequentially with a Kalman filter (Evensen 2009), refining model parameters on the fly (Yu et al. 2020). The second approach assimilates 10 snapshots simultaneously with a Bayesian ensemble of Deep Neural Networks (BayNNE) (Pearce et al. 2020). This gives almost the same results as the Kalman filter but is around 10^6 times faster. Both approaches assimilate data into physics-based models and obtain the expected values and uncertainties of the model parameters.

The third example is the assimilation of experimental data into physics-agnostic models. The models are trained to recognize how close a thermoacoustic system is to instability from the noise that it emits (Sengupta et al. 2021; Waxenegger-Wilfing et al. 2021; McCartney et al. 2022). As for the first two examples, a Bayesian approach is used so that the model can output its certainty about its prediction. This physics-agnostic approach is compared with model-based approaches quantified by the Hurst exponent (Nair et al. 2014), the permutation entropy (Kobayashi et al. 2017), and the autocorrelation decay (Lieuwen and Banaszuk 2005), which are based on a priori assumptions of how the noise signal will change as instability approaches.

Other examples of the application of Machine Learning to Thermoacoustics are in learning the nonlinear flame response with Neural Networks (Jaensch and Polifke 2017; Tathawadekar et al. 2021), identifying nonlinear flame describing functions (McCartney et al. 2020), modelling the flame impulse response from LES with a Gaussian Process surrogate model (Kulkarni et al. 2021), and the use of Gaussian Processes for Uncertainty Quantification (Guo et al. 2021a).

2 Physics-Based Bayesian Inference Applied to a Complete System

Physics-based Bayesian inference starts from a set of physics-based candidate models $\mathcal{H}_i$, each of which has a set of model parameters $\mathbf{a}$. For thermoacoustic systems, typical model parameters would be physical dimensions, temperatures, reflection coefficients, and a flame transfer function. Data, D, arrive and, at the first level of inference, we find the parameters of each model that are most likely to explain the data (MacKay 2003, Sect. 2.6). For thermoacoustic systems, typical data would be temperatures, pressure fluctuations, or natural emission fluctuations. We start from the product rule of probability:

$$P(\mathbf{a}, D|\mathcal{H}_i) = P(\mathbf{a}|D, \mathcal{H}_i)P(D|\mathcal{H}_i) = P(D|\mathbf{a}, \mathcal{H}_i)P(\mathbf{a}|\mathcal{H}_i) \tag{1}$$

where $P(\mathbf{a}|\mathcal{H}_i)$ is our prior assumption about the probability of the parameters, $\mathbf{a}$, given the model $\mathcal{H}_i$. Bayesian inference requires us to impose prior values for the model parameters and their uncertainties. This is appropriate because we usually know the model parameters approximately from previous experiments and will become increasingly certain about them as an experimental campaign progresses. The term $P(D|\mathbf{a}, \mathcal{H}_i)$ contains the data, D, which is fixed by the experiment, and the parameters, $\mathbf{a}$, which we wish to obtain for model $\mathcal{H}_i$. For given D, the term $P(D|\mathbf{a}, \mathcal{H}_i)$ defines the *likelihood* of the parameters, $\mathbf{a}$, of model $\mathcal{H}_i$ (MacKay 2003, p. 29). This likelihood does not have to sum to 1 because the proposed models $\mathcal{H}_i$ are not mutually exclusive or exhaustive. On the other hand, for a given model $\mathcal{H}_i$ and parameters, $\mathbf{a}$, the term $P(D|\mathbf{a}, \mathcal{H}_i)$ defines the *probability* of the data, which does have to sum to 1. This distinction becomes important when incorporating measurement noise.

The term $P(D|\mathcal{H}_i)$ is the evidence for the model. This is the RHS of (1) integrated (also known as marginalized) over all parameter values:

$$P(D|\mathcal{H}_i) = \int_{\mathbf{a}} P(D|\mathbf{a}, \mathcal{H}_i)P(\mathbf{a}|\mathcal{H}_i)\, \mathrm{d}\mathbf{a} \tag{2}$$

which is known as the *marginal likelihood*. At the first level of inference, this quantity has no significance because we simply find $\mathbf{a}$ that maximizes $P(\mathbf{a}|D, \mathcal{H}_i)$ for a given model $\mathcal{H}_i$. It is used in the second level of inference, in which we compare the marginal likelihoods of different candidate models.

The experiments in this section are performed on a vertical Rijke tube containing an electric heater, which is moved through 19 different positions from the bottom end of the tube (Juniper and Yoko 2022; Garita et al. 2021; Garita 2021). The heater power is set to eight different values until the system reaches steady state. Then a loudspeaker at the base of the tube forces the system close to its resonant frequency and probe microphones measure the response throughout the tube. We assimilate the decay rates, S_r, frequencies, S_i, and relative pressures of the microphones, (P_r, P_i) into a thermoacoustic network model.

2.1 Laplace's Method

The most likely parameters, **a**, and their uncertainties can be found with sampling methods such as Markov Chain Monte Carlo (Metropolis et al. 1953; MacKay 2003) or Hamiltonian Monte Carlo. These sample the posterior probability distribution through a random walk. They can be applied to this thermoacoustic problem (Garita 2021) but are quite slow. The assimilation process can be accelerated greatly by assuming that all the probability distributions are Gaussian (MacKay 2003, Chap. 27). The prior probability distribution, which must integrate to 1, is then:

$$P(\mathbf{a}|\mathcal{H}_i) = \frac{1}{\sqrt{(2\pi)^{N_a}|\mathbf{C}_{aa}|}} \exp\left\{-\frac{1}{2}(\mathbf{a} - \mathbf{a}_p)^T \mathbf{C}_{aa}(\mathbf{a} - \mathbf{a}_p)\right\} \tag{3}$$

where N_a is the number of parameters, $\mathbf{a}_p$ are their prior expected values and $\mathbf{C}_{aa}$ is their prior covariance matrix. We assume that, for a given model $\mathcal{H}_i$ with parameters **a**, the measurements D are normally-distributed around the model predictions $\mathcal{D}(\mathbf{a})$:

$$P(D|\mathbf{a}, \mathcal{H}_i) = \frac{1}{\sqrt{(2\pi)^{N_D}|\mathbf{C}_{DD}|}} \exp\left\{-\frac{1}{2}(\mathcal{D}(\mathbf{a}) - D)^T \mathbf{C}_{DD}(\mathcal{D}(\mathbf{a}) - D)\right\} \tag{4}$$

where N_D is the number of datapoints and $\mathbf{C}_{DD}$ is a diagonal matrix containing the variance of each measurement. In this example, epistemic uncertainty such as model error and systematic measurement error is included within $\mathbf{C}_{DD}$.

We define $\mathcal{J}$ to be the negative log of the RHS of (1):

$$\mathcal{J} = -\log\{P(D|\mathbf{a}, \mathcal{H}_i)P(\mathbf{a}|\mathcal{H}_i)\} \tag{5}$$

so that the most probable parameter values, $\mathbf{a}_{mp}$, are found by minimizing $\mathcal{J}$ using an optimization algorithm. The RHS of (1) is the product of two Gaussians (3), (4), meaning that the posterior likelihood of the parameters, $P(\mathbf{a}|D, \mathcal{H}_i)$, is a Gaussian centred around $\mathbf{a}_{mp}$:

$$-\log\{P(\mathbf{a}|D, \mathcal{H}_i)\} = \frac{1}{2}(\mathbf{a} - \mathbf{a}_{mp})^T \mathbf{A} (\mathbf{a} - \mathbf{a}_{mp}) + \text{constant} \tag{6}$$

where **A** is the inverse of the posterior covariance matrix which, by inspection, is the Hessian of $\mathcal{J}$:

$$A_{ij} = \frac{\partial^2 \mathcal{J}}{\partial a_i a_j} \tag{7}$$

The posterior uncertainty in the parameters, $\mathbf{A}^{-1}$, is therefore calculated cheaply. The integral (2), which can be prohibitively expensive to calculate without the Gaussian assumption, is now simply:

$$P(D|\mathcal{H}_i) = P(D|\mathbf{a}_{mp}, \mathcal{H}_i) P(\mathbf{a}_{mp}|\mathcal{H}_i) (\det(\mathbf{A}/2\pi))^{-1/2} \tag{8}$$

This integral allows us to rank different models, $\mathcal{H}_i$. By the product rule of probability $P(\mathcal{H}_i|D)P(D) = P(D|\mathcal{H}_i)P(\mathcal{H}_i)$. If the prior probability, $P(\mathcal{H}_i)$, is the same for each model then the models can be ranked by $P(D|\mathcal{H}_i)$. The fact that (8) is proportional to $\det(\mathbf{A})^{-1/2}$ penalizes models for which $\det(\mathbf{A})$ is large. This tends to favour models with fewer parameters (hence smaller $\mathbf{A}$) even if they do not fit the data as well as models with more parameters. This does not, of course, prevent a model with many parameters from being the highest ranked, as long as the model fits the data well and the measurement uncertainty is small.

2.2 Accelerating Laplace's Method with Adjoint Methods

If all probability distributions are assumed to be Gaussian then $\mathcal{J}$ is the sum of the squares of the discrepancies between the model predictions and the experimental measurements, weighted by our confidence in the experimental measurements, added to the sum of the squares of the discrepancies between the model parameters and their prior estimates, weighted by our confidence in the prior estimates:

$$\begin{aligned} \mathcal{J} &= -\log\{P(D|\mathbf{a}, \mathcal{H}_i)P(\mathbf{a}|\mathcal{H}_i)\} \\ &= (S_r(\mathbf{a}) - S_r)^T C_{Sr}^{-1}(S_r(\mathbf{a}) - S_r)\dots \\ &+ (S_i(\mathbf{a}) - S_i)^T C_{Si}^{-1}(S_i(\mathbf{a}) - S_i)\dots \\ &+ (\mathcal{P}_r(\mathbf{a}) - P_r)^T C_{Pr}^{-1}(\mathcal{P}_r(\mathbf{a}) - P_r)\dots \\ &+ (\mathcal{P}_i(\mathbf{a}) - P_i)^T C_{Pi}^{-1}(\mathcal{P}_i(\mathbf{a}) - P_i)\dots \\ &+ (\mathbf{a} - \mathbf{a}_f)^T C_{aa}^{-1}(\mathbf{a} - \mathbf{a}_f) + \dots \end{aligned} \tag{9}$$

By inspection, the Jacobian and Hessian of $\mathcal{J}$ contain $\partial\square/\partial a_i$ and $\partial^2\square/\partial a_i a_j$ respectively, where $\square$ refers to $S_\star(\mathbf{a})$ and $\mathcal{P}_\star(\mathbf{a})$. These first and second derivatives can be found cheaply with first (Magri and Juniper 2013) and second (Tammisola et al. 2014; Magri et al. 2016) order adjoint methods. The remaining terms in $\mathcal{J}$ contain the normalizing factors in (3), (4). The derivatives w.r.t. the measurement uncertainties can also be calculated and one can then optimize to find the measurement noise that maximizes the posterior likelihood. In this example, the epistemic uncertainty is embedded within the measurement noise, so assimilating the measurement noise also assimilates the epistemic uncertainty.

Adjoint codes require a careful code structure and must avoid non-differentiable operators. The code used here consists of a low level thermoacoustic network model that contains floating parameters to quantify all possible local feedback mechanisms Juniper (2018). The gradients of $(\mathcal{S}_\star, \mathcal{P}_\star)$ w.r.t. all possible feedback mechanisms are calculated. These mechanisms are then ascribed physical meaning by candidate models and the gradients w.r.t. each model's parameters are extracted. The low level function is called by a mid-level function that calculates $\mathcal{J}$ and all its gradients. In turn this is called by a high level function that converges to $\mathbf{a}_{mp}$ and then calculates the likelihoods and marginal likelihoods using Laplace's method. A separate high level function performs Markov Chain Monte Carlo by calling the same mid-level and low-level functions. The code is available at Juniper (2022).

2.3   Applying Laplace's Method to a Complete Thermoacoustic System

Matveev (2003) set out to create a quantitatively-accurate model of the hot wire Rijke tube by compiling quantitatively-accurate models of its components from the literature. Despite being tuned to be quantitatively correct at one heater position, this carefully-constructed model is only qualitatively correct at nearby heater positions (Matveev 2003, Figs. 5-5 to 5-8). This demonstrates the danger of relying on quantitative models from the literature: these models may have been quantitatively correct for the reported experiment, but they are probably only qualitatively correct for other experiments. The Bayesian inference demonstrated in this section uses qualitative models from the literature but, crucially, allows their parameters to float in order to match the new experiment at all operating points. As will be shown later, this creates a quantitatively-accurate model over the entire range studied and, if the model is physically-correct, it can extrapolate beyond the range studied.

Developing a quantitatively accurate model of the hot wire Rijke tube is challenging because the heat release rate is small and therefore the thermoacoustic driving mechanism is weak. For the experiment shown here, which is taken from Juniper and Yoko (2022), the thermoacoustic mechanism contributes around ± 10 rad s^{-1} to the growth rate and ± 100 rad s^{-1} to the frequency. For comparison, Fig. 1 shows the decay rate (negative growth rate) and frequency of acoustic oscillations in the cold Rijke tube (i) when empty, (ii) with the heater prongs in place, (iii) with the heater and prongs in place, and (iv) with the heater, prongs, and thermocouples in place. The growth rate and frequency drifts caused by these elements of the rig, even when the heater is off, are a similar size to the thermoacoustic effect and cannot be ignored in a quantitative model. These elements must be modelled but, even after reading the extensive literature on the Rijke tube such as Feldman (1968); Raun et al. (1993); Bisio and Rubatto (1999) and the references within them, it is not evident *a priori* which physical mechanisms must be included and which can be neglected. Instead, we propose several physics-based models, assimilate the data into those models using Laplace's method combined with adjoint methods, and then select the model with the highest marginal likelihood because it is the one that is best supported by the experi-

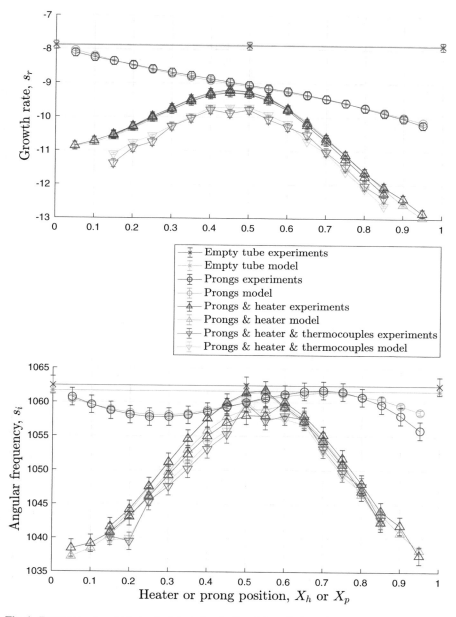

Fig. 1 Expected values (±2 standard deviations) of model predictions $\mathcal{D}(\mathbf{a})$ verses experimentally measured values (±2 standard deviations) D of the growth rates and frequencies of the cold Rijke tube in four configurations: (i) empty tube; (ii) tube containing heater prongs; (iii) tube containing heater prongs and heater; (iv) tube containing heater prongs, heater, and thermocouples. Image adapted from Juniper and Yoko (2022)

Table 1 log(Best Fit Likelihood) per datapoint and log(Marginal Likelihood) per datapoint for seven models of the heater prongs in the cold Rijke tube. The second column contains the number of parameters in each model. The third column describes how the viscous boundary layer on the prongs is modelled: it is the viscous dissipation in the tube's boundary layer multiplied by a real number, a complex number, or zero. The fourth column is the equivalent for the thermal boundary layer. If the third and fourth columns are joined then the same factor is used for both the viscous and thermal boundary layers. The fifth column notes whether the blockage of the prongs is included in the model. Model 4 gives the best fit to the data but is not the most likely model. Model 6 is the most likely model (highest marginal likelihood) because it achieves a good data fit with just two model parameters. (Table adapted from Juniper and Yoko 2022)

Model	Params	Viscous b.l.	Thermal b.l.	Blockage	log(BFL)	log(ML)
1	1	Real		No	−0.3622	−0.7183
2	2	Complex		No	−0.3549	−0.9117
3	2	Real	Real	No	−0.3529	−0.7683
4	4	Complex	Complex	No	+0.9360	−0.2949
5	1	Zero	Zero	Yes	−3.6955	−3.8696
6	2	Real		Yes	+0.6781	+0.1010
7	3	Real	Real	Yes	+0.7099	−0.1096

mental data. For example, Table 1 shows the best fit likelihood, $P(D|\mathbf{a}_{MP}, \mathcal{H}_i)$, and the marginal likelihood, $P(D|\mathcal{H}_i)$, for seven candidate models of the heater prongs. These models contain various combinations of the viscous boundary layer, the thermal boundary layer, and the blockage caused by the prongs, as described in the caption. The best data fit is achieved by model 4 but the highest marginal likelihood is achieved by model 6, which fits the data well with just two parameters. Model 6 contains the blockage caused by the prongs and the visco-thermal drag of the prong's boundary layers, which is expressed as a real multiple of the visco-thermal drag of the tube's boundary layers. It is re-assuring that the model with the highest marginal likelihood contains all the expected physics, but remains simple.

This process is repeated for the heater itself and the thermocouples (Juniper and Yoko 2022) until a quantitatively-accurate model of the cold Rijke tube has been created. Figure 1 shows the model predictions and experimental measurements for the final model. This model is quantitatively accurate across the entire operating range with just a handful of parameters (Juniper and Yoko 2022). Using Laplace's method, accelerated by first and second order adjoint methods, this data assimilation takes a few seconds on a laptop. Using MCMC takes around 1000 times longer on a workstation (Garita 2021). Although time-consuming, MCMC can be useful in order to confirm that the posterior likelihood distributions are close to Gaussian, which justifies the use of Laplace's method.

The fluctuating heat release rate at the wire cannot be measured directly. Analytical relationships between velocity fluctuations and heat release rate fluctuations have been developed (King 1914; Lighthill 1954; Carrier 1955; Merk 1957) but subsequent numerical simulations (Witte and Polifke 2017) have shown that numerically-calculated relationships have a more intricate dependence on Re and St than can be

Table 2 log(Best Fit Likelihood) per datapoint and log(Marginal Likelihood) per datapoint for nine models of the heater in the hot Rijke tube. Model parameters are denoted as k with a numerical index. k_c are the model parameters from the cold experiments, which are fixed. The second column contains the number of parameters in each model. The third and fourth columns describe how the magnitude and phase of the fluctuating heat release rate are modelled. Q_h is the heater power and Q_{King} is adjusted for King's law King (1914); Juniper and Yoko (2022). The fifth and sixth columns describe how the visco-thermal drag at the heater is modelled, where is is the angular frequency and τ_L is Lighthill's time delay Lighthill (1954). (Table adapted from Juniper and Yoko 2022)

Model	Params	Magnitude	Phase	Viscous (k_c = cold value)	Thermal (k_c = cold value)	log(BFL)	log(ML)
1	2	$k_1 \times Q_h$	k_2	k_c	k_c	−4.6018	−4.7039
2	2	$k_1 \times Q_h$	$k_2 \times is$	k_c	k_c	−4.4942	−4.5976
3	3	$k_1 \times Q_h^{k_3}$	k_2	k_c	k_c	−4.5567	−4.6960
4	2	$k_1 \times Q_{King}$	k_2	k_c	k_c	−4.5926	−4.6991
5	2	$k_1 \times Q_{King}$	$k_2 \times is$	k_c	k_c	−4.5670	−4.6750
6	2	$k_1 \times Q_{King}$	$k_2 \times is\tau_L$	k_c	k_c	−5.6770	−5.7794
7	4	$k_1 \times Q_h$	$k_2 \times is$	$k_c + (k_3 + ik_4) \times Q_h$	k_c	−3.3439	−3.6113
8	6	$k_1 \times Q_h$	$k_2 \times is$	$k_c + (k_3 + ik_4) \times Q_h$	$k_c + (k_5 + ik_6) \times Q_h$	−3.1981	−3.5952
9	6	$k_1 \times Q_h$	k_2	$k_c + (k_3 + ik_4) \times Q_h$	$k_c + (k_5 + ik_6) \times Q_h$	−3.5735	−3.9589

derived analytically. Since the 1970s (Bayly 1986) therefore, researchers have tended to use CFD simulations or simple relations that are tuned to a particular operating point (Witte 2018, Table 1; Ghani et al. 2020).

Here we propose six candidate models for the heat release rate and two candidate models for how the thermo-viscous drag of the heater changes with the heater power. We then calculate the marginal likelihoods of these models, allowing the measurement noise to float in order to accommodate epistemic uncertainty such as systematic measurement error and model error. Table 2 shows the candidate models, their assimilated parameters, their log best fit likelihood (BFL) per datapoint, and their log marginal likelihood per datapoint. Model 8 has the highest Marginal Likelihood. In this model, the fluctuating heat release rate is proportional to the steady heat release rate; the time delay between velocity perturbations and subsequent heat release rate perturbations is the same for all configurations, and the thermo-viscous drag of the heater element is proportional to the heater power. There is, of course, no limit to the number of models that can be tested. The interested reader is encouraged to generate and test their own models using the code (Juniper 2022).

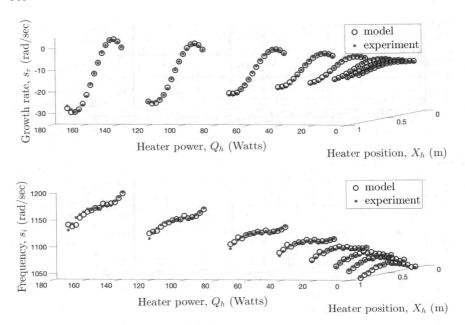

Fig. 2 Expected values of model 8's predictions $\mathcal{D}(\mathbf{a})$ verses experimental measurements D of the growth rates and frequencies of the hot Rijke tube, as a function of heater power and heater position. The model parameters are obtained by assimilating data from all 105 experimental configurations. The model is quantitatively-accurate over the entire operating range. (Image adapted from Juniper and Yoko 2022)

Figure 2 shows the experimental measurements verses the predictions of model 8 for the growth rates and frequencies when assimilating data from all 105 experiments. The agreement is excellent, particularly for the growth rate, which is more practically important than the frequency. Figure 3 is the same as Fig. 2 but is obtained by assimilating data from just 8 of the 105 experiments. The results are almost indistinguishable, which shows that, once a good physics-based model has been identified, very little data is required to tune its parameters. This model can then extrapolate to other operating points, even if they are far from those already examined. This is a desirable feature of any model and shows the advantage of assimilating data into physics-based models with a handful of parameters, rather than physics-agnostic models with many parameters, which would not be able to extrapolate.

As a final comment, this assimilation of experimental data with rigorous Bayesian inference forces the experimentalist to design informative experiments. Firstly, without an excellent initial guess for the parameter values, it is almost impossible to assimilate all the parameters simultaneously. This encourages the experimentalist to assimilate the parameters sequentially with an experimental campaign in which some of the parameters take known values (usually zero) in some of the experiments. Secondly, this process reveals systematic measurement error that was previously

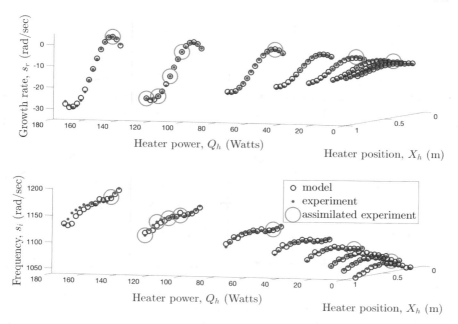

Fig. 3 As for Fig. 2 but when the model parameters are obtained by assimilating data from the eight circled configurations. This model is also quantitatively accurate over the entire operating range, showing that this model can extrapolate beyond the assimilated datapoints. (Image adapted from Juniper and Yoko 2022)

unknown to the experimentalist. This epistemic error is revealed when the parameters shift to absorb the error and seem to uncover impossible physical behaviour.[3] Once this systematic measurement error becomes known, the experimentalist is forced to remove it or avoid it with good experimental design.

3 Physics-Based Statistical Inference Applied to a Flame

The most influential element of any thermoacoustic system is the response of the flame to acoustic forcing. This is also the hardest element to model. In this section, experimental images of forced flames are assimilated into a physics-based model using the first level of inference described in Sect. 2. The physics-based model can then be used in thermoacoustic analysis for example (i) in nonlinear simulations, (ii) to create a nonlinear flame describing function (FDF), or (iii) to create a linear flame transfer function (FTF).

---

[3] As the OPERA team found in 2012, it is wise to search for systematic error before publishing results, however eye-catching they seem (Brumfiel 2012).

3.1 Assimilating Experimental Data with an Ensemble Kalman Filter

We take our model, $\mathcal{H}$, to be a partial differential equation (PDE) discretized onto a grid, with unknown parameters $\mathbf{a}$. As before, we wish to infer the unknown parameters, $\mathbf{a}$, by assimilating data, D, from an experiment. The model, which has state ψ, is marched forwards in time from some initial condition to produce a model prediction, $\mathcal{D}(\psi)$, that can be compared with the experimental measurements, D, over some time period T. In principle, it is possible to use the same method as in Sect. 2.1 to iterate to the values of $\mathbf{a}$ that minimize an appropriate $\mathcal{J}$ for all the data simultaneously. This requires the model predictions, $\mathcal{D}(\psi)$, and their gradients w.r.t. all parameters, a_i, to be stored at all moments at which they are compared with the data D. This is not practical because it would require too much storage. This section describes an alternative approach that requires less storage.

We consider a level set model of a premixed laminar flame, taken from Yu et al. (2020). The state, ψ, is the flame position, and the parameters, $\mathbf{a}$, are the flame aspect ratio β, the Markstein length L, the ratio, K, between the mean flow speed and the phase speed of perturbations down the flame, the amplitude, ϵ, of velocity perturbations, and the parabolicity parameter, α of the base flow, where $U/\overline{U} = 1 + \alpha(1 - 2(r/R)^2)$. The parameters β, L, and α are inferred from an image of an unforced steady premixed bunsen flame. This flame is then forced at 200, 300, and 400 Hz, and the data, D, are experimental images taken at 2800 Hz. The state, ψ, is marched forward in time by the model, $\mathcal{H}$, with parameters $\mathbf{a}$ to an assimilation step. At the assimilation step, the model prediction $\mathcal{D}(\psi)$ is compared with the data D, and the state ψ and remaining parameters $\mathbf{a}$ are both updated to statistically optimal estimates, as described in the next paragraph. The state, ψ, is then marched forward to the next assimilation step and the process is repeated until the parameters $\mathbf{a}$ have converged.

If the evolution were linear or weakly nonlinear then a Kalman filter or extended Kalman filter would be appropriate. The evolution is highly nonlinear, however, with wrinkles and cusps forming at the flame. We therefore use an ensemble Kalman filter (EnKF) in which we generate an ensemble of N states ψ_i from the model $\mathcal{H}$ with different parameter values $\mathbf{a}_i$ (Evensen 2009). At each assimilation step, we append each parameter vector $\mathbf{a}_i$ to its state vector ψ_i to form an augmented state Ψ_i. The expected value $\bar{\Psi}$ and covariance $\mathbf{C}_{\Psi\Psi}$ of the augmented state Ψ are then derived from the ensemble:

$$\bar{\Psi} = \frac{1}{N} \sum_{i}^{N} \Psi_i \tag{10}$$

$$\mathbf{C}_{\Psi\Psi} = \frac{1}{N-1} \sum_{i}^{N} (\Psi_i - \bar{\Psi})(\Psi_i - \bar{\Psi})^T \tag{11}$$

The expected value $\bar{\Psi}$ becomes the prior expected value and replaces $\mathbf{a}_p$ in (3). The covariance $\mathbf{C}_{\Psi\Psi}$ becomes the prior expected covariance and replaces $\mathbf{C}_{aa}$ in (3). The predicted flame position $\mathcal{D}(\bar{\psi})$ is found from the expected state, $\bar{\psi}$. The discrepancy between the experimental flame position D and the model prediction $\mathcal{D}(\psi)$ is then combined with an estimate of the measurement error $\mathbf{C}_{DD}$ in (4). The posterior augmented state Ψ_{mp} and its inverse covariance $\mathbf{A}$ is calculated to be that which maximizes the RHS of (1), as in Sect. 2.1. The state ψ and parameters $\mathbf{a}$ are extracted from the expected value of the posterior augmented state. N states are created with this posterior expected value and covariance, and the process is repeated.

Figure 4 shows the RMS discrepancy between the experiments, D, and the expected value of the simulations, $\mathcal{D}(\psi)$, for flames forced at three different frequencies. The EnKF is switched on from time periods 10 to 15. The RMS discrepancy drops by more than one order of magnitude during this time, to a floor set by the model error. The largest drops in discrepancy occur when the EnKF is assimilating data just as a bubble of unburnt gases is pinching off from the flame. During these moments, which are relatively rare, the parameters converge rapidly towards their final values. This shows that relatively rare events contain more information than relatively common events, as is quantified, for example, through the Shannon information content of an event (MacKay 2003, Eq. (2.34)). After 5 time periods the EnKF is switched off and the tuned models evolve for a further 3 periods without assimilating data. Figure 5 shows the models' expected values and uncertainties (yellow) and the experimental measurements (black) for one further period. This shows that the EnKF has successfully assimilated the model parameters from the experi-

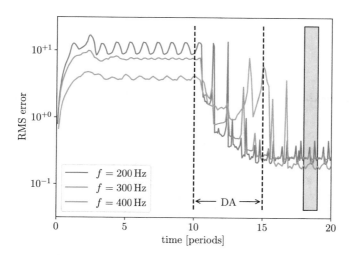

Fig. 4 Root-mean-square (RMS) discrepancy between experimental data, D, and model predictions, $\mathcal{D}$, for a conical bunsen flame forced at 200, 300 and 400 Hz (blue/orange/green, respectively). Data is assimilated from the experiments into the model (DA) between 10 and 15 periods. The snapshots shown in Fig. 5 are taken from the grey window. Image taken from Yu et al. (2020)

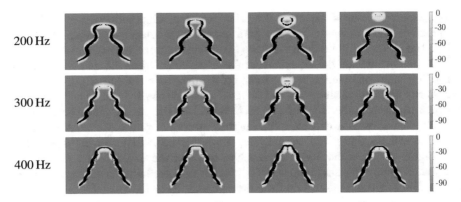

Fig. 5 Snapshots of log-normalized likelihood over one forcing period after combined state and parameter estimation for 200, 300 and 400 Hz (top/middle/bottom row, respectively). Highly likely positions of the flame surface are shown in yellow; less likely positions in green. The flame surface from experimental images is shown as black dots. Image taken from Yu et al. (2020)

mental images and that simulations with these parameters remain accurate beyond the assimilation period.

The EnKF has the advantages that (i) no calculations are required before the assimilation process begins, (ii) it can assimilate any experimental flame that can be represented by the model $\mathcal{H}$. It has the disadvantages that (i) it cannot run in real time because the computational time of the simulations, $O(10^1)$ seconds, exceeds the time between assimilation steps, $O(10^{-3})$ seconds; (ii) if the ensemble starts far from the data, the ensemble tends to diverge rather than converge to the experimental results.

3.2 Assimilating with a Bayesian Neural Network Ensemble

The two disadvantages of the EnKF can be overcome, while retaining uncertainty estimates, by assimilating data, D, with a Bayesian Neural Network ensemble (BayNNE) (Pearce et al. 2020; Gal 2016; Sengupta et al. 2020). Each Neural Network, $\mathcal{M}_i$, in the ensemble is a repeated composition of the function $f(\mathbf{W}_i\mathbf{x} + \mathbf{b}_i)$ where f is a nonlinear function, $\mathbf{x}$ are the inputs, $\mathbf{W}_i$ is a matrix of weights, and $\mathbf{b}_i$ is a vector of biases. Together $\mathbf{W}_i$ and $\mathbf{b}_i$ comprise the parameters θ_i of each neural network. The set of all parameters in the ensemble is denoted $\{\theta_i\}$. The posterior state, $\Psi(D, \{\theta_i\})$, contains the predicted parameters (e.g. β, L, K, ϵ, α) of the numerical simulation. The true targets, $\mathbf{a}$, are the actual parameters of the simulations. The distribution of the prediction is assumed to be Gaussian: $P(\Psi|D, \{\theta_i\}) = \mathcal{N}(\bar{\Psi}, C_{\Psi\Psi})$. Creating this prediction means learning the mean $\bar{\Psi}(D, \{\theta_i\})$ and the covariance $C_{\Psi\Psi}(D, \{\theta_i\})$ of the ensemble.

Each NN in the ensemble produces the expected value, $\mu_i(D, \theta_i)$, and covariance, $\sigma_i^2(D, \theta_i)$, of a Gaussian distribution by minimising the loss function:

$$\mathcal{J}_i = (\mathbf{a} - \mu_i)^T \Sigma_i^{-1}(\mathbf{a} - \mu_i) + \log(|\Sigma_i^{-1}|) \tag{12}$$

$$+(\theta_i - \theta_{i,anc})^T \Sigma_{prior}^{-1}(\theta_i - \theta_{i,anc}) \tag{13}$$

where

$$\Sigma_i^{-1} = \text{diag}(\sigma_i^2) \tag{14}$$

and $\theta_{i,anc}$ are the initial weights and biases of the ith NN. These are sampled from the prior distribution $P(\theta) = \mathcal{N}(\mathbf{0}, \Sigma_p)$, where $\Sigma_p = \text{diag}(1/N_H)$, where N_H is the number of units in each hidden layer. The above task is time-consuming but is performed just once.

The ensemble therefore contains a set of Gaussians, each with their own means, μ_i, and covariances, σ_i^2. These are approximated by a single Gaussian with mean $\bar{\Psi}(D, \{\theta_i\})$ and covariance $C_{\Psi\Psi}(D, \{\theta_i\})$ using (Lakshminarayanan et al. 2017):

$$\bar{\Psi}(D, \{\theta_i\}) = \frac{1}{N} \sum_{i=1}^{N} \mu_i(D, \theta_i) \tag{15}$$

$$C_{\Psi\Psi}(D, \{\theta_i\}) = \text{diag}(\mathbf{c}_{\Psi\Psi}(D, \{\theta_i\})) \tag{16}$$

where N is the number of NNs in the ensemble and

$$\mathbf{c}_{\Psi\Psi}(D, \{\theta_i\}) = \frac{1}{N} \sum_{i=1}^{N} \sigma_i^2(D, \theta_i) + \frac{1}{N} \sum_{i=1}^{N} \mu_i^2(D, \theta_i) - \left(\frac{1}{N} \sum_{i=1}^{N} \mu_i(D, \theta_i)\right)^2 \tag{17}$$

The uncertainty of the ensemble therefore contains the average uncertainty of its members, combined with uncertainty arising from the distribution of the means of the ensemble members. If this uncertainty is large, the observed data is likely to have been outside the training data. This task is quick and is performed at each operating condition.

The BayNNE is trained on 8500 simulations of the level set solver used in Sect. 3.1. The parameters varied are the flame aspect ratio β, the Markstein length L, the ratio, K, between the mean flow speed and the phase speed of perturbations down the flame, the amplitude of velocity perturbations, ϵ, the mean flow parabolicity, α, and the Strouhal number, St. The parameters are sampled using quasi-Monte Carlo in order to obtain good coverage of the parameter space within fixed ranges. For each simulation, 200 evenly-spaced snapshots of a forced periodic solution are stored. The data, D, used for training takes the form of 10 consecutive snapshots extracted from these images. The total library of data therefore consists of $8500 \times 200 = 1.7 \times 10^6$ sets of data, D, each with known parameters $\mathbf{a}$. The neural networks are trained to

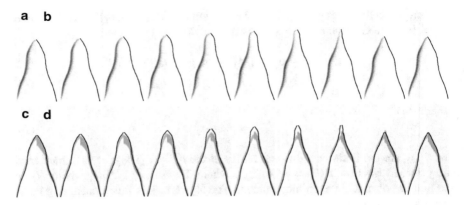

Fig. 6 Top row: experimental images of one cycle of an acoustically forced conical Bunsen flame; the left half **a** shows the raw image while the right half **b** shows the detected edge. Bottom row: the flame edge and its uncertainty when assimilated into a G-equation model with an EnKF (**c**) and a BayNNE (**d**). With this model, propagation of perturbations down the flame is captured well but the pinch-off event is not. Image adapted from Croci et al. (2021)

recognize the parameter values **a** from the data D. Training takes around 12 hours per NN on an NVIDIA P100 GPU. Recognizing the parameter values takes $O(10^{-3})$ seconds on an Intel Core i7 processor on a laptop, which is sufficiently fast to work in real time.

The top row of Fig. 6 shows 10 snapshots of a forced bunsen flame experiment alongside the automatically-detected flame edge. The bottom row shows the modelled flame edge and its variance, assimilated with the EnKF (left) and the BayNNE (right). The flame edge is shown in black. As expected, the expected values found with both assimilation methods are almost identical. The prediction is close to the experiments but, because of model error, the EnKF and the BayNNE both struggle to fit the most extreme pinch off event at $0.6T$. The uncertainty in the BayNNE is greater than that of the EnKF because it assimilates just 10 flame images, while the EnKF has assimilated over 500 images by the time this sequence is generated. Alternative NN architectures, such as long-short term memory networks may be able to reduce this uncertainty.

The fact that the BayNNE assimilates just 10 snapshots is a disadvantage when the flame behaviour is periodic over many cycles, as in the previous example, but an advantage when the flame behaviour is intermittent, as in the next example. Intermittency is commonly observed in thermoacoustic systems, particularly when they are close to a subcritical bifurcation to instability (Juniper and Sujith 2018; Nair et al. 2014). Bursts of periodic behaviour are interspersed within moments of quasi-stochastic behaviour and, while these can be identified by eye and with recurrence plots (Juniper and Sujith 2018), they are not sufficiently regular to be assimilated with the EnKF.

In the next example, images of a bluff-body stabilized turbulent premixed flame Paxton et al. (2019, 2020) are recorded at 10 kHz using OH PLIF, and the flame edge is extracted and smoothed to remove the turbulent wrinkles. A BayNNE trained on 10 snapshots of G-equation simulations with 2400 combinations of parameters, **a**, then identifies the most likely parameters from 10 observed snapshots. In this example the model contains an extra parameter: the spatial growth rate, η, of perturbations,

Figure 7 shows the five assimilated parameters, $(K, \epsilon, \eta, St, \beta)$ and their uncertainties during 430 timesteps of an experimental run imaged at 2.5 kHz. During this run, there are four to five oscillation cycles. The BayNNE successfully identifies the G-equation parameters that match the experimental results and, importantly, estimates their uncertainties. At four moments during the run, Fig. 7 shows snapshots of the experimental image (top left quadrant) alongside the expected value and uncertainty from the G-equation simulations. Because the G-equation simulation is physics-based, it can extrapolate beyond the window viewed in the experiments, as shown in the images. The distribution of fluctuating heat release rate, with its uncertainty, can be calculated from the model. This can then be expressed as a spatial distribution of the flame interaction index, n, and the flame time delay, τ, as in Fig. 8, which can then be entered into a thermoacoustic network model or Helmholtz solver.

4 Identifying Precursors to Thermoacoustic Instability with BayNNEs

The noise from a thermoacoustically-stable turbulent combustor has broadband characteristics and is often assumed to be stochastic (Clavin et al. 1994; Burnley and Culick 2000). This assumption is a reasonable starting point for stochastic analysis (Clavin et al. 1994) but does not exploit the fact that combustion noise contains useful information about the system's proximity to thermoacoustic instability (Juniper and Sujith 2018, Sect. 4). Analysis of this noise usually involves a statistical measure to detect transition away from stochastic behaviour. This can be a measure of departure from chaotic behaviour, using techniques for analysing dynamical systems (Gotoda et al. 2012; Sarkar et al. 2016; Murugesan and Sujith 2016), or the detection of precursors such as intermittency (Juniper and Sujith 2018; Nair et al. 2014; Scheffer et al. 2009).

These methods quantify the behaviour that a researcher thinks should be important, based on observation of similar systems. This approach is generally applicable but has the disadvantage that it will miss information that the researcher does not think is important, and cannot extract information that is peculiar to a particular engine. Given that this research is motivated by industrial applications in which several nominally-identical models of the same engine are deployed, it makes sense to extract as much information as possible from that particular engine model. In other words, we ask whether machine learning techniques can learn to recognize

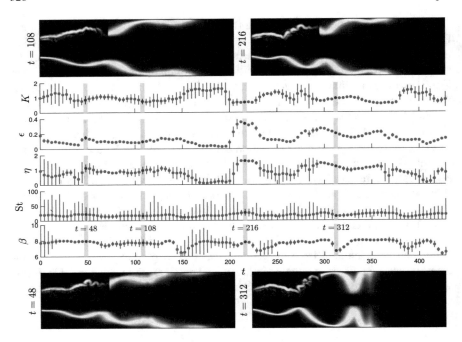

Fig. 7 Assimilated parameters $(K, \epsilon, \eta, \text{St}, \beta)$ of a G-equation model of a bluff-body-stabilized premixed flame during a sequence of 428 snapshots. The parameters are assimilated with a Bayesian Neural Network Ensemble (BayNNE), which also estimates the uncertainty in the assimilated values. The four flame images show (top-left of each frame) the detected flame edge from the experimental OH PLIF image and (remainder of each frame) the expected values and uncertainties in the G-equation model prediction. Image adapted from Croci et al. (2021)

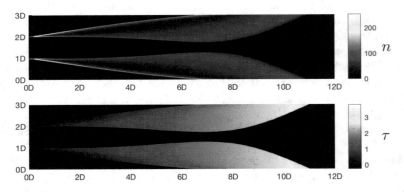

Fig. 8 Spatial distribution of n and τ derived from the G-equation model of the bluff-body-stabilized premixed flame shown in Fig. 7

precursors on one set of engines and then identify precursors on another set of nominally-identical engines. Further we ask whether the machine learning approach is better than techniques that use a statistical measure. In this section, we examine a laboratory scale combustor to develop the method, then three aeroplane engine fuel injector nozzles in an intermediate pressure rig, and then 15 full scale commercial aeroplane engines.

4.1 Laboratory Combustor

In the first study we place a 1 kW turbulent premixed flame inside a steel tube with length 800 mm and diameter 80 mm (Sengupta et al. 2021). The system is run at 900 different operating conditions varying power, equivalence ratio, fuel composition, and the tube exit area. All operating points are thermoacoustically stable, but the thermoacoustic mechanism is active and some points are close to thermoacoutic instability.

For each operating point, the combustion noise is recorded at 10, 000 Hz. The system is then forced for 50 ms at 230 Hz, which is close to the natural frequency of the first longitudinal mode. The decay rate of the acoustic oscillations is extracted from the microphone signal. We then train a Bayesian Neural Network ensemble (BayNNE) to identify the decay rate from 300 ms clips of combustion noise before the acoustic excitation. The decay rate changes from negative to positive at the point of thermoacoustic instability, so is a good measure of the proximity to thermoacoustic instability. The BayNNE returns the uncertainty in its predictions, ensuring that the model does not make overconfident predictions from inputs that differ significantly from those on which it was trained. If the priors are specified correctly, this technique can work with smaller amounts of data and be more resistant to over-fitting (Pearce et al. 2020).

Before training, all the input variables are normalized in order to remove the amplitude information. The parameters $\mathbf{a}_i$ of each ensemble member are initialized by drawing from a Gaussian prior distribution with zero mean and variance equal to $1/N_H$, where N_H is the number of hidden nodes in the previous layer of the NN. This initialization means that the distribution of predictions made by the untrained prior neural network will be approximately zero-centred with unit variance. Each ensemble member is trained normally, but with a modified loss function that anchors the parameters to their initial values. This procedure approximates the true posterior distribution for wide neural networks (Pearce et al. 2020). We train on 80% of the operating points, retain 20% for testing, and train ten different models using ten random test-train splits. This ensures the stability of our algorithm's performance with respect to different train-test splits.

Figure 9a shows the decay timescale (the reciprocal of the decay rate) predicted by the BayNNE, compared with the decay timescale measured from the subsequent response to the pulse. The grey bars show the uncertainty outputted by the BayNNE. The decay timescales are predicted reasonably accurately. The grey uncertainty bars

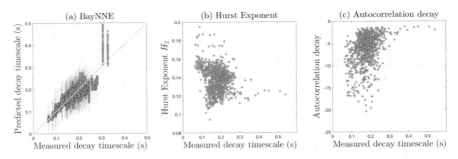

Fig. 9 a Decay timescale, ±2 standard deviations, predicted with a BayNNE, **b** Hurst exponent **c** autocorrelation decay, as functions of the measured decay timescale for thermoacousic oscillations of a turbulent premixed Bunsen flame in a tube. The BayNNE provides the most reliable indicator of proximity to thermoacoustic instability. This figure recreated is based on the data in Sengupta et al. (2021)

widen for the operating points closer to instability because there are only a few operating points close to instability; the decay timescale exceeds 0.3 s for just 13 operating points in the training set. This shows that the BayNNE can successfully predict how far the system is from instability while also indicating how confident it is in that prediction.

Figure 9b, c show the generalized Hurst exponent and the Autocorrelation decay of the combustion noise as functions of the measured decay timescale. As expected, the Hurst exponent drops and the autocorrelation decay increases as the decay timescale increases, showing that these measurements are working as precursors of combustion instability. They are not as accurate, however, as the BayNNE and contain no measure of uncertainty. It is clear therefore that, when trained on this specific combustor, the BayNNE out-performs the Hurst exponent and autocorrelation decay. This outcome would be reversed, of course, if the BayNNE were applied to a different combustor, without retraining.

We also trained the BayNNEs to recognize the equivalence ratio and burner power from 300 ms of combustion noise. The BayNNe could recognize the equivalence ratio with a rms error of 3.5% and the power with a rms error of 2%. This shows that each operating condition has a unique acoustic signature that the BayNNE can learn. The experimentalist in the room can hear that all operating conditions sound slightly different, but cannot recognize the operating condition to the accuracy that the BayNNE can achieve.

4.2 Intermediate Pressure Industrial Fuel Spray Nozzle

The second study is on an industrial intermediate pressure combustion test rig, which is equipped with three pressure transducers, sampling at 50 kHz. Experiments are performed on three different fuel injectors over a range of operating points in order

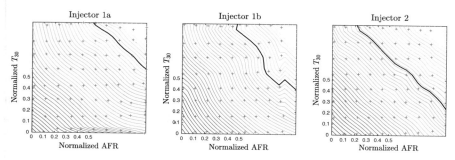

Fig. 10 The black line shows the thermoacoustic instability threshold as a function of air-fuel ratio (AFR) and exit temperature T_{30} for three aeroplane engine fuel injectors in an intermediate pressure rig. The coloured lines show the distance to the black line. Injectors $1a$ and $1b$ are nominally identical

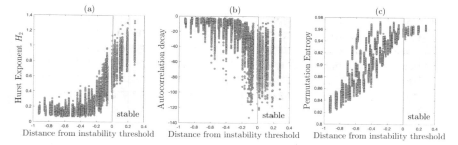

Fig. 11 **a** Hurst exponent, **b** autocorrelation decay, **c** permutation entropy calculated from the pressure signal of injector $1a$ in the intermediate pressure rig, as a function of the distance to the instability threshold in Fig. 10a. A positive (negative) distance indicates stable (unstable) thermoacoustic behaviour

to identify operating points that are thermoacoustically unstable. The injectors are labelled $1a$, $1b$, and 2. Injectors $1a$ and $1b$ are nominally identical. The operating points are identified by their air-fuel ratio (AFR) and their exit temperature (T_{30}). The threshold of thermoacoustic instability is defined as the operating points at which the acoustic amplitude exceeds 0.5% of the static pressure. The black lines in Fig. 10 show this threshold in (AFR, T_{30})–space. Despite being nominally identical, injectors $1a$ and $1b$ have instability thresholds at slightly different positions in (AFR, T_{30})–space.

We normalize the ranges of AFR and T_{30} to run from 0 to 1 and then train a BayNNE to recognize the Euclidian distance to the instability threshold, based on 500 ms of normalized pressure measurements. Stable points are assigned positive distances and unstable points are assigned negative distances. We compare the predictions from the BayNNE with those from the autocorrelation decay, the permutation entropy, and the Hurst exponent.

Figure 11a–c show the Hurst exponent, the autocorrelation decay, and the permutation entropy for injector $1a$. The Hurst exponent reduces significantly as the system

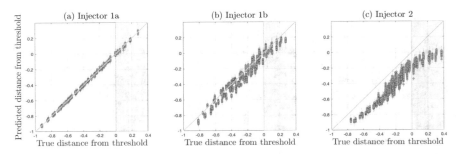

Fig. 12 Predicted distance to the instability threshold ±2 s.d. as a function of measured distance to instability threshold for **a** injector 1a, **b** injector 1b, **c** injector 2 when the prediction is obtained from a BayNNE trained on injector 1a. Injectors 1a and 1b are nominally identical

becomes unstable and this is a useful indicator of the instability threshold, albeit with significant unquantified uncertainty. The autocorrelation decay tends towards zero as the system becomes more unstable but, for this data, barely changes across the instability threshold and therefore does not provide a useful indicator for the threshold. The permutation entropy drops after the system has crossed the threshold from stable to unstable operation, meaning that it is not useful as a precursor in these experiments.

The BayNNE is trained on the training points of 1a and then applied to test points of 1a, 1b, and 2. Figure 12a shows the distance from instability threshold predicted by the BayNNE compared with the true distance. Uncertainty bands of the BayNNE are shown in grey. The BayNNE provides a remarkably accurate prediction of the distance to instability from the pressure signal alone. Figure 12b shows the distance from the instability threshold predicted by the BayNNE trained on injector 1a when applied to the pressure data from the nominally identical injector 1b. The BayNNE performs well when the system is unstable (distance less than 0) but performs less well, and assigns itself greater uncertainty, when the system is stable (distance greater than 0). As mentioned above, 1b is unstable over a different range of (AFR,/T30)–space than 1a, and, despite this difference, the BayNNE has successfully identified the distance to instability on the new injector. The prediction is most inaccurate and uncertain, however, when the system is stable, which is the most useful scenario because it then acts as a precursor to instability. Figure 12c shows the distance from the instability threshold predicted by the BayNNE trained on 1a when applied to the pressure data from injector 2. The BayNNE performs badly, particularly when the system is stable. This confirms that a BayNNE trained on one thermoacoustic system is a good indicator of thermoacoustic precursors on nominally-identical thermoacoustic systems, out-performing statistical measures, but is not useful for different systems. This is not surprising, given that the BayNNE is using all available information from the pressure signal of this particular system, while the statistical methods are quantifying general features in all systems.

4.3 Full Scale Aeroplane Engine

The third study is on 15 full scale prototype aeroplane engines operating at sea level (McCartney et al. 2022). The engines are equipped with two dynamic pressure sensors upstream of the combustor, sampling at 25 kHz. The compressor exit temperature, compressor exit pressure, fuel flow rate, primary/secondary fuel split, and core speed are sampled at 20 Hz. The core speed is increased (known as a *ramp acceleration*) such that the engines deliberately enter a thermoacoustically-unstable operating region. The instability threshold is defined by the point at which the peak to peak amplitude exceeds a certain value. Although the engines are nominally identical, the instability threshold is exceeded at a different core speed for each engine. Here, we investigate whether a BayNNE trained on the operating points and pressure signals from some of the engines can provide a useful warning of impending instability during a ramp acceleration in the other engines.

Previously we used BayNNEs to predict the decay rate of acoustic oscillations (Sect. 4.1) or the distance to instability in parameter space (Sect. 4.2). Now we consider a more practical quantity: the probability that the combustor will become thermoacoustically unstable within the next Δt seconds during a ramp acceleration. We assume that this probability depends on the current operating point of the system, the future operating point, and the time it will take to reach the future operating point. In line with Sects. 4.1 and 4.2 we also assume that the current pressure signal contains useful information about how close the combustor is to thermoacoustic instability. We downsample the signal from a single sensor to 25 kHz, extract 4096 datapoints, which corresponds to around 160 ms, and then process it: (i) into a binary indication of whether the peak to peak pressure threshold has been exceeded; (ii) with de-trended fluctuation analysis (DFA) (Gotoda et al. 2012). The BayNNE is trained to learn the binary signal at time Δt in the future, based on the operating conditions at time Δt in the future and the pressure signal in the present. The future time, Δt, is varied from 100 ms to 1000 ms in steps of 100 ms. For comparison, a BayNNE is trained to learn the binary signal at time Δt in the future, based on the operating conditions alone (i.e. without the pressure data).

There are three stages: tuning, training, and testing. In the tuning stage the number of hidden layers (2–10) and number of neurons in each layer (10–100) are optimized by performing a random search over these hyperparameters. For each combination, a BayNNE is trained on the training data and evaluated on the tuning data. We then select the hyperparameters and number of training epochs that perform best. In the testing stage, the BayNNE with optimal hyperparameters is applied to the testing data. This outputs the log likelihood of the BayNNE model, $\mathcal{M}$, given the data D. The different BayNNEs can then be ranked by the relative sizes of their log likelihoods. (The absolute value is not important.)

Figure 13 shows the log likelihoods of the BayNNE trained on the operating point (OP) alone and the BayNNE trained on the operating point and the DFA pressure signal (DFA). The OP BayNNE is the baseline against which to compare the DFA BayNNE. For future times below 400 ms, the tuned DFA BayNNE model fits the

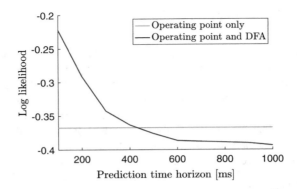

Fig. 13 The log likelihood of observing this data, given (i) the BayNNE trained on the operating point alone (OP) and (ii) the BayNNE trained on the operating point and DFA pressure signal (OP & DFA). For prediction horizons lower than 400 ms, inclusion of the pressure signal renders the model more likely and therefore more predictive. This figure is recreated based on the data in McCartney et al. (2022)

binary signal at that future time better than the tuned OP BayNNE model. In other words, the inclusion of pressure data gives smaller errors in the predicted probability that the threshold will be exceeded at that future time. For future times above 400 ms, the tuned DFA BayNNE model is marginally less likely than the OP BayNNE. This shows that the current pressure signal contains information that is useful up to 400 ms into the future, but no longer.

Figure 14 shows the error in the predicted core speed at which the system becomes unstable. The OP BayNNE knows only the future operating point. The error in the predicted onset core speed arises from differences between the engines being tested. If

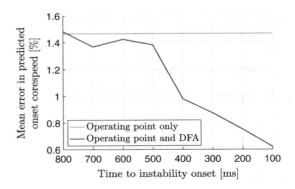

Fig. 14 Mean error in the predicted core-speed at which the engine will become thermoacoustically unstable as a function of time to instability onset as predicted by the BayNNE trained on the OP alone, and the BayNNE trained on the OP and the DFA pressure signal. This figure is recreated based on the data in McCartney et al. (2022)

all the engines were to behave identically, this error would be zero. The DFA BayNNE knows the future operating point and the current pressure signal. As expected from Fig. 13, the error in the predicted onset core speed drops around 400 ms before the instability starts. In other words, in this ramp acceleration, the pressure signal becomes informative around 400 ms before an instability starts but is not informative before then.

5 Conclusion

In the late 1990s, we were promised that the internet would change everything. Three decades later, very few internet-only companies have survived. The winners have been the companies who integrated the internet into what they did well already. If Machine Learning is to science what the internet was to business then the fields that thrive will be those that integrate machine learning into what they do well already. Fluid Dynamics in general, and Thermoacoustics in particular, is well placed to do this because the methods work well and the industrial motivation is strong.

Machine learning is successful because of its relentless focus on data, rather than on models, correlations, and assumptions that the research community has become used to. These models are not badly wrong, but they are rarely quantitatively accurate, and are therefore of limited use for design. It is particularly powerful to combine these physics-based models with one of the tools of probabilistic machine learning: Bayesian inference. By assimilating experimental or numerical data, we can turn qualitatively-accurate models into quantitatively accurate models, quantify their uncertainty, and rank the evidence for each model given the data. This should become standard practice at the intersection between low order models and experiments (numerical or physical). The days of sketching a line by eye through a cloud of points on a 2D plot should be over. This should be replaced by rigorous Bayesian inference, with all subjectivity well-defined, and in as many dimensions as required.

For low order models, assimilation with Laplace's method combined with first and second order adjoints of those models is fast and powerful. For models with more than a few hundred degrees of freedom, this method becomes cumbersome. Nevertheless, it is still possible to assimilate data into larger physics-based models and to estimate the uncertainty in their parameters using iterative methods such as the Ensemble Kalman Filter, or parameter recognition with Bayesian Neural Network Ensembles. This is a powerful way to combine the practical aspects of Machine Learning with the attractive aspects of physics-based models. It is demonstrated here for a simple level set solver but, with enough simulations, could be extended to CFD.

Sometimes, however, we must accept that we do not recognise or cannot model the influential physical mechanisms in a system we are observing. In these circumstances, physics-agnostic neural networks are an ideal tool because they can learn to recognise features that humans will miss. Perhaps the most striking conclusion of the experiment reported in Sect. 4.1 is that every operating point had a different sound and that a Neural Network could recognise the operating point just from that sound. A human may suspect this but would be unable to remember them all. This is an interesting feature for aircraft engines because fleets contains thousands of nominally-identical but slightly different engines. The signs of impending thermoacoustic instability can therefore be learned from the sound on a handful of engines and applied confidently to the others. This gives a way to avoid thermoacoustic instability, even if it has been impossible to design it out.

For thermoacoustics, this chapter shows some promising ways to combine 30 years of machine learning with 200 years of physics-based learning. If we continue to fly long distance or send rockets into space, we will need to continue to avoid thermoacoustic instability. With novel research methods and continual industrial motivation, the field of thermoacoustics looks set to be interesting for many decades to come.

Acknowledgements The work presented in this chapter arose out of collaborations with Ushnish Sengupta, Max Croci, Hans Yu, Michael McCartney, Matthew Yoko, Francesco Garita, and Luca Magri. The author would particularly like to thank Ushnish Sengupta, Max Croci, and Michael McCartney for their important contributions to the manuscript. The computer code "DATAN: Data Assimilation Thermo-Acoustic Network model" may be obtained from the link https://doi.org/10.17863/CAM.84141.

References

Aguilar JG, Juniper MP (2020) Thermoacoustic stabilization of a longitudinal combustor using adjoint methods. Phys Rev Fluids 5:083902

Bayly BJ (1986) Onset and equilibration of oscillations in general Rijke devices. J Acoust Soc Am 79(3):846–851

Bisio G, Rubatto G (1999) Sondhauss and Rijke oscillations - thermodynamic analysis, possible applications and analogies. Energy 24(2):117–131

Brumfiel G (2012) Neutrinos not faster than light. Nature 2012.10249

Burnley VS, Culick FEC (2000) Influence of random excitations on acoustic instabilities in combustion chambers. AIAA J 38(8):1403–1410

Carrier GF (1955) The mechanics of the Rijke tube. Q Appl Math 12(4):383–395

Clavin P, Kim JS, Williams FA (1994) Turbulence-induced noise effects on high frequency combustion instabilities. Combust Sci Technol 96(1–3):61–84

Crocco L, Cheng S-I (1956) Theory of combustion instability in liquid propellant rocket motors. Butterworths, London

Croci ML, Sengupta U, Juniper MP (2021) Data assimilation using heteroscedastic Bayesian neural network ensembles for reduced-order flame models. In: Paszynski M, Kranzlmüller D, Krzhizhanovskaya VV, Dongarra JJ, Sloot MP (eds) Computational Science - ICCS 2021, vol 12746. Lecture Notes in Computer Science. Springer, Cham., pp 408–419

Croci M, Sengupta U, Juniper MP (2021) Bayesian inference in physics-based nonlinear flame models. In: 35th conference on neural information processing systems (NeurIPS)

Culick F (2006) Unsteady motions in combustion chambers for propulsion systems. AGARD AG-AVT-039

Dowling AP, Morgans AS (2005) Feedback Control of Combustion Oscillations. Ann Rev Fluid Mech 37(1):151–182

Dyson F (2004) A meeting with Enrico Fermi. Nature 427:297

Evensen G (2009) Data assimilation. Springer, Berlin

Feldman KT (1968) Review of the literature on Rijke thermoacoustic phenomena. J Sound Vib 7(1):83–89

Gal Y (2016) Uncertainty in deep learning. PhD thesis, University of Cambridge, Cambridge, UK

Garita F (2021) Physics-based statistical learning in thermoacoustics. PhD thesis, University of Cambridge, Cambridge, UK

Garita F, Yu H, Juniper MP (2021) Assimilation of experimental data to create a quantitatively accurate reduced-order thermoacoustic model. J Eng Gas Turb Power 143(2):021008

Ghani A, Boxx I, Noren C (2020) Data-driven identification of nonlinear flame models. J Eng Gas Turb Power 142(12):121015

Gotoda H, Amano M, Miyano T, Ikawa T, Maki K, Tachibana S (2012) Characterization of complexities in combustion instability in a lean premixed gas-turbine model combustor. Chaos 22(4):043128

Guo S, Silva CF, Yong KJ, Polifke W (2021a) A Gaussian-process-based framework for high-dimensional uncertainty quantification analysis in thermoacoustic instability predictions. Proc Combust Inst 38:6251–6259

Guo S, Silva CF, Polifke W (2021b) Robust identification of flame frequency response via multi-fidelity Gaussian process approach. J Sound Vib 502:116083

Hemchandra S (2009) Dynamics of turbulent premixed flames in acoustic fields. PhD thesis, Georgia Tech, Atlanta, GA, USA

Higgins B (1802) On the sound produced by a current of hydrogen gas passing through a tube. J Nat Philos Chem Arts 1:129–131

Huang Y, Yang V (2009) Dynamics and stability of lean-premixed swirl-stabilized combustion. Prog Energy Combust Sci 35(4):293–364

Jaensch S, Polifke W (2017) Uncertainty encountered when modelling self-excited thermoacoustic oscillations with artificial neural networks. Int J Spray Combust Dyn 9:367–379

Juniper MP (2018) Sensitivity analysis of thermoacoustic instability with adjoint Helmholtz solvers. Phys Rev Fluids 3:110509

Juniper MP (2022) Datan: data assimilation thermo-acoustic network model. https://doi.org/10.17863/CAM.84141

Juniper MP, Sujith RI (2018) Sensitivity and nonlinearity of thermoacoustic oscillations. Ann Rev Fluid Mech 50:661–689

Juniper MP, Yoko M (2022) Generating a physics-based quantitatively-accurate model of an electrically-heated Rijke tube with Bayesian inference. J Sound Vib 535:117096. https://doi.org/10.1016/j.jsv.2022.117096

King LV (1914) On the convection of heat from small cylinders in a stream of fluid: determination of the convection constants of small platinum wires with applications to hot-wire anemometry. Philos Trans R Soc A 214:373–432

Kobayashi H, Gotoda H, Tachibana S, Yoshida S (2017) Detection of frequency-mode-shift during thermoacoustic combustion oscillations in a staged aircraft engine model combustor. J Appl Phys 122:224904

Kulkarni S, Guo S, Silva CF, Polifke W (2021) Confidence in flame impulse response estimation from large eddy simulation with uncertain thermal boundary conditions. J Eng Gas Turb Power 143(12):121002

Lakshminarayanan B, Pritzel A, Blundell C (2017) Simple and scalable predictive uncertainty estimation using deep ensembles. In: 31st conference on neural information processing systems

Lieuwen TC (2012) Unsteady combustor physics. Cambridge University Press

Lieuwen TC, Banaszuk A (2005) Background noise effects on combustor stability. J Propul Power 21(1):25–31

Lieuwen TC, McManus KR (2003) Combustion dynamics in lean-premixed prevaporized (lpp) gas turbines. J Propul Power 19:721

Lighthill MJ (1954) The response of laminar skin friction and heat transfer to fluctuations in the stream velocity. Proc R Soc A 224(1156):1–23

MacKay DJC (2003) Information theory, inference, and learning algorithms. Cambridge University Press

Magri L, Juniper MP (2013) Sensitivity analysis of a time-delayed thermo-acoustic system via an adjoint-based approach. J Fluid Mech 719:183–202

Magri L, Bauerheim M, Juniper MP (2016) Stability analysis of thermo-acoustic nonlinear eigenproblems in annular combustors. Part I Sensitivity. J Comput Phys 325:395–410

Matveev KI (2003) Thermoacoustic instabilities in the Rijke tube: experiments and modeling. PhD thesis, California Institute of Technology

McCartney M, Haeringer M, Polifke W (2020) Comparison of machine learning algorithms in the interpolation and extrapolation of flame describing functions. J Eng Gas Turb Power 142(6):061009

McCartney M, Sengupta U, Juniper MP (2022) Reducing uncertainty in the onset of combustion instabilities using dynamic pressure information and Bayesian neural networks. J Eng Gas Turb Power 144(1):011012

Merk HJ (1957) Analysis of heat-driven oscillations of gas flows II on the mechanism of the Rijke-Tube phenomenon. App Sci Res A 6:402–420

Metropolis N, Rosenbluth AW, Rosenbluth MN, Teller AH, Teller E (1953) Equation of state calculations by fast computing machines. J Chem Phys 21:1087–1092

Mongia HC, Held TJ, Hsiao GC, Pandalai RP (2003) Challenges and progress in controlling dynamics in gas turbine combustors introduction. J Propul Power 19(5):822–829

Murugesan M, Sujith RI (2016) Detecting the onset of an impending Thermoacoustic instability using complex networks. J Propul Power 32(3):707–712

Nair V, Thampi G, Sujith RI (2014) Intermittency route to thermoacoustic instability in turbulent combustors. J Fluid Mech 756:470–487

Oefelein JC, Yang V (1993) Comprehensive review of liquid-propellant combustion instabilities in F-1 engines. J Propul Power 9(5):657–677

Paxton BT, Fugger CA, Rankin BA, Caswell AW (2019) Development and characterization of an experimental arrangement for studying bluff-body-stabilized turbulent premixed propane-air flames. In: AIAA Scitech Forum, 2019. American Institute of Aeronautics and Astronautics

Paxton BT, Fugger CA, Tomlin AS, Caswell AW (2020) Experimental investigation of fuel chemistry on combustion instabilities in a premixed bluff-body combustor. In AIAA Scitech, Forum. American Institute of Aeronautics and Astronautics

Pearce T, Leibfried F, Brintrup A (2020) Uncertainty in neural networks: Approximately Bayesian ensembling. In: Chiappa S, Calandra R (eds) Proceedings of 23rd international conference on artificial intelligence and statistics, vol 108, pp 234–244. PMLR 108:234–244

Raun RL, Beckstead MW, Finlinson JC, Brooks KP (1993) A review of Rijke tubes, Rijke burners and related devices. Prog Energy Combust Sci 19:313–364

Rayleigh JWSB (1878) The explanation of certain acoustical phenomena. Nature 18(455):319–321

Rigas G, Jamieson NP, Li LKB, Juniper MP (2016) Experimental sensitivity analysis and control of thermoacoustic systems. J Fluid Mech 787(R1):1–11

Rijke PL (1859) Notice of a new method of causing a vibration of the air contained in a tube open at both eneds. The Lond Edinburgh, and Dublin Philos Mag J Sci 17(4):419–422

Saito T (1965) Vibrations of air columns excited by heat supply. Bull Jpn Soc Mech Eng 8(32):651–659

Sarkar S, Chakravarthy SR, Ramanan V, Ray A (2016) Dynamic data-driven prediction of instability in a swirl-stabilized combustor. Int J Spray Combust Dyn 8:235–253

Scheffer M, Bascompte J, Brock WA, Brovkin V, Carpenter SR, Dakos V, Held H, van Nes EH, Rietkerk M, Sugihara G (2009) Early-warning signals for critical transitions. Nature 461:53–59

Sengupta U, Rasmussen CE, Juniper MP (2021) Bayesian machine learning for the prognosis of combustion instabilities from noise. J Eng Gas Turb Power 143(7):071001

Sengupta U, Amos M, Hosking J, Rasmussen C, Juniper MP, Young P (2020) Ensembling geophysical models with Bayesian neural networks. Adv Neural Inf Proc Syst (Neurips)

Tammisola O, Giannetti F, Citro V, Juniper MP (2014) Second-order perturbation of global modes and implications for spanwise wavy actuation. J Fluid Mech 755:314–335

Tathawadekar N, Doan AKN, Silva CF, Thuerey N (2021) Modeling of the nonlinear flame response of a Bunsen-type flame via multi-layer perceptron. Proc Combust Inst 38:6261–6269

Waxenegger-Wilfing G, Sengupta U, Martin J, Armbruster W, Hardi J, Juniper MP, Oschwald M (2021) Early detection of thermoacoustic instabilities in a cryogenic rocket thrust chamber using combustion noise features and machine learning. Chaos: J Nonlinear Sci 31:063128

Williams FA (1985) Combustion theory. Avalon

Witte A (2018) Dynamics of unsteady heat transfer and skin friction in pulsating flow across a cylinder. PhD thesis, T U Munich, Munich

Witte A, Polifke W (2017) Dynamics of unsteady heat transfer in pulsating flow across a cylinder. Int J Heat Mass Trans 109:1111–1131

Yu H, Juniper MP, Magri L (2021) A data-driven kinematic model of a ducted premixed flame. Proc. Combust. Inst. 38:6231–6239. Also in Yu H (2020) Inverse problems in thermoacoustics. PhD thesis, Cambridge

Summary

The increasing availability of data is a shared trait of several research fields. It opens up great opportunities to advance our understanding of physical processes and lead to disruptive technological innovations. Machine learning methods are becoming an essential resource in combustion science to deal with previously unmet challenges in the field, associated with the number of species involved in combustion processes, the small scales and the non-linear turbulence-chemistry interactions characterising the behaviour of combustion devices. Turbulent reacting flows are inherently multi-scale and multi-physics and involve a broad range of scales, both for chemistry and fluid dynamics. Unlike typical machine learning applications that rely on inexpensive system evaluations, combustion involves experiments that may be difficult to repeat (especially at the scale of interest) and simulations on high-performance computing infrastructures. Contrary to common intuition, available combustion data are very sparse: massive datasets are available, but for very few operating conditions (in terms of chemical composition, turbulence level, turbulence/chemistry interactions, etc.), resulting in generalisation of machine learning algorithms to be a challenging task. This leads to specialised needs that have pushed the research into developing hybrid physics-based, data-driven methods for combustion applications. This book stems from this observation to present current trends for ML methods in combustion research, in particular:

1. The use of machine learning to understand and learn reaction rates and reaction mechanisms and accelerate chemistry integration.
2. The use of linear and non-linear dimensionality reduction approaches for feature extraction, classification and development of reduced-order models (ROMs).
3. The combination of advanced neural network architectures and physics-based models to parametrise the unresolved quantities of interest in reacting flow simulations and to model, forecast, and avoid thermoacoustic instabilities.
4. The development of data-based frameworks designed to detect spatial and temporal events of interest during high-performance computing (HPC) simulations.

N. Swaminathan and A. Parente (eds.), *Machine Learning and Its Application to Reacting Flows*, Lecture Notes in Energy 44, https://doi.org/10.1007/978-3-031-16248-0

    This book explored the growing intersection of machine learning methods with physics-based modelling for turbulent combustion problems. Without the ambition of being exhaustive, it gathered contributions from international experts in the field, covering a variety of problems and application areas. As such, it offers a snapshot of the current trends in the community and discusses potential developments forward. Looking ahead, the main challenge for data-driven approaches applied to combustion will be to demonstrate the interpretability, explainability, and generalizability of the proposed modelling strategies in practical applications. This is critical to implementing major technological modifications and leading the technological transformation towards sustainable combustion technologies based on renewable fuels, including E-fuels. We are certain that this field will advance rapidly in the near future and we hope that the information presented in this volume would contribute towards that development and specifically help curious readers.

Index

Symbols

0D ignition data, 140

2D simulations, 131

A

Acoustic oscillation, 308, 331

Activation function, 121, 155, 160, 180, 181,
188, 203, 222–225

Adaptive chemistry, 118, 138

Adaptive reduced chemistry, 130

Aircraft engines, 308

Air-fuel ratio, 329

Air quality, 136

Algebraic models, 150

Ammonia, 149

Analysis partition in event detection, 57

Analysis partitioning mask, 68

Anomaly detection, 55

A posteriori validation, 167, 168

Approximate deconvolution method, 98

Approximate methods, 98

A priori evaluation, 150, 168

Arrhenius model, 119

Artificial intelligence, 134

Artificial neural network, 99, 102, 106, 118,
119, 127, 176, 180, 183, 187, 190,
193, 199, 202, 204, 210, 211, 220,
222, 247, 260

Autoencoders, 25, 129

Auto-ignition detection in combustion, 82

AutoKeras, 124

Automated machine learning, 124

Auto-PyTorch, 124

AVBP, 160

B

Back-propagation algorithm, 176, 180

Backscatter, 93, 102

Backward propagation, 222

Bayesian inference, 311, 314, 333

Bayesian neural network ensemble, 310,
322, 333

Bayesian optimization, 138, 261

Best Fit Likelihood, 317

β-function, 212, 213, 228, 238

Bias, 180, 188

Bias neurons, 120

Bias-variance trade-off, 156

Bluff-body stabilized flame, 325

Bond order, 18

Born-Oppenheimer approximation, 17

C

Chemical kinetics, 210

Chemical mechanism, 118

Chemical reaction neural networks, 127

Chemical reactions, 118

Chemical source terms, 118

Chemistry acceleration, 117, 119

Chemistry integration, 118, 119, 123, 133

Chemistry reduction, 117, 118, 131

Chemistry regression, 122

Chemistry tabulation, 118, 123, 138, 187,
203

CHG, 2

Clark model, 96

Classification, 92, 119, 157

Classifier, 131

Clustering, 22, 119, 123, 125, 130

N. Swaminathan and A. Parente (eds.), *Machine Learning and Its Application to Reacting Flows*, Lecture Notes in Energy 44, https://doi.org/10.1007/978-3-031-16248-0

Printed in the United States
by Baker & Taylor Publisher Services